D1035319

PROCEEDINGS OF SYMPOSIA
IN PURE MATHEMATICS
Volume XXII

Algebraic Topology

American Mathematical Society
Providence, Rhode Island
1971

Proceedings of the Seventeenth Annual
Summer Research Institute of the
American Mathematical Society

Held at the University of Wisconsin
Madison, Wisconsin
June 29–July 17, 1970

Prepared by the American Mathematical Society under National Science Foundation Grant GP-19276

Edited by
ARUNAS LIULEVICIUS

International Standard Book Number 0-8218-1422-2
Library of Congress Catalog Number 72-167684

Printed in the United States of America

CONTENTS

PREFACE

The American Mathematical Society held its Seventeenth Summer Research Institute at the University of Wisconsin, Madison, Wisconsin, from June 29 to July 17, 1970. The topic for the institute was algebraic topology. The organizing committee for the institute consisted of William Browder, Edward R. Fadell, Edwin E. Floyd, Peter J. Hilton, Richard K. Lashof, Arunas L. Liulevicius (chairman), Mark E. Mahowald, R. James Milgram, Franklin P. Peterson, James D. Stasheff, and P. Emery Thomas.

The program of the institute was divided into four parts: (1) survey talks on recent developments in the field of algebraic topology; (2) invited one-hour talks on important recent work (these tended to be more technical than the survey talks); (3) sessions on problems; (4) seminars organized by the participants in the institute. The survey talks were given by J. Frank Adams, William Browder, Edgar H. Brown, Jr., Samuel Gitler, Richard K. Lashof, Franklin P. Peterson, Larry Smith, and James D. Stasheff. Lecture notes for these series appear in this volume, the only exception being William Browder's "Homotopy and manifolds" (to appear in the proceedings of the Amsterdam conference on topology of manifolds). The remainder of this volume consists of the texts of the invited one-hour talks and a list of research problems. There were four sessions on problems in algebraic topology: a general session (R. James Milgram), immersions and embeddings (Samuel Gitler), *H*-spaces (James D. Stasheff), homotopy theory (Franklin P. Peterson); the resulting list of research problems was edited by R. James Milgram. Seminars at the summer institute were organized by the participants themselves. A partial list of the more stable seminars included categorical methods, cobordism, cohomology of Postnikov systems, *H*-spaces, foliations, and group actions. No seminar talks are included in this volume.

The summer institute was sponsored by the National Science Foundation under Grant GP 19276. The organizing committee wishes to thank Lillian R. Casey of the American Mathematical Society for her help in organizing the conference. The institute was fortunate in having Edward R. Fadell as its administrative director who made sure that things ran smoothly. To him and the other friendly people at the University of Wisconsin go the thanks of all of the participants.

Arunas Liulevicius

ALGEBRAIC TOPOLOGY IN THE LAST DECADE

J. F. ADAMS

1. **Introduction.** I am grateful to the American Mathematical Society for offering me this rather challenging assignment of reporting on the progress of algebraic topology in the past decade. I should tell you how I will interpret this rather large subject. I will say little or nothing about the applications of algebraic topology. There will be little or nothing about the Atiyah-Singer Index Theorem. There will be little or nothing about the recent dramatic progress in the topology of manifolds stemming from the work of Kirby and Siebenmann. I know that such pieces of work deserve to be recorded in letters of gold; but on the one hand I expect that we will hear all about them at the International Congress of Mathematicians, Nice 1970; and on the other hand, I do not know them well enough to do them justice.

I will be concerned mainly with progress inside algebraic topology; and here I will be more concerned with methods which are likely to be useful in the future than with particular results. I hope that everything I say will come into the class of things which can usefully be remembered by those who have the care of graduate students.

Since algebraic topologists need to apply algebra, I will also mention certain pieces of algebra which may be useful to topologists. If I had been reporting on the progress of the previous decade, 1950–1960, I would surely have had to say something about homological algebra. Similarly for the last decade. So I will begin this lecture on the boundary between algebra and algebraic topology. In my second lecture I will take the theme of generalized homology and cohomology theories. In my third lecture I will discuss various more miscellaneous topics.

What, then, are the pieces of algebra which we have learnt in the past decade, which topologists should remember?

AMS 1970 *subject classifications.* Primary 18F25, 55-02, 55B20, 55E50; Secondary 18G10, 55B45, 55C05, 55E05, 55E10, 55G20, 55H05, 55H15, 55H20.

2. **Algebraic K-theory.** Homological algebra starts from the regrettable fact that not all modules are projective. Similarly, algebraic K-theory starts from the regrettable fact that not all projective modules are free.

This subject began in the previous decade, with a most germinal seminar by J.-P. Serre [1]. In this seminar, Serre exploits an analogy between vector bundles and projective modules. Let X be a compact Hausdorff space. Let $A = R(X)$ be the ring of continuous real-valued functions $f: R \to X$. Let ξ be a real vector-bundle over X. Let $\Gamma(\xi)$ be the set of continuous sections of ξ; then $\Gamma(\xi)$ is a module over A. Moreover, it is a finitely generated projective module; for we can find a complementary bundle η so that $\xi \oplus \eta$ is trivial of dimension n, say $\xi \oplus \eta \cong n$; then we have

$$\Gamma(\xi) \oplus \Gamma(\eta) \cong \Gamma(n) \cong A^n.$$

This construction sets up a (1-1) correspondence between isomorphism classes of vector bundles over X and isomorphism classes of finitely generated projective modules over A. The proofs here were carefully written up by R. G. Swan [2].

By means of this analogy, we can use our knowledge of vector bundles to suggest results, proofs and constructions for projective modules.

For example, suppose given any ring A. We consider functions f which assign to each finitely generated projective A-module P a value $f(P)$. Each function f is supposed to map into some abelian group, and to satisfy the following axioms:

(i) If $P \cong Q$, then $f(P) = f(Q)$.

(ii) $f(P \oplus Q) = f(P) + f(Q)$.

Among such functions f there is one which is universal; the group in which it takes its values is written $K_0(A)$ and called the projective class group of A. If we impose the further axiom

(iii) $f(A) = 0$,

we obtain the reduced group $\tilde{K}_0(A)$.

If we substitute $A = R(X)$, then $K_0(A)$ becomes the group $K^0(X)$ of Grothendieck, Atiyah, and Hirzebruch. However, for other applications we take smaller rings A, and here we rely on the algebraists to compute the group $K_0(A)$ for us. Similar remarks apply to the group $K_1(A)$, when I get so far. For an account of the theorems which allow you to compute these groups, see H. Bass [3].

As an application, I want to cite the work of C. T. C. Wall on finiteness obstructions; see C. T. C. Wall [4], [5].

Let X be a CW-complex, and let A be the integral group ring of its fundamental group, $A = Z(\pi_1(X))$. Let $\tilde{X}$ be the universal cover of X, and let us consider the chain groups $C_*(\tilde{X})$. (It does not matter much what sort of chain groups we use; let us use cellular chains.) Then $C_*(\tilde{X})$ is a chain complex of A-modules. Let us suppose, to begin with, that we can replace $C_*(\tilde{X})$ by another chain complex C', chain-equivalent over A to $C_*(\tilde{X})$, with the following properties.

(i) Each component C'_n is a finitely generated projective module over A.

(ii) The groups $H_n(C')$ and $H^n(C'; M)$ (where M is any A-module) are zero for n sufficiently large, say $n \geqq \nu$.

I will not discuss how to find such a C'; but if we cannot find such a C', then it is not even plausible to suppose that X is equivalent to a finite complex.

Suppose, then, that we have such a C'. By a further reduction, we can obtain a chain complex C chain equivalent over A to C' with the following properties.

(i) Each component C_n is a finitely generated projective module over A.

(ii) $C_n = 0$ for n sufficiently large, say $n \geqq \nu$.

This is very easy; you just take the chain complex C' and chop it off at a suitable point. Observe, however, that even if all the modules C'_n are free (which is the usual case), the last nonzero module C_n need not be free; we can only say that it is projective.

We can now form Wall's invariant

$$\chi(C) = \sum_i (-1)^i [C_i] \in \tilde{K}_0(A).$$

This depends only on the chain-equivalence class of C, and so it depends only on the homotopy type of X.

If X is homotopy-equivalent to a finite complex K then we can take $C = C_*(\tilde{K})$; $C_n(\tilde{K})$ is free, so $\chi(C) = 0$. On the other hand, there exist examples in which $\chi(C) \neq 0$. In this way Wall was able to answer an outstanding problem raised by J. H. C. Whitehead: if X is dominated by a finite CW-complex, does it follow that X is homotopy-equivalent to some finite complex? The answer is "no".

This work probably went along with Wall's work on surgery for non-simply-connected manifolds M. In this case also A is the group ring $Z(\pi_1(M))$. Wall's surgery obstructions take their values in a group G which can be considered as a Grothendieck group. To construct G, one considers projective A-modules P equipped with extra structure maps, such as a bilinear pairing $\varphi: P \times P \to A$. Until the appearance of Wall's book [**6**], the reference is C. T. C. Wall [**7**].

Another sort of module with extra structure is a finitely generated projective module P provided with an automorphism $f: P \to P$. If we think a bit about the properties of the determinant in the classical case, we see that we should consider functions d which assign to each pair (P, f) an element $d(f)$ in some multiplicative abelian group, and satisfy the following axioms.

(i) If

$$P \xrightarrow{f} P \xrightarrow{g} P,$$

then

$$d(gf) = d(g)d(f).$$

(ii) If

$$\begin{array}{ccccccccc} 0 & \longrightarrow & P & \xrightarrow{i} & Q & \xrightarrow{j} & R & \longrightarrow & 0 \\ & & \Big\downarrow f & & \Big\downarrow g & & \Big\downarrow h & & \\ 0 & \longrightarrow & P & \xrightarrow{i} & Q & \xrightarrow{j} & R & \longrightarrow & 0 \end{array}$$

(where the row is exact), then

$$d(g) = d(f)d(h).$$

(The classical case is that

$$\det \begin{bmatrix} B & C \\ 0 & D \end{bmatrix} = \det(B) \cdot \det(D).)$$

Among such functions d there is one which is universal; the group in which it takes its values is written $K_1(A)$.

Suppose that we are given an acyclic chain complex

$$0 \leftarrow C_0 \leftarrow C_1 \leftarrow \cdots \leftarrow C_n \leftarrow 0$$

in which the C_i are finitely generated free A-modules with given bases, and the boundary maps are A-maps. (Data a little weaker than "given bases" would actually suffice.) Then it turns out that we can associate to such an acyclic chain complex an invariant in $K_1(A)$. This way of looking at things has led to a much clearer understanding of Reidermeister-Franz torsion, and of J. H. C. Whitehead's theory of "simple homotopy type". For a good exposition, see J. Milnor [**8**].

The groups $K_i(A)$ for $i \geqq 2$ are so far more interesting to number theorists than to topologists; but for the latest material, see R. G. Swan [**9**].

3. **Derived functors of the inverse-limit functor.** The next piece of algebra which I think deserves comment concerns the derived functors of the inverse-limit functor. Let I be a directed set of indices α; then we can consider inverse systems of abelian groups $\mathbf{G} = \{G_\alpha, g_{\alpha\beta}\}$. These inverse systems form the objects of a category. The inverse limit is representable in this category; for let $\mathbf{Z}$ be the inverse system with $Z_\alpha = Z$, $z_{\alpha\beta} = 1$; then

$$\mathrm{Hom}\,(\mathbf{Z}, \mathbf{G}) \cong \varprojlim_\alpha \mathbf{G}.$$

Moreover, this category is one in which we can do homological algebra; in particular, there are enough injectives. We can therefore form the functors

$$\lim{}^i\, \mathbf{G} = \mathrm{Ext}^i\,(\mathbf{Z}, \mathbf{G}).$$

We have $\lim^0 = \varprojlim_\alpha$.

The use of these functors goes back to J. Milnor [**10**]. Let H^* be a generalized cohomology theory satisfying the wedge axiom, which says that the canonical map

$$\tilde{H}^*\Big(\bigvee_\alpha X_\alpha\Big) \to \prod_\alpha \tilde{H}^*(X_\alpha)$$

is an isomorphism. (One can use H^* instead of $\tilde{H}^*$ if one uses the disjoint union instead of the wedge.) Suppose given an increasing sequence of CW-pairs (X_n, A_n), and set

$$X = \bigcup_n X_n, \qquad A = \bigcup_n A_n.$$

Then Milnor shows that we have an exact sequence

$$0 \to \lim^1 H^{q-1}(X_n; A_n) \to H^q(X; A) \to \lim^0 H^q(X_n; A_n) \to 0.$$

In particular, substituting X for X_n and X_n for A_n, we have

$$\lim^1 H^*(X; X_n) = 0, \qquad \lim^0 H^*(X; X_n) = 0.$$

In this situation, when I is countable, we have $\lim^i \mathbf{G} = 0$ for $i \geqq 2$.

The careful use of these derived functors enables one to overcome many of the difficulties which arise in the use of inverse limits. For example, suppose that we wish to construct a map $\mu: BU \wedge BU \to BU$ corresponding to the tensor product of virtual bundles of virtual dimension zero. Let X_n be an increasing sequence of finite CW-complexes whose union is BU. We easily construct maps $\mu_n: X_n \wedge X_n \to BU$ and verify that they give an element of

$$\lim^0 [X_n \wedge X_n, BU].$$

Since

$$[BU \wedge BU, BU] \to \lim^0 [X_n \wedge X_n, BU]$$

is an epimorphism, we see that there is a map $\mu: BU \wedge BU \to BU$ which restricts to each μ_n (up to homotopy). Now suppose we wish to check the associativity of μ, that is, to check that $\mu(\mu \wedge 1) \sim \mu(1 \wedge \mu): BU \wedge BU \wedge BU \to BU$. We easily check that $\mu(\mu \wedge 1)$ and $\mu(1 \wedge \mu)$ have the same image in

$$\lim^0 [X_n \wedge X_n \wedge X_n, BU].$$

In order to check that $\mu(\mu \wedge 1)$ and $\mu(1 \wedge \mu)$ are homotopic, we simply check that

$$\lim^1 [X_n \wedge X_n \wedge X_n, U] = 0.$$

This "$\lim^1 = 0$" argument, which is commonplace today, was unknown before 1960; I attribute it to John Milnor.

For some applications it is important to know how inverse limits work in spectral sequences. For the first examples, see D. W. Anderson's thesis [**11**] and M. F. Atiyah [**12**]. Suppose, for example, that we take a generalized cohomology theory H satisfying the wedge axiom and a CW-complex X containing an increasing sequence of subcomplexes $\emptyset = X_{-1} \subset X_0 \subset X_1 \subset \cdots \subset X_n \subset \cdots \subset X$. Suppose also that

$$\lim^0 H^*(X; X_n) = 0, \qquad \lim^1 H^*(X; X_n) = 0.$$

(For example, we might have $X = \bigcup_n X_n$.) Applying H^*, we obtain a half-plane spectral sequence with

$$E_1^{p,q} = H^{p+q}(X_p; X_{p-1}).$$

In what sense does this spectral sequence converge? We may be interested in three conditions.

(i) Observe that $E_{r+1}^{p,q} \to E_r^{p,q}$ is monomorphic for $r > p$. So we can ask that the map

$$E_\infty^{p,q} \to \lim_r^0 E_r^{p,q}$$

should be an isomorphism.

(ii) Similarly, we can ask that

$$\lim_r^1 E_r^{p,q} = 0.$$

(iii) Let $F^{p,q}$ be the filtration quotients of $H^{p+q}(X)$, so that we have exact sequences

$$0 \to E_\infty^{p,q} \to F^{p,q} \to F^{p-1,q+1} \to 0$$

and $F^{-1,q} = 0$. We can ask that the map

$$H^n(X) \to \lim_p^0 F^{p,n-p}$$

should be an isomorphism.

THEOREM 3.1. *Condition* (ii) *is equivalent to* (i) *plus* (iii).

In practice we verify condition (ii); it is equivalent to check that $E_{p+1}^{p,q}$ is complete for the topology defined by the subgroups $E_r^{p,q}$, $r > p$. (Here we use words so that "complete" does not imply "Hausdorff"; it means that each Cauchy sequence has a limit, perhaps not unique.) We then use the theorem to deduce that conditions (i) and (iii) hold.

This result is stronger than previous results which rely on the Mittag-Leffler condition for spectral sequences. I have taken most of it from a seminar by J. M. Boardman, but it may also be included in the work of Eilenberg and Moore [**13**].

We can also generalize the result of Milnor. For convenience I consider the absolute case. Let X be any CW-complex which is the union of a directed set of subcomplexes X_α. Then we have a spectral sequence

$$\lim_\alpha^p H^q(X_\alpha) \underset{p}{\Rightarrow} H^{p+q}(X).$$

And if you want to know in what sense this spectral sequence is convergent, it is convergent in the sense that Theorem 3.1 holds.

It has been claimed by Jensen that there is an example of an inverse system $\mathbf{G}$ of abelian groups such that $\lim^i \mathbf{G} \neq 0$ for all $i > 0$, but I have not seen the proof. The reference is C. U. Jensen [**14**].

4. **Coalgebraic structures.** The next piece of progress which I think deserves comment is the increasing use of coalgebraic structures. Suppose I have in sight some ground ring R. Then an algebra A over R is an R-module provided with an R-linear product map $\varphi: A \otimes_R A \to A$. Of course φ has to satisfy suitable axioms. Similarly, an A-module M is an R-module provided with an R-linear product map $\varphi': A \otimes_R M \to M$. Again, φ' has to satisfy suitable axioms.

Dually, a coalgebra C over R is an R-module provided with an R-linear coproduct map $\psi : C \to C \otimes_R C$. Of course ψ has to satisfy suitable axioms. Similarly, a C-comodule N is an R-module provided with an R-linear map $\psi' : N \to C \otimes_R M$. Again, ψ' has to satisfy suitable axioms.

These definitions, of course, were current long before 1960. I was writing about the cobar construction in 1956, and coalgebras were not new then. But I do claim that in the past decade people have become increasingly used to thinking in coalgebraic terms. Probably topologists were much influenced by the work of Milnor and Moore [**15**].

However, I want to illustrate the point from the homology of fibre spaces. At the end of the last decade it appeared that if you wanted to do homological calculations in fiberings, and if you wanted to work at the chain level, so as to get results more precise than those afforded by the Serre spectral sequence, then the best prospect for the future was some form of twisted tensor product; see E. H. Brown [**17**].

Of course one had the work of Cartan on "constructions", from the previous decade; and it was not long before Eilenberg put matters in the following form. The ordinary theory of Tor and Ext refers to modules over a ring A. However, it does not take much trouble to replace the ring A by a differential graded algebra. Let G be (say) a topological group, and BG its classifying space; then we have

$$H_*(BG; R) \cong \mathrm{Tor}_*^{C_*(G;R)}(R, R), \qquad H^*(BG; R) \cong \mathrm{Ext}^*_{C_*(G;R)}(R, R).$$

It follows that we have spectral sequences

$$\mathrm{Tor}_*^{H_*(G;R)}(R, R) \Rightarrow H_*(BG; R), \qquad \mathrm{Ext}^*_{H_*(G;R)}(R, R) \Rightarrow H^*(BG; R).$$

The canonical joke at this point is that one defines these spectral sequences by filtering according to the number of bars used by Eilenberg and Mac Lane in their bar construction.

These spectral sequences were obtained in a much more geometrical way by Rothenberg and Steenrod [**18**]. Their work remains usable when one wishes to replace ordinary homology by generalized homology.

So far we have been talking about going from G to BG. The converse problem is to go from X to ΩX. The results, at any rate, can be formulated by turning the arrows around; the reference is S. Eilenberg and J. C. Moore [**19**].

First we dualize the definition of the tensor product. Suppose that A is a differential graded algebra, and L, M are differential graded algebras over A, L being a right module and M a left module. Then the definition of $L \otimes_A M$ says that the following sequence is exact:

$$L \otimes A \otimes M \xrightarrow{\varphi\otimes 1 - 1\otimes\varphi} L \otimes M \to L \otimes_A M \to 0.$$

(Here $\otimes$ means $\otimes_R$.) Now let C be a differential graded coalgebra, and let L, M be differential graded comodules over C, L being a right comodule and M a left comodule. We define the cotensor product $\square_C$ so that the following sequence is exact:

$$0 \to L \,\square_C\, M \to L \otimes M \xrightarrow{\psi\otimes 1 - 1 \otimes \psi} L \otimes C \otimes M.$$

The cotensor product has derived functors, and they are written Cotor.

Now suppose given the following diagram.

$$\begin{array}{ccc} E' & \longrightarrow & E \\ {\scriptstyle \pi'}\downarrow & & \downarrow{\scriptstyle \pi} \\ B' & \xrightarrow{\;b\;} & B \end{array}$$

We assume that π is a fibering and that π' is the fibering induced by b. In particular, if B' is a point, then E' will be the fibre F of π. Then $C_*(B; R)$ can be made into a coalgebra, and $C_*(B'; R)$, $C_*(E; R)$ can be made into comodules over this coalgebra. One theorem of Eilenberg and Moore states that under mild restrictions we have

$$H_*(E'; R) \cong \mathrm{Cotor}_*^{C_*(B;R)}(C_*(B'; R), C_*(E; R)).$$

With this viewpoint, the cobar construction is regarded as a canonical resolution for computing Cotor.

Under suitable assumptions, such as the assumption that R is a field, we get a spectral sequence

$$\mathrm{Cotor}_*^{H_*(B;R)}(H_*(B'; R), H_*(E; R)) \Rightarrow H_*(E'; R).$$

(Filter according to the number of cobars used by Coeilenberg, etc.) This is the Eilenberg-Moore spectral sequence, which has already found applications.

5. **Generalized homology and cohomology theories.** A generalized homology or cohomology theory is a functor which satisfies the first six axioms of Eilenberg and Steenrod, but does not necessarily satisfy the seventh axiom, the dimension axiom. As long as we are dealing with finite complexes these axioms suffice. If we have to deal with infinite complexes we have to impose further axioms. In homology, it is appropriate to assume that the obvious map

$$\operatorname*{dir\,lim}_{\alpha} H_n(X_\alpha) \to H_n(X)$$

is an isomorphism; here X_α runs are the finite subcomplexes of X. In cohomology we assume the wedge axiom: the obvious map

$$\tilde{H}^n\Big(\bigvee_\alpha X_\alpha\Big) \to \prod_\alpha \tilde{H}^n(X_\alpha)$$

is an isomorphism.

In the past decade we have seen a lot of progress in this area. It may be that the progress did not come earlier because we did not have convincing examples of such functors; for the beginning of K-theory, see Atiyah and Hirzebruch [**20**]. But by now we have several sorts of K-theory and several sorts of bordism or cobordism, and it is clear that they work; for example, they allow you to solve geometric problems.

The first theorem in this subject is E. H. Brown's Representability Theorem (see E. H. Brown [21], [22]). Suppose given a contravariant functor H. It should be defined on the category in which the objects are CW-complexes with base-point, and the morphisms are homotopy classes of maps; both maps and homotopies are to preserve the base-point. H should take values in the category of sets. Then Brown's Theorem gives very simple necessary and sufficient conditions in order that we have a natural isomorphism $H(X) \cong [X, Y]$ for some fixed CW-complex Y. The first condition is the wedge axiom: the obvious map $H(\bigvee_\alpha X_\alpha) \to \prod_\alpha H(X_\alpha)$ is an isomorphism of sets. The second condition is the Mayer-Vietoris axiom. Consider the following diagram.

$$\begin{array}{ccc} H(U \cup V) & \longrightarrow & H(U) \\ \downarrow & & \downarrow \\ H(V) & \longrightarrow & H(U \cap V) \end{array}$$

We ask that if $u \in H(U)$, $v \in H(V)$ have the same image in $H(U \cap V)$, then they should be the images of a single class $w \in H(U \cup V)$.

Brown's Theorem, then, says that if a functor $H(X)$ satisfies these two conditions, then it has the form $[X, Y]$ up to natural isomorphism. This is a very useful theorem, because we can use it to construct CW-complexes Y.

(i) For the most trivial example, suppose given any space Z and consider the functor $[X, Z]$; we get a CW-complex Y and a natural isomorphism $[X, Y] \cong [X, Z]$. That is, we get a CW-complex weakly equivalent to Z.

(ii) Take $H(X)$ to be the ordinary homology group $\tilde{H}^n(X; \pi)$. We obtain a complex Y, and we easily check that we have obtained an Eilenberg-Mac Lane complex of type (π, n); for

$$\pi_r(Y) = [S^n, Y] \cong \tilde{H}^n(S^r; \pi) = \begin{cases} \pi & (n = r), \\ 0 & (n \neq r). \end{cases}$$

(iii) Suppose given a topological group G. We want to construct the classifying space BG. So we define a functor $H(X)$ by taking the G-bundles over X and classifying them into isomorphism classes. (Actually one has to modify this definition a little to take account of the base-point, but this is just a technicality.) We check that $H(X)$ satisfies the axioms, and we receive a CW-complex Y with the property that G-bundles over X are classified by maps $f: X \to Y$.

It may sometimes happen that one's functor H is defined not on all CW-complexes, but only on some subcategory. For example, suppose we define the Grothendieck-Atiyah-Hirzebruch functor $\tilde{K}(X)$ in terms of vector-bundles, in the classical way. This definition is good only for finite-dimensional complexes X; if we tried to use it for infinite-dimensional complexes, it would not satisfy the wedge axiom. All the same, the theorem of E. H. Brown remains true if the functor H is given only on finite-dimensional complexes. Of course, the complex Y which is constructed will not be finite dimensional, in general.

Similarly, suppose that the functor H is given only on finite CW-complexes. In this case I will assume that the functor H takes values in the category of groups. If so, then E. H. Brown's Theorem is still true. (Again, the complex Y which is constructed will not be a finite complex, in general.)

Suppose now that we have a generalized cohomology theory H. Then we have a sequence of functors $\tilde{H}^n$ satisfying the conditions of Brown's Theorem. So we get a sequence of representing complexes Y_n such that $\tilde{H}^n(X) \cong [X, Y_n]$. Also we have isomorphisms $\tilde{H}^n(X) \cong \tilde{H}^{n+1}(SX)$; these correspond to weak equivalences $Y_n \simeq \Omega Y_{n+1}$; by taking the adjoint we obtain maps $SY_n \to Y_{n+1}$.

In this way we reach the notion of a *spectrum*, first published by Lima in the previous decade. According to G. W. Whitehead, a spectrum is a sequence of CW-complexes E_n provided with maps $e_n: SE_n \to E_{n+1}$.

EXAMPLES. (i) The Eilenberg-Mac Lane spectrum, in which $E_n = K(\Pi, n)$.

(ii) The sphere spectrum, in which $E_n = S^n$.

(iii) The MU-spectrum of Thom and Milnor, in which $E_{2n} = MU(n)$.

(iv) The BU-spectrum, in which E_{2n} is the space BU.

Now we come to the work of G. W. Whitehead [**23**]. According to G. W. Whitehead, a spectrum determines not only a generalized cohomology theory, but also a generalized homology theory. The definition is

$$\tilde{E}_n(X) = \lim_{n \to \infty} \pi_{m+n}(E_m \wedge X).$$

For example, taking $E = MU$ we obtain a homology theory called complex bordism. It was already known that bordism theories were important; see Atiyah [**24**].

THEOREM 5.1 *Every generalized homology theory defined on* CW-*complexes can be obtained by G. W. Whitehead's construction from a suitable spectrum E.*

It is convenient to use the same letter for a spectrum and for its associated homology and cohomology theory. For this reason I will use K for the BU-spectrum and KO for the BO-spectrum.

Now I want to go on and point out that it is highly desirable to have a good category of spectra in which we can work and do stable homotopy theory. For one thing, we would like to make the following definition for the generalized cohomology of a spectrum:

$$E^n(X) = [S^{-n}X, E].$$

Here $S^{-n}X$ means a suitable desuspension of X, and $[A, B]$ means homotopy classes of maps from A to B in our category of spectra. Unfortunately we cannot do this if we use G. W. Whitehead's definition of a "map of spectra".

Again, it would be very convenient if we had smash products of spectra in our category. For example, when G. W. Whitehead considers products in generalized cohomology theories, he has to introduce the notion of a pairing of spectra from E and F to G. It would be very convenient if this were simply a map $\mu: E \wedge F \to G$ in our category.

Let me give an example to show how easy life is when you have the right formalism.

PROPOSITION 5.2. *Assume that* E *is a ring-spectrum,* F *is a module-spectrum over* E, *and* $E_*(X) = 0$. *Then* $F^*(X) = 0$.

This is both a special case of the Universal Coefficient Theorem and a lemma for use in setting up the general case.

PROOF. Take any map $f: X \to F$. We can factor it in the following way.

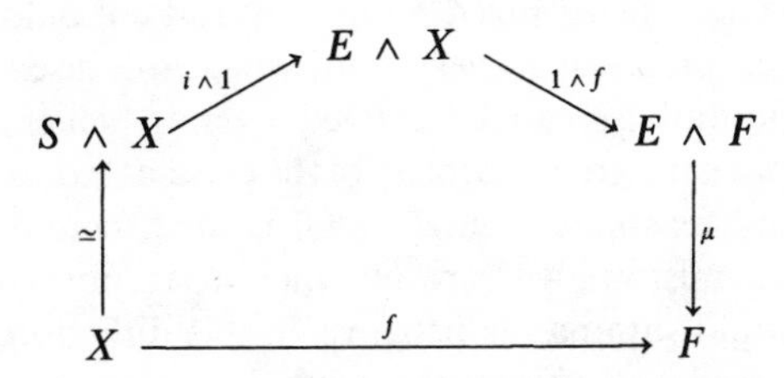

(Here S is the sphere-spectrum, which acts as a unit for the smash-product; and $i: S \to E$ is the unit map for the ring-spectrum E.) We are given $\pi_*(E \wedge X) = 0$, so $E \wedge X$ is contractible and $f \sim 0$. Similarly with X replaced by $S^{-n}X$.

This proof is certainly very easy; but you try doing it without a category in which you can make this argument.

The good category in which to do stable homotopy theory was constructed by Boardman. We all hope that he will publish a readable and definitive account of it; till then, I recommend the exposition by R. Vogt [**25**].

Let me give a quick description of a category equivalent to Boardman's. I restrict attention to spectra E_n in which each map $e_n: SE_n \to E_{n+1}$ is an isomorphism from the complex SE_n to a subcomplex of E_{n+1}. This is no great restriction. On occasion we neglect the maps e_n and regard S^rE_n as embedded in E_{n+r}. A subspectrum $E' \subset E$ is said to be *cofinal* if for each cell $e \subset E_n$ there exists an r such that $S^re \subset E'_{n+r}$. That is, each cell of E gets into E' after enough suspensions.

The maps which we now take from E to F are sequences $f_n: E'_n \to F_n$, where E'_n is a cofinal subspectrum of E and the following diagram is strictly commutative for each n.

$$\begin{array}{ccc} SE'_n & \longrightarrow & E'_{n+1} \\ {\scriptstyle Sf_n}\downarrow & & \downarrow{\scriptstyle f_{n+1}} \\ SF_n & \longrightarrow & F_{n+1} \end{array}$$

An example will illustrate the reason we take this definition. Take two spectra with

$$E_n = S^{n+3} \vee S^{n+7} \vee S^{n+11} \vee \cdots, \qquad F_n = S^n.$$

We would like to make a map from E to F whose component from S^{n+4k-1} to S^n is a generator for the image of J in the stable $(4k - 1)$-stem. But there is no single value of n for which all the requisite maps exist as maps into S^n; we have to concede that for the different cells of E the maps come into existence for different values of n. In other words, the maps exist on a cofinal subspectrum of E. The slogan is "cells now—maps later".

The notion of homotopy is now obvious; a homotopy has to be given on a cofinal subspectrum of the spectrum $\{E_n \wedge (I/\emptyset)\}$.

This gives a category which has all the good properties of the category of CW-complexes.

Let me come back to generalized homology and cohomology. It is now more or less accepted that practically everything which one used to do with ordinary homology and cohomology can be carried over to generalized homology and cohomology. We have the great majority of the classical tools, and we can calculate as well as one could reasonably expect. Of course, before you calculate $E_*(X)$ for any other X, you must know the result when X is a point, that is, the coefficient groups $\pi_*(E)$. I should summarize progress in this direction.

For real and complex K-theory the coefficient groups are determined by the Bott periodicity theorem. I should perhaps cite the proof of this theorem by Atiyah and Bott **[26]**. We have also the generalization to the real case by R. Wood **[27]**, and its subsequent further generalization by Atiyah **[28]**, and by M. Karoubi **[29]**.

For the various bordism theories, $\pi_*(MO)$ and $\pi_*(MSO)$ were computed before the decade started. So was $\pi_*(MU)$, although Milnor did not immediately publish his paper **[30]**. A very firm hold on $\pi_*(MU)$ was provided by the Hattori-Stong theorem. See R. Stong **[31]**, Hattori **[32]**. Although this theorem was originally presented as determining the image of $\pi_*(MU)$ in $H_*(MU)$, it seems now to be accepted that it is best stated in terms of $K_*(MU)$. Further enlightenment was provided by the work of Quillen **[33]**. I have tried to give an exposition in my lecture notes, University of Chicago, 1970. On SU-bordism and spin-bordism I refer you to the following sources: Conner and Floyd **[34]**; Anderson, Brown, and Peterson **[35]**, **[36]**; Wall **[37]**. The structure of $\pi_*(MSp)$ is still not known.

Let me now go on to talk about products. The basic products are defined in G. W. Whitehead's paper, cited above; they are four in number.

(i) An external product in cohomology, say

$$u \otimes v \to u \barwedge v : E^p(X) \otimes F^q(Y) \to (E \wedge F)^{p+q}(X \wedge Y).$$

Of course, if you have a pairing of spectra, that is, a map of spectra $E \wedge F \to G$, you can apply the resulting induced homomorphism to obtain an answer in $G^{p+q}(X \wedge Y)$.

(ii) An external product in homology,

$$u \otimes v \to u \,\underline{\wedge}\, v : E_p(X) \otimes F_q(Y) \to (E \wedge F)_{p+q}(X \wedge Y).$$

(iii) and (iv). Two slant products

$$u \otimes v \to u/v : E^p(X \wedge Y) \otimes F_q(Y) \to (E \wedge F)^{p-q}(X),$$
$$u \otimes v \to u\backslash v : E^p(X) \otimes F_q(X \wedge Y) \to (E \wedge F)_{q-p}(Y).$$

Note. If you wish to make the notation convenient, give a fraction the same variance as its numerator, the opposite variance from its denominator.

The definitions of the products are all easy if one can work in a suitable category.

Apart from obvious naturality properties, the products satisfy two anti-commutative laws and eight associative laws and have a unit. The external products are associative and anticommutative. The remaining six associative laws are easily interpreted as obvious rules for manipulating fractions; for example,

$$u \mathbin{\overline{\wedge}} (v/w) = (u \mathbin{\overline{\wedge}} v)/w.$$

Unfortunately the eight associative laws are not given in G. W. Whitehead's paper; any student of the subject should certainly make a list of them. For the case of ordinary homology and cohomology, six out of the eight can be found scattered through the pages of Spanier's book.

We can now summarize the position about duality in manifolds. Let E be a ring-spectrum, and suppose that M is a manifold for which you have chosen an orientation class in, say, E-cohomology of the tangent bundle. Then the account of duality in Spanier's book goes over word for word; the statements are all true, and the proofs are formally the same once you have the toolkit.

What has to be done for any particular E, then, is to construct good orientations for some useful class of bundles. This can be nontrivial. For example, consider Atiyah, Bott, and Shapiro [**38**]. In this paper they construct a KO-orientation for Spin-bundles. They take particular care that their orientation behaves well on a bundle which happens to be the Whitney sum of two others; in my terminology, they take care to construct a map of ring-spectra

$$M\,\mathrm{Spin} \to KO.$$

One can also orient U-bundles over complex K-theory, and of course G-bundles over MG for any of the usual choices for G (take the identity map $MG \to MG$).

I understand from Wall that Sullivan has shown that if you ignore the prime 2, then PL-bundles can be oriented over KO; but I myself cannot answer for the statement or the proof.

I would now like to cite the theorem of Conner and Floyd [**39**]. I just said that you can orient U-bundles over K. More precisely, we have a map of ring-spectra from MU to K. This leads to a remarkably simple sort of Universal Coefficient Theorem; we have an isomorphism

$$MU_*(X) \otimes_{\pi_*(MU)} \pi_*(K) \xrightarrow{\cong} K_*(X).$$

Similarly for

$$MSp_*(X) \underset{\pi_*(MSp)}{\otimes} \pi_*(KO) \xrightarrow{\cong} KO_*(X).$$

The theory of characteristic classes with values in a generalized cohomology theory is now fairly well understood, at least if you stick to any of the usual sorts of bundle and any of the usual cohomology theories. For example, take $U(n)$ bundles; the Chern classes in ordinary cohomology have analogues in complex K-theory (Grothendieck, Atiyah) and in complex cobordism (see the work by Conner and Floyd cited above).

The theory of formulae of "Riemann-Roch type", involving a multiplicative factor, is also well understood; indeed it goes back to the previous decade. The most elementary situation here is as follows. Take a bundle ξ (over X, say) which has been oriented over two theories E^*, F^*; and suppose given a map of ring-spectra $\alpha: E \to F$. Then we have the following diagram, but it is not commutative.

$$\begin{array}{ccc} \tilde{E}^*(X^\xi) & \xrightarrow{\alpha} & \tilde{F}^*(X^\xi) \\ \uparrow{\scriptstyle \varphi_E} & & \uparrow{\scriptstyle \varphi_F} \\ E^*(X) & \xrightarrow{\alpha} & F^*(X) \end{array}$$

(Here X^ξ means the Thom complex of ξ, and φ_E, φ_F are Thom isomorphisms.) We have

$$\varphi_F^{-1}\alpha\varphi_E x = \alpha x(\varphi_F^{-1}\alpha\varphi_E 1).$$

Here $\varphi_F^{-1}\alpha\varphi_E 1$ is a characteristic class of ξ, which can be evaluated in practical cases. Similarly if we study manifolds and consider duality isomorphisms (instead of φ), homomorphisms induced by maps of manifolds (instead of α) or the "Umkehrungshomomorphismus".

For the more interesting applications of generalized homology and cohomology we require cohomology operations in E-cohomology, or something similar. Here one can obviously consider $E^*(E)$, which is the algebra of stable cohomology operations on E-cohomology. If we take E to be the Eilenberg-Mac Lane spectrum $K(Z_p)$, we get the Steenrod algebra. We can also form $E_*(E)$. If we take E to be the Eilenberg-Mac Lane spectrum $K(Z_p)$, we get the dual of the Steenrod algebra, which was studied by Milnor. I have shown that under mild restrictions on E, $E_*(E)$ is a Hopf algebra in a good sense, and $E_*(X)$ is a comodule over $E_*(E)$ **[40]**.

EXAMPLES. (i) $E = K$ and KO; real and complex K-theory. $E^*(E)$ is very bad. There are however unstable operations in K-theory. Atiyah has related them to Steenrod's method for constructing operations in ordinary cohomology: see Atiyah **[41]**. In this case $E_*(E)$ is very good. It has been computed by Adams, Harris, and Switzer **[42]**.

(ii) $E = MU$. $E_*(E)$ and $E^*(E)$ are strictly dual. They are both very good, and have been computed by Novikov **[43]**. The Steenrod approach was carried out by tom Dieck **[44]**.

(iii) $E = BP$, the Brown-Peterson spectrum. $E_*(E)$ and $E^*(E)$ are strictly dual. They are both very good. $E^*(E)$ has been computed by Quillen [**33**]. The interpretation in terms of $E_*(E)$ is given in my lecture notes, Chicago 1970.

(iv) $E = MSp$. The analogues of Novikov's results on MU are true, but until we know $\pi_*(MSp)$ they are not so much use.

(v) $E = bu$, connective K-theory. Some progress both on $E_*(E)$ and on cohomology operations with suitable coefficients has been made by D. W. Anderson.

I have left myself with little space to mention applications. We have had quite a lot of fun with K-theory; I suspect that the quick profits have probably been reaped, except maybe for some of the fancy forms of K-theory, like the equivariant K-theory of Atiyah and Segal. As for bordism functors like complex bordism, Conner and Floyd have already showed us that it can be very profitable in studying manifolds. I think some of the recent work of Larry Smith, Toda, and others on the spaces $V(n)$ shows that it will probably soon be profitable in homotopy theory.

6. **The groups $J(X)$.** Historically, the subject to be considered in this section arose from two sources. The first was the study of the stable J-homomorphism

$$J: \pi_r(SO) \to \pi_{n+r}(S^n) \qquad (n > r + 1);$$

this is of interest to homotopy-theorists, and also to differential geometers. The second was a **method** which occurred to Atiyah for proving some results of I. M. James about the problem of vector-fields on spheres. This led to Atiyah's celebrated paper [**45**].

Let us take a finite CW-complex X, say connected. We can take the real vector-bundles over X, and proceeding in the usual way we obtain the Grothendieck group $KO(X)$. We can also define a quotient group of $KO(X)$ by considering an equivalence relation between bundles which is cruder than isomorphism, **namely,** fibre homotopy equivalence of the associated sphere bundles. This quotient group is called $J(X)$. Equivalently, $J(X)$ is the image of the following homomorphism:

$$[X, Z \times BO] \to [X, Z \times BG].$$

Here BG is the classifying space for spherical fibrations. This homomorphism is of some interest to differential topologists.

For the usual reason, we have $J(X) \cong Z + \tilde{J}(X)$. One of the central observations in Atiyah's paper was the fact that $\tilde{J}(X)$ is a finite group. If we substitute $X = S^{r+1}$, then $\tilde{J}(S^{r+1})$ is the image of the stable J-homomorphism

$$J: \pi_r(SO) \to \pi_{n+r}(S^n) \qquad (n > r + 1).$$

In order to compute $J(X)$, we must do two things. First, if two sphere-bundles are not fibre homotopy equivalent, we must construct invariants to prove it. The best-known fibre homotopy invariants are the Stiefel-Whitney classes. The reason they are invariant is as follows. Take an $(n - 1)$-sphere bundle E_0 over X, and regard it as the boundary of the associated disc bundle E. We have the following diagram.

$$\begin{array}{ccc} H^n(E, E_0; Z_2) & \xrightarrow{Sq^i} & H^{n+i}(E, E_0; Z_2) \\ \uparrow \varphi & & \uparrow \varphi \\ H^0(X; Z_2) & & H^i(X; Z_2) \end{array}$$

Here φ is the Thom isomorphism, and the Stiefel-Whitney class w_i is given by $w_i = \varphi^{-1}Sq^i\varphi 1$. It is clear that a fibre homotopy equivalence will give a homotopy equivalence from the pair E, E_0 to another such pair, and this will not alter anything. So w_i is a fibre homotopy invariant.

We can copy this argument, replacing $H^*(X; Z_2)$ by $KO^*(X)$ and Sq^i by the operation Ψ^k. However the proof of invariance needs to be changed a little, because a fibre homotopy equivalence $E, E_0 \to E', E'_0$ will generally carry the orientation class $\varphi'1$ not into $\varphi 1$, but into some other class. We have to allow for this in formulating our invariant. In fact, we have to consider the vector $\{\varphi^{-1}\Psi^k\varphi 1\}$ not as an element of $\prod_k KO(X)$, but as an element of a suitable quotient.

We can now define a quotient $J'(X)$ of $KO(X)$ as follows: two bundles ξ and η are to become equal in $J'(X)$ if the fibre homotopy invariant indicated above takes the same value on ξ as on η. It follows immediately that the quotient map $KO(X) \to J'(X)$ factors through $J(X)$; we have the following commutative diagram.

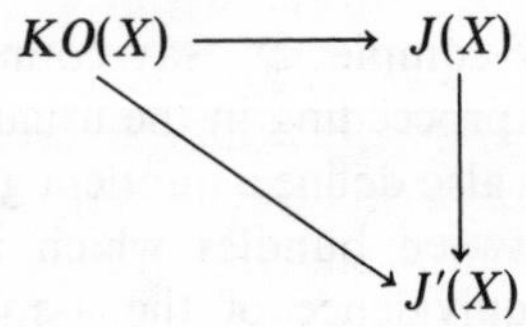

The second thing we have to do to compute $J(X)$ is as follows. If two sphere-bundles are fibre homotopy equivalent, we must give a method for proving it. For this purpose I got interested some time ago in the following formulation.

(6.1) Suppose given a finite complex X, an element $x \in KO(X)$ and an integer k. Then there is an integer n such that $k^n(\Psi^k - 1)x$ maps to zero in $J(X)$.

I proved a few simple special cases of this statement, and explored its relation to other statements. Everybody now seems agreed to call it the Adams conjecture, so I suppose I had better go along with that.

The first piece of progress with it was that Quillen found a line of argument, using algebraic geometry, which would have been a proof if the algebraic geometers had proved one theorem which he needed; unfortunately they had not. The reference is Quillen [**46**]. Sullivan also did a good deal of work on the problem, but I am not too well informed about it. He may well have a proof. Finally Quillen came up with a proof of the Adams conjecture which is now circulating in preprint form. Atiyah has given Quillen's proof in lectures at Oxford. It now seems safe to assume that Quillen's proof is correct, so we can now make use of the Adams conjecture (6.1).

The obvious way to make use of it is to define a group $J''(X)$ so that (6.1) provides us with the following commutative diagram.

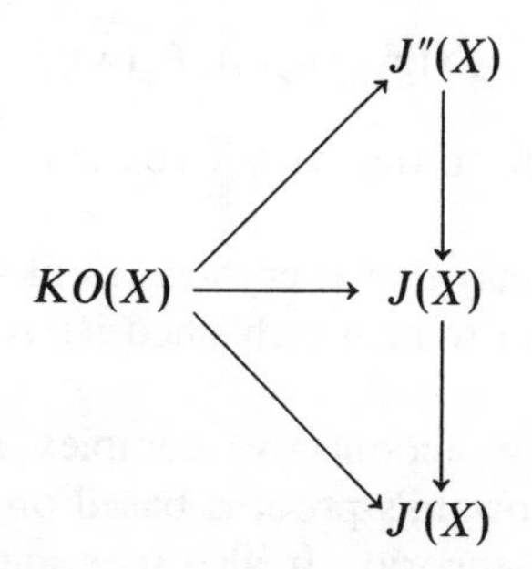

But I had already proved that the quotient map $J''(X) \to J'(X)$ is an isomorphism; see [**47**]. So we have $J''(X) = J(X) = J'(X)$, and we can compute $J(X)$.

For example, the proof of the Adams conjecture should allow one to compute $J(QP^n)$, and so settle the problem of the cross-sections of the symplectic Stiefel fiberings.

As a more classical example, it follows that the image of the stable J-homomorphism

$$J : \pi_r(SO) \to \pi_{n+r}(S^n)$$

has precisely the order suggested by the work of Milnor and Kervaire [**48**], as improved by Atiyah and Hirzebruch [**20**].

I should also mention that if we restrict attention to $X = S^{r+1}$, then there is another proof in the pipeline about the behaviour of the classical stable J-homomorphism. It comes from Mark Mahowald, and I will sketch some of the ideas behind it.

If we want to compute the stable homotopy groups $\pi_*^S(X)$, or at least their p-components, we have a spectral sequence

$$\mathrm{Ext}_{A_*}^{s,t}(Z_p, \tilde{H}_*(X; Z_p)) \underset{s}{\Rightarrow} \pi_{t-s}^S(X).$$

Here Z_p and $\tilde{H}_*(X; Z_p)$ are considered as comodules over the coalgebra A_* dual to the Steenrod algebra. We would like to construct an analogous spectral sequence in which ordinary homology theory $\tilde{H}_*(X; Z_p)$ is replaced by some generalized homology theory. So we take a ring-spectrum E, and form cofibre sequences by induction:

$$X = X_0 \simeq S \wedge X_0 \xrightarrow{i \wedge 1} E \wedge X_0 \longrightarrow X_1,$$

$$X_1 \simeq S \wedge X_1 \xrightarrow{i \wedge 1} E \wedge X_1 \longrightarrow X_2, \quad \text{etc.}$$

Mapping the sphere-spectrum S into these X_{i-1} we obtain a spectral sequence. It is easy to prove that it converges to $\pi_*^S(X)$ if we assume suitable hypotheses on

E; for example, assume

$$\pi_r(E) = 0 \quad \text{for } r < 0, \qquad H_0(E) = Z, \qquad H_1(E) = 0.$$

If $E_*(E)$ is flat over $\pi_*(E)$ then the E_2 term of the spectral sequence is

$$\mathrm{Ext}^{s,t}_{E_*(E)}(E_*(S), E_*(X));$$

but you do not have to stick to this case if you can compute the E_2 term or the E_1 term some other way.

This construction is so easy that everybody thinks it's his idea. In particular, I think it is my idea. I even think I published it; see "Lectures on generalized cohomology" **[40]**.

The case $X = S$, $E = bu$ (connective complex K-theory) has been studied by D. W. Anderson. Mahowald's proof is based on a study of the case $X = S$, $E = bo$ (connective real K-theory). It also uses ideas from his previous paper **[49]**, in which he proved the result up to the dimension 2^{12}. I understand that an announcement by Mahowald is to appear in the Bulletin of the American Mathematical Society **[50]**.

7. **Stable homotopy theory.** The material on the J-homomorphism leads me on to stable homotopy theory.

Well, now, some topologists do have a certain love-hate relationship with the stable homotopy groups of spheres. By using their preferred methods they have, in the past decade, about doubled the dimensional range in which they are intimate with these creatures. However, the philosophy does seem to be gaining ground that for real interest one should try for theorems which give an infinite amount of information. I would like to refer to one method which seems new and worthwhile. It was first published by Toda for the case of an odd prime p; see **[51]**, **[52]**. He has used it to settle some quite subtle questions in stable homotopy theory.

I will describe the construction for the case $p = 2$; the generalization to the case of other primes is fairly obvious. Let X be, say, a finite CW-complex with base-point x_0. Form

$$\frac{X \times X \times S^n}{(x_0 \times X \times S^n) \cup (X \times x_0 \times S^n)}$$

and let Z_2 act on it by

$$(x, y, s) \to (y, x, -s).$$

Let the quotient be $Q^n(X)$.

If $X = S^r$ we can work out what $Q^n(X)$ is; it is obtained by suspending a suitable quotient of real projective spaces RP^a/RP^b.

The functor Q^n does not send cofiberings into cofiberings; it is not "linear", but "quadratic". Nevertheless, the deviation from linearity is small, and if X is a small complex like $S^r \cup_f e^t$ we can work out a good deal about the cell-structure of $Q^n(X)$. This leads to relations between elements of homotopy groups, suggests the construction of secondary homotopy operations, and is otherwise useful.

For an application which I think is quite worthwhile, see D. S. Kahn [**53**].

8. **Unstable homotopy theory.** Here I would like to refer to various points. There has been some work on unstable analogues of the Adams spectral sequence. First we had the work of Massey and Peterson [**54**]. Their method does impose certain restrictions on X, but these are satisfied if $X = S^n$. Then we had a good deal of work by the MIT school, using CSS methods; see in particular, Bousfield, Curtis, Kan, Quillen, Rector, and Schlesinger [**55**], and Rector [**56**].

I would also like to cite the paper by Boardman and Steer [**57**]. It seems to me that this paper says the last word on the formula which replaces the distributive law for $(f + g)h$.

I would also like to mention the work of Barratt and Mahowald [**58**]. This is a very clear and satisfactory result, but most people seem to feel that they do not quite know how to fit it into their general picture of homotopy theory.

To end this section, I would like to mention the work of Mahowald [**59**]. Certainly the results here contain more than a finite amount of information. However, the methods lead me on to the subject of cohomology operations, which is the next topic.

9. **Secondary operations in ordinary cohomology.** There are some questions in homotopy theory which one naturally reduces to questions about secondary operations, either by obstruction theory or by the method of killing homotopy groups. If we dismiss the questions which we can easily answer, we are left with various more subtle questions. For example, one may want to know a Cartan formula for $\Phi(xy)$, including those terms which involve primary operations but no secondary operations. Or one may know that a secondary operation Φ applied in a particularly low dimension gives the same result as some primary operation, and then one wants to know which primary operation it is.

For such questions the most powerful method and the most precise information is contained in the papers of Kristensen. See Kristensen [**60**], [**61**], [**63**]; Kristensen and Madsen [**62**].

Outsiders should be warned, however, that the method is hard work. Although painless general methods like acyclic models are used whenever possible, the basic method is one which relies on definite cochain operations.

10. **Iterated loop-spaces and infinite loop-spaces.** One of the basic papers here, of course, came in the previous decade: Kudo and Araki [**64**]. Kudo and Araki set up the basic operations in homology with mod 2 coefficients. Early in this decade we had the corresponding operations in homology with mod p coefficients: Dyer and Lashof [**65**]. In general, however, the subject was unjustly neglected until later in the decade. Then we had some very successful work on G and BG, the classifying space for spherical fibrations. Some of the references are: A. Tsuchiya [**66**]; R. J. Milgram [**67**]; J. Peter May [**68**]; Ib Madsen [**69**].

You have heard Peter May explain how a fairly clean way of setting up some of the basic machinery is provided by the work of Boardman and Vogt [**70**]; see also mimeographed notes with the same title, University of Warwick, 1968. The work of Boardman and Vogt, in its turn, may perhaps be made even smoother by forthcoming work of G. Segal.

NOTE ADDED IN PROOF. This subject has been making rapid progress recently. See Part IV of the survey by J. Stasheff in this symposium [**16**].

11. ***H*-spaces.** I will treat this topic in the most sketchy fashion, since I am sure that Stasheff will give you a much better account of it later. I would like to mention two points about H-spaces. First the well-known work of Peter Hilton and Joseph Roitberg [**71**]. J. Hubbuck [**72**] succeeded in classifying finite complexes which are homotopy-commutative H-spaces.

REFERENCES

1. J.-P. Serre, *Modules projectifs et espaces fibrés à fibre vectorielle*, Séminaire P. Dubreil, Dubreil-Jacotin et C. Pisot, 1957/58, Exposé 23, Secrétariat mathématique, Paris, 1958. MR **31** #1277.

2. R. G. Swan, *Vector bundles and projective modules*, Trans. Amer. Math. Soc. **105** (1962), 264–277. MR **26** #785.

3. H. Bass, *Algebraic K-theory*, Benjamin, New York, 1968. MR **40** #2736.

4. C. T. C. Wall, *Finiteness conditions for* CW-*complexes*, Ann. of Math. (2) **81** (1965), 56–69. MR **30** #1515.

5. ———, *Finiteness conditions for* CW-*complexes*. II, Proc. Roy. Soc. Ser. A **295** (1966), 129–139.

6. ———, *Surgery on compact manifolds*, Academic Press, New York, 1970.

7. ———, *Surgery of non-simply-connected manifolds*, Ann. of Math. (2) **84** (1966), 217–276. MR **35** #3692.

8. J. Milnor, *Whitehead torsion*, Bull. Amer. Math. Soc. **72** (1966), 358–426. MR **33** #4922.

9. R. G. Swan, *Non-abelian homological algebra and K-theory*, Proc. Sympos. Pure Math., vol. 17, Amer. Math. Soc., Providence, R.I., 1970.

10. J. Milnor, *On axiomatic homology theory*, Pacific J. Math. **12** (1962), 337–341. MR **28** #2544.

11. D. W. Anderson, *The real K-theory of classifying spaces*, Thesis, Univ. of California, Berkeley, Calif., 1964.

12. M. F. Atiyah, *Characters and cohomology of finite groups*, Inst. Hautes Études Sci. Publ. Math. No. 9 (1961), 23–64. MR **26** #6228.

13. S. Eilenberg and J. C. Moore, *Limits and spectral sequences*, Topology **1** (1962), 1–23.

14. C. U. Jensen, *On the vanishing of* $\varprojlim^{(i)}$, J. Algebra **15** (1970), 151–166.

15. J. W. Milnor and J. C. Moore, *On the structure of Hopf algebras*, Ann. of Math. (2) **81** (1965), 211–264. MR **30** #4259.

16. James D. Stasheff, *H-spaces and classifying spaces*: *Foundations and recent developments*, Proc. Sympos. Pure Math., vol. 22, Amer. Math. Soc., Providence, R.I., 1971.

17. E. H. Brown, Jr., *Twisted tensor products*. I, Ann. of Math. (2) **69** (1959), 223–246. MR **21** #4423.

18. M. Rothenberg and N. E. Steenrod, *The cohomology of classifying spaces of H-spaces*, Bull. Amer. Math. Soc. **71** (1965), 872–875. MR **34** #8405.

19. S. Eilenberg and J. C. Moore, *Homology and fibrations*. I. *Coalgebras, cotensor product and its derived functors*, Comment. Math. Helv. **40** (1966), 199–236. MR **34** #3579.

20. M. F. Atiyah and F. Hirzebruch, *Riemann-Roch theorem for differentiable manifolds*, Bull. Amer. Math. Soc. **65** (1959), 276–281. MR **22** #989.

21. E. H. Brown, Jr., *Cohomology theories*, Ann. of Math. (2) **75** (1962), 467–484; correction, ibid. (2) **78** (1963), 201. MR **25** #1551; MR **27** #749.

22. ———, *Abstract homotopy theory*, Trans. Amer. Math. Soc. **119** (1965), 79–85. MR **32** #452.

23. G. W. Whitehead, *Homology theories and duality*, Trans. Amer. Math. Soc. **102** (1962), 277–283.

24. M. F. Atiyah, *Bordism and cobordism*, Proc. Cambridge Philos. Soc. **57** (1961), 200–208. MR **23** #A4150.

25. R. Vogt, *Boardman's stable category*, Lecture Note Series, no. 21, Aarhus Universitet, Aarhus, 1970.

26. M. F. Atiyah and R. Bott, *On the periodicity theorem for complex vector bundles*, Acta Math. **112** (1964), 229–247. MR **31** #2727.

27. R. Wood, *Banach algebras and Bott periodicity*, Topology **4** (1965), 371–389. MR **32** #3062.

28. M. F. Atiyah, *K-theory and reality*, Quart. J. Math. Oxford Ser. (2) **17** (1966), 367–386. MR **34** #6756.

29. M. Karoubi, *Algèbres de Clifford et K-theorie*, Ann. Sci. École Norm. Sup. (4) **1** (1968), 161–270. MR **39** #287.

30. J. W. Milnor, *On the cobordism ring* Ω^* *and a complex analogue*. I, Amer. J. Math. **82** (1960), 505–521. MR **22** #9975.

31. R. E. Stong, *Relations among characteristic numbers*. I, Topology **4** (1965), 267–281. MR **33** #740.

32. A. Hattori, *Integral characteristic numbers for weakly almost complex manifolds*, Topology **5** (1966), 259–280. MR **33** #742.

33. D. Quillen, *On the formal group laws of unoriented and complex cobordism theory*, Bull. Amer. Math. Soc. **75** (1969), 1293–1298. MR **40** #6565.

34. P. E. Conner and E. E. Floyd, *The* SU-*bordism theory*, Bull. Amer. Math. Soc. **70** (1964), 670–675. MR **29** #5253.

35. D. W. Anderson, E. H. Brown, Jr., and F. P. Peterson, SU-*cobordism*, KO-*characteristic numbers, and the Kervaire invariant*, Ann. of Math. (2) **83** (1966), 54–67. MR **32** #6470.

36. ———, *The structure of the spin cobordism ring*, Ann. of Math. (2) **86** (1967), 217–298. MR **36** #2160.

37. C. T. C. Wall, *Addendum to a paper of Conner and Floyd*, Proc. Cambridge Philos. Soc. **62** (1966), 171–175. MR **32** #6472.

38. M. F. Atiyah, R. Bott and A. Shapiro, *Clifford modules*, Topology **3** (1964), suppl. 1, 3–38. MR **29** #5250.

39. P. E. Conner and E. E. Floyd, *The relation of cobordism to K-theories*, Lecture Notes in Math., no. 28, Springer-Verlag, Berlin and New York, 1966. MR **35** #7344.

40. J. F. Adams, *Lectures on generalized cohomology*, Lecture Notes in Math., no. 99, Springer-Verlag, Berlin and New York, 1966.

41. M. F. Atiyah, *Power operations in K-theory*, Quart. J. Math. Oxford Ser. (2) **17** (1966), 165–193. MR **34** #2004.

42. Adams, Harris, and Switzer, *Hopf algebras of cooperations for real and complex K-theory*, Proc. London Math. Soc. (to appear).

43. S. P. Novikov, *The methods of algebraic topology from the viewpoint of cobordism theory*, Izv. Akad. Nauk SSSR Ser. Mat. **31** (1967), 855–951 = Math. USSR Izv. **1** (1967), 827–913. MR **36** #4561.

44. T. tom Dieck, *Steenrod-Operationen in Kobordismen-Theorien*, Math. Z. **107** (1968), 380–401. MR **39** #6302.

45. M. F. Atiyah, *Thom complexes*, Proc. London Math. Soc. (3) **11** (1961), 291–310. MR **24** #A1727.

46. D. G. Quillen, *Some remarks on étale homotopy theory and a conjecture of Adams*, Topology **7** (1968), 111–116. MR **37** #3572.

47. J. F. Adams, *On the groups* $J(X)$. III, Topology **3** (1965), 193–222. MR **33** #6627.

48. J. W. Milnor and M. W. Kervaire, *Bernoulli numbers, homotopy groups, and a theorem of Rohlin*, Proc. Internat. Congress Math. 1958, Cambridge Univ. Press, New York, 1960, pp. 454–458. MR **22** #12531.

49. M. Mahowald, *On the order of the image of J*, Topology **6** (1967), 371–378. MR **35** #3663.

50. ———, *On the order of the image of the J-homomorphism*, Bull. Amer. Math. Soc. **76** (1970), 1310–1313.

51. H. Toda, *An important relation in homotopy groups of spheres*, Proc. Japan Acad. **43** (1967), 839–842. MR **37** #5872.

52. ———, *Extended p-th powers of complexes and applications to homotopy theory*, Proc. Japan Acad. **44** (1968), 198–203. MR **37** #5873.

53. D. S. Kahn, *Cup-i products and the Adams spectral sequence*, Topology **9** (1970), 1–9. MR **40** #6552.

54. W. S. Massey and F. P. Peterson, *The* mod 2 *cohomology structure of certain fibre spaces*, Mem. Amer. Math. Soc. No. 74 (1967). MR **37** #2226.

55. A. K. Bousfield, E. B. Curtis, D. M. Kan, D. G. Quillen, D. L. Rector and J. W. Schlesinger, *The* mod-*p lower central series and the Adams spectral sequence*, Topology **5** (1966), 331–342. MR **33** #8002.

56. D. L. Rector, *An unstable Adams spectral sequence*, Topology **5** (1966), 343–346.

57. J. M. Boardman and B. Steer, *On Hopf invariants*, Comment. Math. Helv. **42** (1967), 180–221. MR **36** #4555.

58. M. G. Barratt and M. E. Mahowald, *The metastable homotopy of $O(n)$*, Bull. Amer. Math. Soc. **70** (1964), 758–760. MR **31** #6229.

59. M. E. Mahowald, *Some Whitehead products in S^n*, Topology **4** (1965), 17–26. MR **31** #2724.

60. L. Kristensen, *On secondary cohomology operations*, Math. Scand. **12** (1963), 57–82. MR **28** #2550.

61. ———, *On a Cartan formula for secondary cohomology operations*, Math. Scand. **16** (1965), 97–115. MR **33** #4926.

62. L. Kristensen and Ib Madsen, *On evaluation of higher order cohomology operations*, Math. Scand. **20** (1967), 114–130. MR **36** #5936.

63. L. Kristensen, *On secondary cohomology operations*. II., Conf. on Algebraic Topology (Univ. of Illinois at Chicago Circle, Chicago, Ill., 1968), Univ. of Illinois at Chicago Circle, Chicago, Ill., 1969, pp. 117–133. MR **40** #3539.

64. T. Kudo and S. Araki, *Topology of H_n-spaces and H-squaring operations*, Mem. Fac. Sci. Kyūsū Univ. Ser. A. **10** (1956), 85–120. MR **19,** 442.

65. E. Dyer and R. K. Lashof, *Homology of iterated loop spaces*, Amer. J. Math. **84** (1962), 35–88. MR **25** #4523.

66. A. Tsuchiya, *Characteristic classes for spherical fiber spaces*, Proc. Japan Acad. **44** (1968), 617–622. MR **40** #2115.

67. R. J. Milgram, *The* mod 2 *spherical characteristic classes*, Ann. of Math. (2) **92** (1970), 238–261.

68. J. Peter May, *Geometry of iterated loop spaces*, in preparation; see also his contribution to this symposium.

69. Ib Madsen, *On the action of the Dyer-Lashof algebra in $H_*(G)$ and $H_*(G/\mathrm{Top})$*, Thesis, Chicago, Ill., 1970.

70. J. M. Boardman and R. M. Vogt, *Homotopy-everything H-spaces*, Bull. Amer. Math. Soc. **74** (1968), 1117–1122. MR **38** #5215. See also: University of Warwick Mimeographed Notes, 1968.

71. P. Hilton and J. Roitberg, *On principal S^3 bundles over spheres*, Ann. of Math. (2) **90** (1969), 91–107. MR **39** #7624.

72. J. Hubbuck, *On homotopy commutative H-spaces*, Topology **8** (1969), 119–126.

UNIVERSITY OF MANCHESTER AND
UNIVERSITY OF CAMBRIDGE, ENGLAND

SPECTRA AND Γ-SETS

D. W. ANDERSON

0. **Introduction.** This is a report on the recent work of Graeme Segal on homotopy everything H-spaces. I shall attempt to give the broad outline of his work, together with those details which seem to be required to make it easier to read his paper "Homotopy everything H-spaces".

I have made one technical change in the presentation, in order to relate it to some work of my own—I have kept the discussion entirely simplicial. This helps to stress the role of multisimplicial complexes—objects which no doubt are worthy of further study.

Segal's technique is to introduce a structure, which he calls a Γ-structure. This is slightly more restrictive than a simplicial structure. Through various devices, it is not difficult to show that many categories give rise to Γ-structures. Also, if G is an abelian group, $\overline{W}(G)$ has not only a simplicial structure, but also a Γ-structure.

From any Γ-set (to be thought of as a simplicial set with further structure), one can, by simple manipulations, extract a sequence of simplicial spaces, which form a spectrum.

More generally, one can consider simplicial Γ-sets. These, just as simply as for Γ-sets, give rise to spectra. One can give simple criteria on the simplicial Γ-structure to insure that this spectrum be an Ω-spectrum.

The functor which assigns to a (special) simplicial Γ-set an Ω-spectrum has an adjoint. More interestingly, the functors which assign to a (special) simplicial Γ-set the 1-term of its associated Ω-spectrum has an adjoint $X \mapsto \Gamma_X$, where X is a simplicial complex, and Γ_X is a simplicial Γ-set. Γ_X suffers from a slight defect, in that the initial term of its associated spectrum may not have a homotopy inverse. A slight modification, Γ'_X, of Γ_X does have an Ω-spectrum associated to it. We direct the interested reader to Segal's paper for technical details.

The spectrum, associated to Γ'_X when X is two points, is the Ω-spectrum associated to the sphere spectrum. This follows immediately from the adjointness

AMS 1970 *subject classifications*. Primary 55B20.

of $X \mapsto \Gamma_X$ and the functor associating to simplicial Γ-sets a spectrum. Since the definition of the 1-term of the spectrum associated to Γ'_X is identical with the Barratt-Quillen construction, it follows that the Barratt-Quillen construction applied to X yields $\Omega^\infty S^\infty X$.

Proofs that BU, BPL, $B\ Top$, and BF are infinite loop spaces can be given easily in this framework. The technique is as follows. First, one constructs a simplicial Γ-category. For example, when considering BU, the initial term of the simplicial Γ-category is the category of finite-dimensional complex vector spaces, together with the singular complex of their group of automorphisms as morphisms. There is a functor M: Simplicial Categories $\to$ Simplicial Sets, which is a form of the $\overline{W}$-construction. When this is applied to simplicial Γ-categories, it produces simplicial Γ-sets. The most helpful feature of M is that it transforms natural transformations of functors into homotopies. Thus, for vector bundles, it suffices to know that direct sum is isomorphism commutative, rather than show that it is in some sense homotopy commutative.

The functor M has an adjoint, the exploded category of a complex. We do not explain this, but it has applications. For example, Segal uses it to show that any representable functor which has a Gysin map (called a trace by Quillen) for finite coverings extends to a cohomology theory. This had been conjectured by Quillen. (Any cohomology theory has a trace.)

If X, Y are simplicial categories, one can define homotopy classes of maps from X to Y. Furthermore, one can define directly what it means for a map $X \to Y$ to be of dimension n. Thus, one could define homology and cohomology for simplicial Γ-spaces by taking $H_n(X; Y) = [\Gamma_{S^0}, X \wedge Y]^{-n}$, $H^n(X; Y) = [X, Y]^n$, where $[\ ,\ ]^n$ denotes homotopy classes of maps of dimension n. If A is a space, $H_n(A; Y) = H_n(\Gamma_A; Y)$, $H^n(A; Y) = H^n(\Gamma_A; Y)$. I hope to expand this discussion in a later paper.

As a final application, Segal shows that if $K_\otimes(X)$ is the K-theory of virtual bundles of dimension 1, with group structure given by tensor product, the classifying space $BU_\otimes$ is an infinite loop space. Since the multiplicative structure on Γ_{S^0} is given by the Cartesian product of sets, this suggests that there is an infinite loop map $F \to BU_\otimes$ whose loop $\Omega F \to U$ is the usual map. Since Segal has shown that there are H-maps for each prime p, $BU \to BU_\otimes$ (using exterior powers) which induce an isomorphism on all homotopy groups of sufficiently high dimension, there are various interesting consequences to this conjecture.

1. **$\mathscr{C}$-sets.** If $\mathscr{C}$ is a category, a $\mathscr{C}$-set is a functor $\Phi:\mathscr{C} \to$ Sets. We assume that Φ is contravariant. Given two $\mathscr{C}$-categories Φ, Ψ, we have the notion of a product $\Phi \times \Psi$, which is the composition of the diagonal functor $\mathscr{C} \to \mathscr{C} \times \mathscr{C}$ with the categorical product of Φ and Ψ.

We can also define a $\mathscr{C}$-set $\mathrm{Hom}(\Phi, \Psi)$ as follows. If C is an object of $\mathscr{C}$, let $\mathscr{C}(C):\mathscr{C} \to$ Sets be given by $\mathscr{C}(C)(C') = \mathrm{Hom}_{\mathscr{C}}(C', C)$. Then $\Lambda = \mathrm{Hom}(\Phi, \Psi)$ is defined by $\Lambda(C) = \mathrm{Hom}_{\mathscr{C}}(\Phi \times \mathscr{C}(C), \Psi)$, the set of natural transformations ($\mathscr{C}$-maps) of $\Phi \times \mathscr{C}(C)$ to Ψ. Λ has the obvious definition on morphisms.

The constructions product and Hom are naturally adjoint. This is well known when $\mathscr{C} = \Delta$, so that we are dealing with simplicial sets. The following proof is left to the reader.

PROPOSITION. *For any* Φ_1, Φ_2, Φ_3, $\mathrm{Hom}(\Phi_1 \times \Phi_2, \Phi_3)$ *is naturally equivalent to* $\mathrm{Hom}(\Phi_1, \mathrm{Hom}(\Phi_2, \Phi_3))$ *as a* $\mathscr{C}$*-set.*

2. **Realization and condensation of multisimplicial sets.** If X is a topological space, one can define the singular n-fold complex $S^n(X)$ as the complex of all continuous maps $|\Delta_{i_1}| \times \cdots \times |\Delta_{i_n}| \to X$, where $|\Delta_i|$ is the topological i-simplex. The geometric realization $|K|$ of a simplicial n-fold complex is defined by the adjointness condition

$$\mathrm{Hom}_{\mathrm{Topological}}(|K|, X) = \mathrm{Hom}_{\Delta^n}(K, S^n(X)).$$

It is a standard exercise to show that such an adjoint exists. Alternatively, one could construct $|K|$ directly by taking a suitable identification space of the disjoint union of copies of spaces $|\Delta_{i_1}| \times \cdots \times |\Delta_{i_n}|$, one for each multisimplex of degree $(i_1, \ldots, i_n)$ of K. In this manner, one can see that if K_1, K_2 are Δ^{n_1}, Δ^{n_2}-sets respectively, $|K_1 \times K_2| = |K_1| \times |K_2|$.

Not all multisimplicial complexes can be decomposed as products, and this can lead to somewhat unfamiliar phenomena. Consider, for example, an n-fold complex K whose only nondegenerate multisimplices are in degrees $(0, \ldots, 0)$ and $(1, \ldots, 1)$. Then $|K|$ is a quotient of the n-cube and can fairly easily be seen to be the n-sphere.

Since each product $\Delta_{i_1} \times \cdots \times \Delta_{i_n}$ has a well-defined structure of a simplicial set, we can define the singular multicomplex S^n for simplicial sets as well as for spaces. If L is a simplicial set, let $S^n(L)(i_1, \ldots, i_n) = \mathrm{Hom}_\Delta(\Delta_{i_1} \times \cdots \times \Delta_{i_n}, L)$. The condensation $K.$ of an n-fold complex K is defined by the adjointness condition

$$\mathrm{Hom}_\Delta(K., L) = \mathrm{Hom}_{\Delta^n}(K, S^n(L)).$$

Clearly, $K. = K$ if $n = 1$. Also, $|K.| = |K|$. If one writes down an explicit formula for $K.$, one can see that $(K' \times K''). = (K'). \times (K'').$, where the left-hand product is the product of a $\Delta^{n'}$-set with a $\Delta^{n''}$-set as a $\Delta^{n'+n''}$-set, and the right-hand product is the product of simplicial sets.

Segal constructs spaces from multisimplicial sets by a process of successive geometric realization. This suffers from a lack of obvious associativity and commutativity. However, it is not difficult to show that if K is a multisimplicial set, the geometric realization of the condensation $K.$ is naturally homeomorphic to Segal's successive geometric realization.

A further simplification has been pointed out to me by Quillen and by Bousfield. It can be proved by direct, if tedious, means that the condensation of a multisimplicial set is naturally isomorphic to the diagonal simplicial set. Therefore, below, the reader can substitute "diagonal" for "condensation" everywhere.

3. **The categories Δ and Γ.** The simplicial category Δ is the category whose objects are the finite ordered sets $\Delta_n = (0, 1, \ldots, n)$. The morphisms $\varphi: \Delta_n \to \Delta_k$ are the order preserving functions from the set Δ_n to the set Δ_k.

The category Γ has as its objects the finite sets $\Gamma_n = (1, \ldots, n)$ for $n > 0$, and $\Gamma_0 = \emptyset$, the empty set. A morphism $\varphi: \Gamma_n \to \Gamma_k$ is a function defined on the set of subsets of Γ_n, $\mathscr{P}(\Gamma_n) \to \mathscr{P}(\Gamma_k)$, to the set of subsets of Γ_k which preserves union and set difference. There is a natural functor $\Delta \to \Gamma$ given by $\varphi \mapsto \varphi'$, $\Delta_n \to \Gamma_n$, where φ' is determined by $\varphi'(\{1, \ldots, r\}) = \{1, \ldots, \varphi(r)\}$. Notice that $\Delta \to \Gamma$ is not faithful.

The opposite category to Γ is the category of finite sets $\Gamma_n^{\mathrm{op}} = (0, \ldots, n)$ and set maps which preserve the basepoint 0. The correspondence is as follows. If $\varphi: \Gamma_n \to \Gamma_k$, let $\varphi^{\mathrm{op}}: \Gamma_k^{\mathrm{op}} \to \Gamma_n^{\mathrm{op}}$ be defined by $\varphi^{\mathrm{op}}(i) = j$ if $i \in \varphi(\{j\})$, $\varphi^{\mathrm{op}}(i) = 0$ if $i \notin \varphi(\Gamma_n)$. Then $\varphi \mapsto \varphi^{\mathrm{op}}$ defines a functor $\Gamma^{\mathrm{op}} \to$ finite pointed sets. The inverse is $\psi \mapsto \psi^{\mathrm{op}}$, where $\psi^{\mathrm{op}}(\{i\}) = \psi^{-1}(\{i\}) - \{0\}$. Notice that in the category of finite pointed sets, Δ^{op} corresponds to those maps which increase up to some point, and are 0 from that point on. Permutations in $\mathrm{Hom}(\Gamma_n, \Gamma_n)$ correspond to basepoint fixing permutations in $\mathrm{Hom}(\Gamma_n^{\mathrm{op}}, \Gamma_n^{\mathrm{op}})$. Clearly, every morphism $\Gamma_n^{\mathrm{op}} \to \Gamma_k^{\mathrm{op}}$ can be written as a morphism $\Delta_n^{\mathrm{op}} \to \Delta_k^{\mathrm{op}}$ composed with a permutation of Γ_k^{op}. Thus the morphisms of Γ are generated by the morphisms of Δ together with the permutations of the simplex which leave the initial vertex fixed.

There is a product functor $\Gamma \times \Gamma \to \Gamma$ given by $(\Gamma_n, \Gamma_k) \mapsto \Gamma_{nk}$ on objects, and by Cartesian product of morphisms. Notice that $\Delta \times \Delta \to \Gamma$ does not factor through $\Delta \to \Gamma$. Notice also that $\Gamma \times \Gamma \to \Gamma$ is associative.

In order to simplify notation, we shall adopt Segal's notation of n for Δ_n, and $\boldsymbol{n}$ for Γ_n.

If we regard $\Gamma(\boldsymbol{1})$ as a simplicial set, notice that $\Gamma(\boldsymbol{1})(n) = \mathrm{Hom}_\Gamma(\boldsymbol{n}, \boldsymbol{1})$ contains $n + 1$ objects. If $n > 1$, all of these are in the image of an element of $\Gamma(\boldsymbol{1})(n - 1)$, since every morphism $\boldsymbol{n} \to \boldsymbol{1}$ factors through $\boldsymbol{n} - \boldsymbol{1}$ (this is easy to see in Γ^{op}). Thus, as a simplicial set, $\Gamma(\boldsymbol{1}) = S^1$, the 1-sphere with one vertex.

PROPOSITION 3.1. *As a simplicial complex*, $\Gamma(\boldsymbol{k})$ *is* $S^1 \times \cdots \times S^1$, *k factors in all.*

PROOF. In Γ, $\boldsymbol{k}$ is the direct product of $\boldsymbol{1}$ with itself k times. Thus $\Gamma(\boldsymbol{k}) = \Gamma(\boldsymbol{1})^k$.

Since the Cartesian product is associative, there is a well-defined functor $\Gamma^n = \Gamma \times \cdots \times \Gamma \to \Gamma$. Thus, we can associate to a Γ-set S its nth power $B^n S$ as a Δ^n-set, by $B^n S(i_1, \ldots, i_n) = S(\boldsymbol{i}_1 \times \cdots \times \boldsymbol{i}_n)$.

Notice that any morphism $\boldsymbol{i}_1 \times \cdots \times \boldsymbol{i}_n \to \boldsymbol{1}$ takes a nonempty value on at most one set containing a single element. Thus, it factors through $\boldsymbol{1} \times \cdots \times \boldsymbol{1}$ via maps $\boldsymbol{i}_i \to \boldsymbol{1}$. Since every Γ-morphism $\boldsymbol{i} \to \boldsymbol{1}$ can easily be seen to lie in Δ, every multisimplex in $B^n\Gamma(\boldsymbol{1})$ is degenerate except for the basepoint and one multisimplex in degree $(1, \ldots, 1)$. Thus $B^n\Gamma(\boldsymbol{1})$. is the n-fold smash product of S^1 with itself and is thus equivalent to S^n.

PROPOSITION 3.2. $B^n\Gamma(\boldsymbol{k})$. *is homotopy equivalent to* $(S^n)^k$.

4. $\Delta^n \times \Gamma$**-sets and the functor** B. If X is a $\Delta^n \times \Gamma$-set, we can define a $\Delta^{n+1} \times \Gamma$-set BX by using the Cartesian product $\times$ in Γ. More explicitly,

$$BX(i_1, \ldots, i_n, i_{n+1}, \boldsymbol{k}) = X(\boldsymbol{i}_1, \ldots, \boldsymbol{i}_n, \boldsymbol{i}_{n+1} \times \boldsymbol{k}).$$

Notice that $BX(i_1, \ldots, i_n, i_{n+1}, \boldsymbol{0}) = X(i_1, \ldots, i_n, \boldsymbol{0})$, so that if $X(-, \boldsymbol{0})$ contains only degenerate multisimplices, so does $BX(-, \boldsymbol{0})$.

Let X_k be the Δ^n-set $X_k(i_1, \ldots, i_n) = X(i_1, \ldots, i_n, \boldsymbol{k})$. We write BX_k for $(BX)_k$. $BX_1(i_1, \ldots, i_n, i_{n+1}) = X_{i_{n+1}}(i_1, \ldots, i_n)$, so that we have a natural identification of sets $X_1 = BX_1(-, 1)$. This gives us a natural map of Δ^{n+1}-complexes $X_1 \times S^1 \rightarrow BX_1$. If X_0 has only degenerate multisimplices, the condensation of this map factors through the suspension $X_1. \wedge S^1 \rightarrow BX_1.$. Thus, we see that X defines a spectrum $X_1., BX_1., B^2X_1., \ldots$.

5. **Special** $\Delta \times \Delta$**-sets.** There is a fairly simple condition given by Segal to insure that the sequence $X_1., (BX)_1., \ldots$ is an Ω-spectrum, at least beyond the first stage. First, we make a diversion.

If G is a simplicial group, we can form a Δ^2-complex BG by letting $(BG)(i, j) = G(i) \times \cdots \times G(i)$, j factors in all. Face operators in the first variable are given by those of G. In the second variable, differentials are defined by $d_0(x_1, \ldots, x_n) = (x_2, \ldots, x_n)$, $d_i(x_1, \ldots, x_n) = (x_1, \ldots, x_i x_{i+1}, \ldots, x_n)$ for $0 < i < n$, $d_n(x_1, \ldots, x_n) = (x_1, \ldots, x_{n-1})$. Then $(BG).$ is a form of the $\overline{W}$-construction. It is the simplicial form of Milgram's construction of classifying spaces for topological groups. There is a fibration $(BG). \rightarrow \overline{W}(G)$ with acyclic fiber. It is well known that the usual map $G \rightarrow \Omega((BG).)$ is a homotopy equivalence.

The constructions $(BG).$ and $\overline{W}(G)$ both can be made for any associative simplicial monoid G which has a unit (which is needed to define degeneracies). $\overline{W}(G)$ is well known to be a Kan complex if and only if G is a group. However, $(BG).$ can be seen to be a Kan complex if G has the right sort of homotopy inverse. This is a small, but pleasant, advantage that B has over $\overline{W}$.

A more important advantage which B has over $\overline{W}$ is that it is particularly easy to formulate a type of homotopy associativity. Notice that there are preferred maps $m_i : \mathit{1} \rightarrow n$ in Δ, given by $m_i(0) = i - 1, m_i(1) = i$. Then, if A is any $\Delta \times \Delta =$ ' complex, $m = \pi(m_i^*) : A(-, n) \rightarrow A(-, \mathit{1})^n$. We call A a special $\Delta \times \Delta$ complex if $A(-, 0)$ contains only one point, and all its simplices are degenerate, and if $A(-, n) \rightarrow A(-, \mathit{1})^n$ is a homotopy equivalence for all $n > 0$.

Notice that if A is special, $A_1 = A(-, \mathit{1})$ is an H-space, via $A_1 \times A_1 \cong A(-, 2) \xrightarrow{d_1} A_1$. Segal proves that if A_1 has a homotopy inverse in this H-space structure, the natural map $A_1 \rightarrow \Omega(A.)$ is a homotopy equivalence if A is a special $\Delta \times \Delta$-set.

If X is a $\Delta^n \times \Gamma$-set, let A be the $\Delta \times \Gamma$ complex obtained by condensing with respect to the first n variables. Because condensation is associative, and the product in Γ is associative, the condensation of BX with respect to the first $(n + 1)$ variables equals the condensation of BA with respect to the first two.

We call X a special $\Delta^n \times \Gamma$-set if A is special as a $\Delta \times \Delta$-set. Notice that if X is special, $(BA_1)(i, j) = A(i, j)$.

Thus $(BA_1)(i, j) \to (BA_1(i))^j = (A(i, 1))^j$ gives a homotopy equivalence $BA_1(-, j) \to BA_1(-)^j$ for all j. Thus BA is special if A is, so BX is special if X is. Thus, the spectrum $\boldsymbol{B}X = X_1., BX_1., \ldots$ is an Ω-spectrum if X is special and X_1 has a homotopy inverse.

We have defined a functor $\boldsymbol{B}$ from special $\Delta^n \times \Gamma$-spaces X with homotopy inverse to spectra. There is a functor from Ω-spectra to special $\Delta \times \Gamma$-spaces with homotopy inverse given by $X \to \boldsymbol{\Omega}X$, where $\boldsymbol{\Omega}X(\boldsymbol{k}) = \mathrm{Hom}(\boldsymbol{B}\Gamma(\boldsymbol{k}), X)$, where we are considering only degree 0 spectral maps. Recall that $B^n\Gamma(\boldsymbol{k}) = (S^n)^k$. The functors $\boldsymbol{B}$ and $\boldsymbol{\Omega}$ are adjoint. If X is a connective spectrum, $\boldsymbol{B\Omega}X \to X$ is a homotopy equivalence.

6. **The morphism complex of a category.** If $\mathscr{C}$ is a category, we can associate to $\mathscr{C}$ a simplicial set, $M(\mathscr{C})$, which we call the morphism complex of $\mathscr{C}$. We define $M = M(\mathscr{C})$ as follows. $M(0)$ is the set of objects of $\mathscr{C}$. If $n > 0$, $M(n)$ is the set of sequences of morphisms of $\mathscr{C}$ of the form $(\alpha_1, \ldots, \alpha_n)$ such that $\alpha_1 \cdots \alpha_n$ is defined. The face operators $d_0, d_1 : M(1) \to M(0)$ are source and target, respectively. The face operators $d_i : M(2) \to M(1)$ are given by $d_0(\alpha_1, \alpha_2) = \alpha_2$, $d_1(\alpha_1, \alpha_2) = \alpha_1\alpha_2$, $d_2(\alpha_1, \alpha_2) = \alpha_1$, and face operators in higher degrees are defined similarly. Degeneracy operators are defined by inserting appropriate identity maps.

Notice that M: Categories $\to$ Δ-sets is a functor and preserves disjoint union and product.

The category J with two objects 0, 1 and one nonidentity morphism $0 \to 1$, has the property that $M(J) = \Delta(1)$, the 1-simplex.

If $\Phi_1, \Phi_2 : \mathscr{C} \to \mathscr{D}$, a natural transformation from Φ_1 to Φ_2 determines and is determined by an extension of $\Phi_1 + \Phi_2 : \mathscr{C} + \mathscr{C} \to \mathscr{D}$ to $\mathscr{C} \times J \to \mathscr{D}$. Thus, if there is a natural transformation from Φ_1 to Φ_2, $M(\Phi_1)$ and $M(\Phi_2)$ are homotopic maps of $M(\mathscr{C})$ to $M(\mathscr{D})$.

Clearly, $M(\mathscr{C}_1 \times \mathscr{C}_2) = M(\mathscr{C}_1) \times M(\mathscr{C}_2)$. Now, if we have $\Phi : \Gamma \to$ Categories, a functor into the category of small categories and functors, $M(\Phi)(i, \boldsymbol{k}) = M(\Phi(\boldsymbol{k}))(i)$ defines $M(\Phi)$ as a $\Delta \times \Gamma$-set. We call Φ special if $\Phi(\boldsymbol{0})$ contains one object and one morphism, and if $\Phi(\boldsymbol{n}) \to \Phi(\boldsymbol{1})^n$ is an equivalence. Then if Φ is special, so is $M(\Phi)$.

A simplicial category has the obvious meaning. If $\mathscr{C}$ is a simplicial category, $M(\mathscr{C})$ is a $\Delta \times \Delta$-set in the obvious way.

If $\mathscr{C}$ is a category with one object, $M(\mathscr{C}) = \overline{W}(\mathrm{Mor}(\mathscr{C}))$, where $\mathrm{Mor}(\mathscr{C})$ is regarded as an associative monoid.

If G is a discrete group which acts on a simplicial complex X, let $\mathscr{C}(G, X)$ be the simplicial category whose objects are the simplices of X, and whose morphisms are the elements of G, so that $g : x \to y$ if $y = g(x)$. Then $M(\mathscr{C}(G, X)) = W(G) \times_G X$, which is sometimes written as X_G. (For those not familiar with simplicial constructions, $W(G)$ is an acyclic complex upon which G acts freely.)

7. **Categories with direct sum.** Let $\mathscr{C}$ be a category with direct sums defined. Segal constructs a commutative composition law $\mathscr{L}(\mathscr{C})$ for $\mathscr{C}$ as follows. Let $\mathscr{P}(n)$ be the category whose objects are the subsets of the set with n elements, and whose morphisms are the inclusions of one subset in another. Then $\mathscr{L}(\mathscr{C})(\boldsymbol{n})$ is the

subcategory of the category of functors $\mathscr{P}(n) \to \mathscr{C}$ defined as follows. If $\Phi: \mathscr{P}(n) \to \mathscr{C}$, and σ_1, σ_2 are two subsets of n, with $\sigma_1 \cap \sigma_2 = \emptyset$, then the natural map $\Phi(\sigma_1) \coprod \Phi(\sigma_2) \to \Phi(\sigma_1 \cup \sigma_2)$, defined by the sum of the values of Φ on the inclusions $\sigma_1, \sigma_2 \to \sigma_1 \cup \sigma_2$, is an isomorphism. The morphisms are all natural transformations η of functors which have the property that $\eta(\sigma_1 \cup \sigma_2) = \eta(\sigma_1) \coprod \eta(\sigma_2)$ for all σ_1, σ_2 with $\sigma_1 \cap \sigma_2 = \emptyset$.

Notice that any such $\Phi: \mathscr{P}(n) \to \mathscr{C}$ is naturally equivalent in an obvious way to Φ', where $\Phi'(\sigma) = \coprod \{\Phi(\{i\}) \mid i \in \sigma\}$, so that $\mathscr{L}(\mathscr{C})(n) \to \mathscr{L}(\mathscr{C})(1)^n = \mathscr{C}^n$ is an equivalence of categories.

Notice that if $\mathscr{C}_0$ is a subcategory of $\mathscr{C}$ which is closed under direct sum, there is a commutative composition law $\mathscr{L}(\mathscr{C}_0) \subset \mathscr{L}(\mathscr{C})$ consisting of those functors $\mathscr{P}(n) \to \mathscr{C}$ which factor through $\mathscr{C}_0$.

If $\mathscr{C}$ is a category with direct products, one can produce a composition law $\mathscr{L}(\mathscr{C})$ using contravariant functors $\mathscr{P}(n) \to \mathscr{C}$ in the same fashion as for direct sum.

8. **The free Γ-complex of a simplicial set.** If X is a Δ-set, and if Y is a special $\Delta \times \Gamma$-set, any map $X \to Y_1$ determines, up to homotopy, maps $X^n \to Y_n$, since $Y_n \cong (Y_1)^n$. There are multiplications $Y_n \to Y_1$, corresponding to $\{1\} \mapsto \{1, \ldots, n\}$ (if $Y = BG$ for a group G, $G^n \to G$ is given by $(g_1, \ldots, g_n) \mapsto g_1 \cdots g_n$). Thus we have also $X^n \to Y_1$ determined up to homotopy. Furthermore, these maps are homotopy equivariant with respect to the action of the symmetric group acting on X^n, since the symmetric group acts on Y_n in such a way that total multiplication $Y_n \to Y_1$ is equivariant.

Thus, we see that if $Z_1 = \coprod_{n \geq 0} X^n$, $Z_n = X_1^n$, X determines a $\Delta \times \Gamma$-space Z, and every map $X \to Y_1$ extends, up to some questions of homotopy, to a map $Z \to Y$. Furthermore, $Z_n \to Y_n$ is homotopy equivariant with respect to the action of the various symmetric groups acting on the various subsets of Z_n.

To clear up these questions of homotopy, we make some auxiliary constructions. From X we construct Γ_X, which replaces Z as a richer object, which allows us to express the action of the symmetric group on X^n in terms of a $\Delta \times \Gamma$-set. Then, we enlarge this to $\Gamma_f \to \Gamma_X$, a fibering over Γ_X depending on f, to account for the fact that Y_n is only homotopy equivalent to Y_1^n, not actually equal to it. There will be a natural projection $\Gamma_f \to Y$. If Y is reasonable, $\Gamma_f \to \Gamma_X$ will be a homotopy equivalence, and thus we will have $\Gamma_X \to Y$ extending $X \to Y_1$.

The $\Delta \times \Gamma$-set Γ_X is called the free Γ-set on X. It is adjoint in the homotopy category to the functor $Y \to Y_1$ from special $\Delta \times \Gamma$-sets to Δ-sets, at least if we make a slight restriction on the type of $\Delta \times \Gamma$-sets which we consider.

Γ_X is defined to be the composition law for the following subcategory of a category with direct product defined as follows. Let $\mathscr{C}(X)$ be the category whose objects are functions $f: [n] \to X$, where $[n]$ is the set of n elements. The set maps $[n] \to [k]$ define the morphisms in $\mathscr{C}(X)$. The disjoint sum $f_1 \coprod f_2 : [n_1 + n_2] \to X$ for $f_i : [n_i] \to X$ is the direct product in $\mathscr{C}(X)$. Let $\mathscr{C}_0(X)$ be the subcategory with the same objects and the isomorphisms as morphisms. Then $\mathscr{C}_0$ is closed under direct product.

The $\Delta \times \Gamma$-category which defines Γ_f for $f: X \to A_1$ has as objects for $\Gamma_f(\mathbf{1})$ the simplices of the fiber products $X^n \times_{A_1^n} A_n$, with morphisms the appropriate permutations. This is a category with direct product, and so has a commutative composition law which defines Γ_f.

Since $A_n \to A_1^n$ is a homotopy equivalence, so is $X^n \times_{A_1^n} A_n \to X^n$. Thus $\Gamma_f(\mathbf{n}) \to \Gamma_X(\mathbf{n})$ is a homotopy equivalence for all n. If the vertical and horizontal homotopy of A are sufficiently independent (that is, A is what Segal calls a $\Delta \times \overline{\Gamma}$-set), there will be a homotopy inverse $\Gamma_X \to \Gamma_f$ which is a $\Delta \times \Gamma$-map.

MASSACHUSETTS INSTITUTE OF TECHNOLOGY

A FREE GROUP FUNCTOR FOR STABLE HOMOTOPY

M. G. BARRATT

Along with sundry other curiosities, an extension Γ of Milnor's functor F, such that $\Gamma(X)$ is equivalent to $\Omega^\infty\Sigma^\infty |X|$, is described.

1. **An exact sequence.** For any pair $A \supset B$ of sets with base points, let $\mu: A \to A/B$ collapse B to the base point. Let P be a functor from pointed sets to groups; then an exact sequence

$$(1.1) \quad \cdots \longrightarrow P_k(A, B) \xrightarrow{P(\mu_k)} \cdots \xrightarrow{P(\mu_1)} P(A) \xrightarrow{P(\mu)} P(A/B) \longrightarrow \{1\}$$

may be described, together with short exact sequences

$$(1.2) \quad P_{k+1}(A, B) \xrightarrow{\subset} P_k(A \vee B', B) \xrightarrow{P(\beta_{k+1})} P_k(A, B) \qquad (B' \approx B).$$

These are defined by means of the sets

$$A_k = A \vee B_1 \vee \cdots \vee B_k = \text{the union of the axes of } A \times (B)^k,$$

and the maps μ_k and β_t $(1 \leqq t \leqq k)$ of A_k to A_{k-1} given by

μ_k maps A_{k-1} identically and folds B_k to $B \subset A$,
β_t maps A_{t-1} identically and B_i to B_{i-1} when $i \geqq t$,
except that β_1 collapses B_1 to a point.

Then $P_k(A, B)$ is the intersection of the kernels of the homomorphisms $P(\beta_t)$ $(t \leqq k)$, and the homomorphisms above and the restrictions of the homomorphisms $P(\mu_k)$, $P(\beta_{k+1})$.

The exactness of (1.2) is obvious, and that of (1.1) may be proved using certain right inverses λ_k to μ_k, given by

$$\lambda_k : A_{k-1} \to A_k \text{ maps } (A_{k-1} - B) \text{ identically, and } B \text{ onto } B_k.$$

AMS 1970 *subject classifications*. Primary 55E10, 55D35; Secondary 20E05.

2. **A spectral sequence.** Now let P be a functor from pointed simplicial sets (CSS complexes) to simplicial groups (CSS group complexes), and suppose P given in each dimension by a functor on (graded) sets. Such a functor is Milnor's functor F.

Then the maps μ_k, β_t of §1 are simplicial (although λ_k is not), and hence there is an exact sequence of simplicial groups

$$\cdots \to P_k(A, B) \to \cdots \to P_1(A, B) \to P(A) \to P(A/B) \to \{1\} \tag{2.1}$$

for any subcomplex B of A, together with short exact sequences (fibrations)

$$P_{k+1}(A, B) \to P_k(A \vee B_{k+1}, B) \to P_k(A, B). \tag{2.2}$$

It follows that there is a spectral sequence $\{E^r_{p,q}, d^r\}$ converging to $\pi_*(P(A/B))$, with $E^1_{p,q} = \pi_q(P_p(A, B))$. When $P = F$ and A is contractible, this is a spectral sequence converging to $\pi_*(\Sigma^2 B)$, with E^1 terms the group of cross terms in bouquets of copies of ΣB.

If, however, P is distributive over wedges in the sense that the natural map

$$P(B_1) \to P_1(A, B) = (\ker P(A \vee B_1) \to P(A)) \tag{2.3}$$

is a homotopy equivalence, for all (A, B), then (2.2) shows that $P_k(A, B)$ is contractible for all $k > 1$, and (2.1) implies

THEOREM 2.4. *Then $P(B)$ is homotopically equivalent to the fibre of*

$$P(\mu): P(A) \to P(A/B),$$

that is, P turns cofibrations into fibrations.

It follows that $\pi_* \circ P$ is an homology theory satisfying the Mayer-Vietoris Theorem, provided it satisfies the homotopy axiom. SP^∞ is an example of such a functor, of course.

REMARK. The spectral sequence above is similar to but quite different from the spectral sequence described in the author's 1962 Aarhus notes. Calculations made with it may be described elsewhere.

3. **On the symmetric groups.** Let $[n]$ denote the ordered set of integers $0 \leqq i \leqq n$, and let G be an abstract group. The minimal contractible G-free complex WG has all functions $[n] \to G$ as n-simplices, the face and degeneracy operators being defined by composition with nondecreasing functions $[m] \to [n]$ in the standard way. G acts on WG by *left* translation:

$$\phi: [n] \to G \quad \text{implies} \quad (g \cdot \phi)(i) = g \cdot \phi(i).$$

Furthermore, WG is a group, by multiplying values:

$$(\phi \circ \psi)(i) = \phi(i) \cdot \psi(i).$$

Let S_n denote the group of permutations of the positive integer $1 \leqq i \leqq n$, embedded in S_{n+k}: then

$$S_{n+1} = S_n \circ \{1, \tau_{n+1,n}, \ldots, \tau_{n+1,1}\} \qquad (\tau_{n+1,k} = (n+1, n, \ldots, k)).$$

Let $T: S_{n+1} \to S_n$ be the left S_n-map which sends each $\tau_{n+1,k}$ to 1, and induces a simplicial S_n-map $T: WS_{n+1} \to WS_n$.

Let $i_m: S_n \to S_{m+n}$ embed S_n as the subgroup of permutations of integers $>m$, so that $1 \times i_m$ embeds $S_m \times S_n$ in S_{m+n}.

LEMMA. *$WS_m \times WS_n = W(S_m \times S_n)$ is an $S_m \times S_n$ equivariant deformation retract of WS_{m+n}.*

(For $WS_m \times WS_n/S_m \times S_n$ is a deformation retract of $W(S_{m+n})/S_m \times S_n$.)

4. **The free monoid $\Gamma^+(X)$.** Let $\mathcal{D}$ be the equivalence relation on $\bigcup_n (X)^n \times WS_n$ (X being a CSS complex) generated by the relations $((x_1, \ldots, x_n), w) \equiv (\sigma(x_1, \ldots, x_n), \sigma w)$ if $\sigma \in S_n$, $((x_1, \ldots, x_n), w) = ((x_1, \ldots, x_{n-1}), Tw)$ if $x_n =$ base point. Let $\Gamma^+(X)$ be the set of equivalence classes, $[x_1, \ldots, x_n, w]$.

THEOREM 4.1. *$\Gamma^+(X)$ is a free simplicial monoid if multiplication is defined by $[a, u] \cdot [b, v] = [a, b, u \cdot i_m v]$ if $u \in WS_m$.*

For a direct calculation shows that the multiplication is well defined and that decomposition into irreducible factors is unique.

The isomorphism $X \cong X \times WS_1$ can be shown to induce an embedding

$$\gamma_X: X \subset \Gamma^+(X) \tag{4.2}$$

which extends to an embedding of the free monoid $F^+(X)$ (which is James' monoid X_∞). Further, if $\Gamma^+(X)$ is filtered by the images of $(X)^n \times WS_n$, the successive quotients are X and the complexes

$$X^{(n)} \rtimes_{S_n} WS_n = ((X \wedge X \wedge \cdots \wedge X) \times_{S_n} WS_n)/\overline{W}S_n, \tag{4.3}$$

the case $n = 2$ being the infinite quadratic construction on X. Hence

LEMMA 4.4. *If X is k-connected, $(\Gamma^+(X), X)$ is $(2k+1)$-connected.*

5. **Homotopy commutativity.** $\Gamma^+(X)$ is strongly homotopy commutative, in the sense that there are higher homotopy commutations of all orders. This follows from the map

$$h: \Gamma^+(\Gamma^+(X)) \to \Gamma^+(X) \tag{5.1}$$

defined as follows. For any k elements $\alpha_1, \ldots, \alpha_k$ of $\Gamma^+(X)$ there is an integer n such that each α_i has a representative α_i' in $(X)^n \times WS_n$. The wreath product with the identity of S_n embeds S_k (as a set) in S_{kn}, and so induces a simplicial map $n^*: WS_k \to WS_{kn}$. The right action of S_{kn} on itself defines a right action of WS_{kn} on itself, and so on the product of the α_i''s. It may be verified that

$$h[\alpha_1, \ldots, \alpha_k, w] = [\alpha_1' \cdots \alpha_k' \cdot n^* w] \tag{5.2}$$

is a well-defined homomorphism of simplicial groups and also that

THEOREM 5.3. *h composed with either $\Gamma^+(\gamma_X)$ or $\gamma_{\Gamma^+(X)}$ is the identity map of $\Gamma^+(X)$. Furthermore, Γ^+ with the maps γ, h, generates a cosimplicial complex*

$$\Gamma^+(X) \underset{\rightarrow}{\overset{\leftarrow}{\rightarrow}} \Gamma^+(\Gamma^+(X)) \underset{\underset{\rightarrow}{\rightarrow}}{\overset{\overset{\leftarrow}{\leftarrow}}{\rightarrow}} \Gamma^+(\Gamma^+(\Gamma^+(X))) \cdots.$$

From h composed with the identification map on $(\Gamma^+(X))^2 \times WS_2$ it follows that

COROLLARY 5.4. *$\Gamma^+(X)$ is homotopy commutative.*

COROLLARY 5.5. *$\Gamma^+(A) \times \Gamma^+(B)$ is a deformation retract of $\Gamma^+(A \vee B)$.*

These involve some computation, and the second uses the Lemma in §3; (5.5) is true even though $\Gamma^+(X)$ need not be a Kan complex.

6. **On $\Gamma(X)$.** Let $\Gamma(X)$ denote the free simplicial group on the same generators as the free simplicial monoid $\Gamma^+(X)$, which is embedded in $\Gamma(X)$. It follows from a theorem of J. C. Moore that $\Gamma(X)$ and $\Gamma^+(X)$ have the same homology if X (and so $\Gamma^+(X)$) is connected; in that case the fundamental groups of $\Gamma^+(X)$ and $\Gamma(X)$ are isomorphic. Therefore, by an extension of a theorem due to J. H. C. Whitehead

THEOREM 6.1. *$\Gamma^+(X) \equiv \Gamma(X)$ if X is connected.*

Now the results of §5 pass over to properties of $\Gamma(X)$, so that by §2, for any cofibration $B \to B \to A/B$, the natural homomorphism

$$\Gamma(B) \to (\ker \Gamma(A) \to \Gamma(A/B))$$

is a homotopy equivalence. In particular, taking A contractible,

THEOREM 6.2. *$\Gamma(B) \equiv \Omega\Gamma\Sigma(B)$.*

By (4.4), $\Sigma^n B$ and $(\Omega^n |\Gamma\Sigma^n B|, \Omega^n\Sigma^n |B|)$ are $(n-1)$-connected, so

COROLLARY 6.3. *$|\Gamma(B)| \equiv \lim_{n\to\infty} \Omega^n\Sigma^n |B| = \Omega^\infty\Sigma^\infty |B|$, for all B.*

Obviously, if B has components $\{B_\alpha \mid \alpha \in S\}$, there is a cofibration $S \to B \to B_\alpha$ and so

$$(6.4) \qquad \begin{aligned} \Gamma^+(B) &\equiv \prod_{\alpha\neq\alpha_0} \Gamma^+(S^0_\alpha) \times \prod_\alpha \Gamma^+(B_\alpha), \\ \Gamma(B) &\equiv \prod_{\alpha\neq\alpha_0} \Gamma(S^0_\alpha) \times \prod_\alpha \Gamma(B_\alpha), \end{aligned} \qquad \text{where } \Gamma^+(B_\alpha) \equiv \Gamma(B_\alpha),$$

(the products being the restricted products if S is infinite).

Now $\Gamma^+(S^0)$ is the disjoint union of the Eilenberg-Mac Lane spaces $\overline{W}S_n$ $(n \geqq 1)$, and right translation by trivial simplices embeds each $\overline{W}S_n$ in later ones in the natural way. It may be shown that every homology class of the component of the identity of $\Gamma(S^0)$ has a representative in a translate of $\overline{W}S_n$ for some n, and so may be deduced the theorem

THEOREM 6.5. *The homologies, with simple coefficients, of* $\Omega^\infty S^\infty$ *and of* $Z \times \overline{W}S_\infty$ *are naturally isomorphic.*

REMARK. This was first proved by Priddy and independently by Quillen. Constructions similar to Γ^+ apparently were known to Dyer and Lashof and also to Quillen. The work of J. Milgram should also be mentioned. G. Segal has recently produced a more sophisticated description.

MANCHESTER UNIVERSITY,
NORTHWESTERN UNIVERSITY, AND
THE UNIVERSITY OF WASHINGTON

HOMOTOPY STRUCTURES AND THE LANGUAGE OF TREES

J. M. BOARDMAN

These preliminary notes offer further glimpses into the theory of trees and H-spaces developed jointly with R. Vogt, beyond what was announced in [Bull. Amer. Math. Soc. **74** (1968), 1117–1122], which we refer to as [BV]. Some proofs are sketched. The trees may be regarded as a convenient disguise for the bar construction. This work is not entirely frivolous; in §8 we give a simple proof of an unpublished theorem of J. F. Adams on monoids, and in §12 we sketch the application to infinite loop spaces. (Some of this material was circulated semi-privately last year from Haverford College.)

1. **Some topology.** To avoid spurious topological difficulties we work entirely in the category of Hausdorff k-spaces or compactly generated spaces, or equivalently in the category of Hausdorff spaces and k-continuous maps. (A function $f: X \to Y$ between topological spaces is said to be *k-continuous* if $f | K: K \to Y$ is continuous whenever K is a compact subspace of X. We call X a *k-space* if for all Y and f, k-continuity coincides with continuity.) For any k-space X, the function space Y^X with the compact-open topology has all the properties expected of it.

Pairs. We consider only closed pairs (X, A), that is, A closed in X. Let us define $(X, A) \times (Y, B) = (X \times Y, X \times B \cup A \times Y)$, and hence

$$(X, A)^n = (X, A) \times (X, A) \times \cdots \times (X, A) = (X^n, \ldots).$$

If (X, A) and (Y, B) have the homotopy extension property (HEP), so does $(X, A) \times (Y, B)$.

We need more. Suppose finite groups G and H act on (X, A) and (Y, B) respectively; then $G \times H$ acts on $(X, A) \times (Y, B)$, and K acts on $(X, A)^n$, where

AMS 1970 *subject classifications.* Primary 18C10, 55D35, 55F35, 57F30; Secondary 55D10.

K is generated by $G \times G \times \cdots \times G$ and the symmetric group S_n (a split extension). The standard proof extends immediately to these cases.

LEMMA 1.1. *If (X, A) and (Y, B) have the equivariant* HEP, *so do the pairs $(X, A) \times (Y, B)$ and $(X, A)^n$. If A is equivariantly a strong deformation retract of X, the same holds for the pairs $(X, A) \times (Y, B)$ and $(X, A)^n$.*

We shall need HELP, the homotopy extension and lifting property of a general homotopy equivalence, which deserves to be better known. It is due to Samelson.

THEOREM 1.2 (HELP). *Let $p: Y \to Z$ be a homotopy equivalence and (X, A) a closed pair with* HEP. *Suppose given a map $g: X \to Z$, a map $f_A: A \to Y$, and a homotopy $H_A: g \mid A \simeq p \circ f_A$. Then there exist a map $f_X: X \to Y$ extending f_A, and a homotopy $H_X: g \simeq p \circ f_X$ extending H_A.*

We need also the equivariant HELP, when a finite group G acts on all the spaces, and the maps and homotopies are equivariant. Suitable hypotheses on p are:

(a) p is an equivariant homotopy equivalence; or

(b) p is an ordinary homotopy equivalence, and X is G-free and paracompact.

2. **Trees of operations.** Suppose given a set X and a collection of operations $X^n \to X$, for various $n \geqq 0$. (Since X^0 is a point, a 0-ary operation simply selects a point in X.) We consider the most general operation $X^k \to X$ obtained from the given operations.

We may obviously compose two maps $X^m \to X^n$ and $X^n \to X^p$ to obtain $X^m \to X^p$.

We have for free the *projection* operations $p_i: X^n \to X$ given by

$$p_i(x_1, x_2, \ldots, x_n) = x_i,$$

in particular the *identity* operation $1 = p_1: X \to X$.

We can construct products. Given k n-ary operations $\theta_1, \theta_2, \ldots, \theta_k: X^n \to X$, we define $\theta = (\theta_1, \theta_2, \ldots, \theta_k): X^n \to X^k$ on $x = (x_1, x_2, \ldots, x_n) \in X^n$ by $\theta(x) = (\theta_1(x), \theta_2(x), \ldots, \theta_k(x)) \in X^k$. Conversely, we recover the θ_i as $\theta_i = p_i \circ \theta$. Hence it is enough to consider operations with range X. In particular, any map of sets $\sigma: \{1, 2, \ldots, m\} \to \{1, 2, \ldots, n\}$ induces the operation

$$\sigma^* = (p_{\sigma 1}, p_{\sigma 2}, \ldots, p_{\sigma m}): X^n \to X^m,$$

by $\sigma^*(x_1, x_2, \ldots, x_n) = (x_{\sigma 1}, x_{\sigma 2}, \ldots, x_{\sigma m})$. It follows that for any

$$\theta = (\theta_1, \theta_2, \ldots, \theta_n): X^r \to X^n$$

we have the formula

$$\sigma^* \circ \theta = (\theta_{\sigma 1}, \theta_{\sigma 2}, \ldots, \theta_{\sigma m}): X^r \to X^m.$$

It is easy enough to see that the general operation $X^k \to X$ is a composite of products of given operations and projections. This description is unsatisfactory, because it is far from unique: for example, given $\alpha: X^2 \to X$ we have

$$\alpha \circ (\alpha, \alpha) = \alpha \circ (\alpha, p_3) \circ (p_1, p_2, \alpha) = \alpha \circ (p_3, \alpha) \circ (p_1, p_2, \alpha).$$

Electrically, we regard a given n-ary operation $\alpha: X^n \to X$ as a black box labeled α, with n inputs numbered $1, 2, \ldots, n$, and one output. When we feed x_i into the ith input for each i, we get out $\alpha(x_1, x_2, \ldots, x_n)$. To obtain the general operation $X^k \to X$, we wire black boxes together, being careful to avoid loops.

We therefore define the tree composite operations.

DEFINITION 2.1. A *0-fold tree composite* is a projection. By induction, a *higher tree composite* operation $X^k \to X$ consists of a *root* operation $\alpha: X^n \to X$, together with previously defined tree composites $\beta_i: X^k \to X$, for $1 \leqq i \leqq n$. Its value on $x = (x_1, x_2, \ldots, x_k) \in X^k$ is $\alpha(\beta_1(x), \beta_2(x), \ldots, \beta_n(x))$. That is, $\alpha \circ (\beta_1, \beta_2, \ldots, \beta_n)$.

This description has the advantage of being unique.

The general tree composite operation $X^k \to X$ is conveniently represented by a certain kind of tree in the plane (plane tree?) with decorations. Each edge has an arrow, but need not have vertices at both ends; we call an edge a *twig* if it has no beginning vertex, the *root* if it has no end vertex, and an *internal edge* otherwise. Each vertex is labeled with one of the given operations; if labeled by $\alpha: X^n \to X$, it has n incoming edges and one outgoing edge, arranged in clockwise order in the plane. Each twig is labeled by a number in $\{1, 2, \ldots, k\}$. The *trivial* tree with one edge, no vertex, and twig label j represents the projection p_j.

Thus the answer to our problem is that the set of all operations from X^n to X is an arboretum, the set of all possible trees with twig labels selected from $\{1, 2, \ldots, n\}$.

EXAMPLE. A ring has five operations: binary operations $+$ and $\times$, a unary operation $-$, and 0-ary operations (or constants) 0 and 1. These satisfy the axioms:

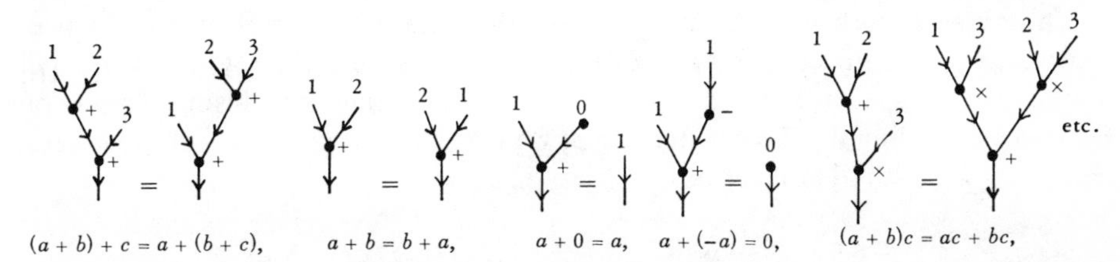

$(a + b) + c = a + (b + c)$, $a + b = b + a$, $a + 0 = a$, $a + (-a) = 0$, $(a + b)c = ac + bc$,

REMARK. It is important to distinguish between twigs and the nonbotanical edges from constant operations; let us call these edges *stumps*. We cannot do without constants.

The general operation from X^m to X^n is a *copse* or ordered system of n trees, with roots labeled in order from 1 to n. To compose two copses $\alpha: X^m \to X^n$ and $\beta: X^n \to X^r$ to form $\beta \circ \alpha: X^m \to X^r$, we take the copse β, and to each twig of β with twig label i we graft a copy of the ith tree in the copse α.

This is not general enough, because it is still not an adequate description of algebra. (In the theory of group representations, for example, one considers groups G acting on vector spaces V, and both G and V may need to be regarded as variable.) What we need instead of one set X is a finite family of objects $X_1, X_2, \ldots, X_n$, which we shall arbitrarily call *colors*, with given operations $\alpha: X_{i_1} \times X_{i_2} \times \cdots \times X_{i_n} \to X_j$. So we need to consider trees in which each edge is also assigned a color, that are correctly colored around each vertex: at a vertex labeled α, the incoming edges are colored $i_1, i_2, \ldots, i_n$ in order, and the outgoing edge is colored j.

3. **Topological-algebraic theories.** Lawvere has formalized the concept of an algebraic theory given by operations and laws without existential quantifiers. As examples we have the theories of monoids, groups, rings, etc. (whose axioms can be put into the required form although they are usually not) but not the theory of fields. He considers the category of *all* operations that can be written down in the theory, instead of selecting certain operations that generate the rest.

We always have the set operations σ^* (see §2). Let $\boldsymbol{S}$ be in effect the category of finite sets, with objects $0, 1, 2, 3, \ldots$, and $\boldsymbol{S}(m, n)$ the set of all maps from $\{1, 2, \ldots, m\}$ to $\{1, 2, \ldots, n\}$.

Since we have to dualize sometime, for technical convenience we dualize now.

DEFINITION 3.1. An *algebraic cotheory* is a category Θ with objects $0, 1, 2, 3, \ldots$ together with a functor $\boldsymbol{S}^{\mathrm{op}} \to \Theta$ that preserves objects and products. (In particular 0 is a terminal object.) We call Θ *topological-algebraic* if each set $\Theta(m, n)$ is a topological space, with continuous composition and products (we require $\Theta(m, n) \cong \Theta(m, 1)^n$ a homeomorphism). A *Θ-space* is a continuous functor $\Theta \to \boldsymbol{T}$, where $\boldsymbol{T}$ is the category of topological spaces, such that $\boldsymbol{S}^{\mathrm{op}} \to \Theta \to \boldsymbol{T}$ preserves products; the image of 1 is called the *underlying space*. A *homomorphism* of Θ-spaces is a natural transformation between such functors. If Θ_1 and Θ_2 are cotheories, a *cotheory functor* is a continuous functor $\Theta_1 \to \Theta_2$ whose composite with $\boldsymbol{S}^{\mathrm{op}} \to \Theta_1$ is $\boldsymbol{S}^{\mathrm{op}} \to \Theta_2$.

In any such cotheory we have the *free* Θ-space FZ on a given topological space Z, with an injection map $i: Z \to FZ$. (That is, given any Θ-space G and map $Z \to G$ of spaces, there is a unique continuous homomorphism $FZ \to G$ of Θ-spaces extending it.) To construct FZ, we take the disjoint union of the spaces $\Theta(n, 1) \times Z^n$, with identifications

$$(\alpha \circ \sigma^*; x_1, x_2, \ldots, x_n) = (\alpha; x_{\sigma 1}, x_{\sigma 2}, \ldots, x_{\sigma m})$$

for any $\sigma \in \boldsymbol{S}(m, n)$ and any $\alpha \in \Theta(m, 1)$. In the category of *based* topological spaces, we need further identifications. (For example, in the cotheory of monoids, FZ is James' reduced product space Z_∞, and in the cotheory of commutative monoids FZ is the infinite symmetric product of Z.)

Now Θ^{op} has the very convenient interpretation as the category of finitely generated free objects: $\Theta(m, n)$ is the space of all Θ-homomorphisms from $Fn = F\{z_1, z_2, \ldots, z_n\}$ to $Fm = F\{z_1, z_2, \ldots, z_m\}$, and $\Theta(n, 1)$ is homeomorphic to the free object Fn.

It is easy to characterize Θ-spaces. For any space Z we have the homomorphism $m: FFZ \to FZ$ which extends the identity map $1: FZ \to FZ$.

THEOREM 3.2. *The space Z is a Θ-space if and only if the injection map $i: Z \to FZ$ admits a retraction $r: FZ \to Z$ that makes the square*

$$\begin{array}{ccc} FFZ & \xrightarrow{Fr} & FZ \\ \downarrow{\scriptstyle m} & & \downarrow{\scriptstyle r} \\ FZ & \xrightarrow{r} & Z \end{array}$$

commute.

PROOF. If Z is a Θ-space, we define $r: FZ \to Z$ as the homomorphism that extends the map $1: Z \to Z$; the square commutes by general theory of adjoint functors. Conversely, given r, we make Z a Θ-space by defining the action of $\alpha \in \Theta(n, 1)$ on Z as the composite

$$Z^n \xrightarrow{i^n} (FZ)^n \xrightarrow{\alpha} FZ \xrightarrow{r} Z.$$

Again we need a generalization to cover cotheories on several objects. Let K be a finite set, whose elements we call *colors* or objects; we have the category $S(-, K)$ of sets over K, whose objects are maps $a: \{1, 2, \ldots, n\} \to K$. In the above definition we simply replace S by $S(-, K)$. Then a Θ-space has many underlying spaces, one for each color $c \in K$. We have the free Θ-space FZ on the space Z, for a given map $q: Z \to K$. Given $a: \{1, 2, \ldots, n\} \to K$, the free object Fa has $\Theta(a, c)$ as its underlying space with color c.

4. **Interchange.** As has been pointed out by Mac Lane and others, the concept of "interchange" is fundamental in this subject. If X and Y are Θ-spaces, then $X \times Y$ is canonically also a Θ-space, componentwise, and in particular X^n becomes a Θ-space.

DEFINITION 4.1. Suppose X is both a $\boldsymbol{U}$-space and a $\boldsymbol{V}$-space. We say these actions *interchange* if each morphism $\alpha: X^m \to X^n$ in the $\boldsymbol{U}$-action is a homomorphism of $\boldsymbol{V}$-spaces, or, equivalently, if each morphism $\beta: X^p \to X^q$ in the $\boldsymbol{V}$-action is a homomorphism of $\boldsymbol{U}$-spaces. We write $\boldsymbol{U} \otimes \boldsymbol{V}$ for the cotheory that acts on X whenever $\boldsymbol{U}$ and $\boldsymbol{V}$ both act and interchange; there are canonical cotheory functors $\boldsymbol{U} \to \boldsymbol{U} \otimes \boldsymbol{V}$ and $\boldsymbol{V} \to \boldsymbol{U} \otimes \boldsymbol{V}$, which need not be inclusions.

The condition is expressed more symmetrically as follows: given $\alpha \in \boldsymbol{U}(m, n)$ and $\beta \in \boldsymbol{V}(p, q)$, the square

$$\begin{array}{ccc} (X^p)^m \cong X^{mp} = (X^m)^p & \xrightarrow{\alpha^p} & (X^n)^p = X^{np} \cong (X^p)^n \\ \downarrow{\scriptstyle \beta^m} & & \downarrow{\scriptstyle \beta^n} \\ (X^q)^m \cong X^{mq} = (X^m)^q & \xrightarrow{\alpha^q} & (X^n)^q = X^{nq} \cong (X^q)^n \end{array}$$

commutes, where the isomorphisms are appropriate shuffle permutations. We write $\alpha \otimes \beta: X^{mp} \to X^{nq}$ for the composite, and call the diagram the shuffle condition for $\alpha \otimes \beta$. Thus $\boldsymbol{U} \otimes \boldsymbol{V}$ may be described as generated by $\boldsymbol{U}$ and $\boldsymbol{V}$, with a shuffle relation for each morphism α in $\boldsymbol{U}$ and each β in $\boldsymbol{V}$, and an appropriate identification topology.

The condition is automatic if α or β is a set operation. If it holds for $\alpha' \otimes \beta$ and $\alpha'' \otimes \beta$, it holds also for $(\alpha' \circ \alpha'') \otimes \beta$, and similarly for the other variable. Hence commutativity $\boldsymbol{U} \otimes \boldsymbol{V} \cong \boldsymbol{V} \otimes \boldsymbol{U}$ and associativity

$$(\boldsymbol{U} \otimes \boldsymbol{V}) \otimes \boldsymbol{W} \cong \boldsymbol{U} \otimes (\boldsymbol{V} \otimes \boldsymbol{W}).$$

Also, the condition need only be verified in the case $n = q = 1$, since it holds for $(\alpha_1, \alpha_2, \ldots, \alpha_n) \otimes (\beta_1, \beta_2, \ldots, \beta_q)$ if it holds for each $\alpha_i \otimes \beta_j$.

EXAMPLE. If $\boldsymbol{U}$ is the cotheory of left R-modules, and $\boldsymbol{V}$ is the cotheory of left S-modules, then $\boldsymbol{U} \otimes \boldsymbol{V}$ is the cotheory of left $(R \otimes S)$-modules; whence our notation.

EXAMPLE [MAC LANE-ECKMANN-HILTON-ETC.]. If Θ_m is the cotheory of monoids, then $\Theta_m \otimes \Theta_m$ is isomorphic to the cotheory of commutative monoids. (That is, two monoid multiplications on X that interchange must coincide and be commutative.)

The structure of $\boldsymbol{U} \otimes \boldsymbol{V}$ is in general anything but clear. Any morphism in $\boldsymbol{U} \otimes \boldsymbol{V}$ can be written as a word $\alpha_1 \circ \beta_1 \circ \alpha_2 \circ \beta_2 \circ \cdots \circ \beta_k$, where $\alpha_i \in \boldsymbol{U}$ and $\beta_i \in \boldsymbol{V}$, with the various shuffle relations. As in §2, the general operation in $\boldsymbol{U} \otimes \boldsymbol{V}$ is better represented by a tree with vertex labels in $\boldsymbol{U}$ or $\boldsymbol{V}$. An edge that joins vertices with labels both in $\boldsymbol{U}$ or both in $\boldsymbol{V}$ may be shrunk away, by composing the labels. The relations between these trees are however very difficult to handle, and in general we have a substantial word problem.

We can and must allow $\boldsymbol{U}$ or $\boldsymbol{V}$ to be a cotheory with colors. If $\boldsymbol{U}$ has colors K and $\boldsymbol{V}$ has colors L, then $\boldsymbol{U} \otimes \boldsymbol{V}$ can still be defined and is a cotheory with colors $K \times L$.

One case we need and can handle is the structure of $\boldsymbol{U} \otimes \boldsymbol{L}_n$, where $\boldsymbol{U}$ is any monochrome cotheory and $\boldsymbol{L}_n$ is the "linear" category whose colors are 0, 1, 2, ..., n, with one morphism $i \to j$ if $i \leqq j$, and none otherwise, and of course the products of these. That is, a $\boldsymbol{L}_n$-space is a sequence of spaces and maps $X_0 \to X_1 \to X_2 \to \cdots \to X_n$, nothing more. A $(\boldsymbol{U} \otimes \boldsymbol{L}_n)$-structure on this $\boldsymbol{L}_n$-space makes the spaces into $\boldsymbol{U}$-spaces and the maps into $\boldsymbol{U}$-homomorphisms.

LEMMA 4.2. *Let $K = \{0, 1, 2, \ldots, n\}$, $a:\{1, 2, \ldots, m\} \to K$, and $c \in K$. Then $(\boldsymbol{U} \otimes \boldsymbol{L}_n)(a, c) \cong \boldsymbol{U}(m, 1)$ if $ai \leqq c$ for all i, and is empty otherwise.*

PROOF. After shuffles, any morphism in $(\boldsymbol{U} \otimes \boldsymbol{L}_n)(a, c)$ has the form

$$(\alpha \otimes 1_c) \circ (\beta_1 \oplus \beta_2 \oplus \cdots \oplus \beta_m),$$

where $\alpha \in \boldsymbol{U}(m, 1)$, $\beta_i : ai \to c$ in $\boldsymbol{L}_n$, and $\oplus : \boldsymbol{L}_n \times \boldsymbol{L}_n \to \boldsymbol{L}_n$ denotes the product functor.

We have one useful general result. Constant operations are in many ways a nuisance.

LEMMA 4.3. *Let $\boldsymbol{U}$ and $\boldsymbol{V}$ be cotheories with at least one constant each. Let $\boldsymbol{U}'$ and $\boldsymbol{V}'$ be the subcotheories of $\boldsymbol{U}$ and $\boldsymbol{V}$ without the constants. Then $\boldsymbol{U} \otimes \boldsymbol{V}$ has* *precisely one constant operation* $0 \to 1$, *and for* $n > 0$, $(\boldsymbol{U} \otimes \boldsymbol{V})(n, 1)$ *is a quotient of* $(\boldsymbol{U}' \otimes \boldsymbol{V}')(n, 1)$.

PROOF. Let e and f be any constants in $\boldsymbol{U}$ and $\boldsymbol{V}$. The shuffle condition for $e \otimes f$ yields $e = f$ in $\boldsymbol{U} \otimes \boldsymbol{V}$. Given any tree in $\boldsymbol{U} \otimes \boldsymbol{V}$, we can prune away all the stumps one by one. If an edge joins a vertex labeled e to a vertex labeled $\beta \in \boldsymbol{V}$, we first replace e by f and then compose in $\boldsymbol{V}$, to remove that edge. We end up with either a tree without stumps or a tree consisting of just one stump and no other vertex.

5. **Split cotheories.** We do not yet know how to handle general cotheories. We restrict attention to the two kinds of cotheory that interest us most, in which we can work satisfactorily. It is clear from examples, and the interpretation of $\Theta(n, 1)$ as the free space Fn, that cotheories tend to be inconveniently large. We need a device for cutting down the number of operations to be considered.

Let Θ_m be the cotheory of monoids, and Θ_{cm} the cotheory of commutative monoids. We have the abelianization functor $\Theta_m \to \Theta_{cm}$.

In Θ_m we pay particular attention to the elements $\lambda_n = z_1 z_2 \cdots z_n \in \Theta_m(n, 1) \cong Fn$, the free monoid on n generators $z_1, z_2, \ldots, z_n$, and their products

$$\lambda_{n_1} \oplus \lambda_{n_2} \oplus \cdots \oplus \lambda_{n_r} \in \Theta_m(n, r),$$

where $\oplus : \Theta_m \times \Theta_m \to \Theta_m$ is the product functor and $n = \sum_i n_i$. Denote by $\boldsymbol{A}$ the subcategory of Θ_m having these as morphisms. Then $\boldsymbol{A}$ has the important property that any morphism in Θ_m has uniquely the form $\lambda \circ \sigma^*$, where λ is a morphism in $\boldsymbol{A}$ and σ is a set operation.

DEFINITION 5.1. We call $\boldsymbol{U}$ a *split cotheory over* Θ_m if we are given a cotheory functor $P : \boldsymbol{U} \to \Theta_m$ such that any morphism in $\boldsymbol{U}$ has uniquely the form $\lambda \circ \sigma^*$, where σ is a set operation and $P\lambda$ is a morphism in $\boldsymbol{A}$, and $\boldsymbol{U}$ has the appropriate topology. We call $P^{-1}(\boldsymbol{A})$ a *cotheory over* $\boldsymbol{A}$.

Let $\boldsymbol{B} = P^{-1}(\boldsymbol{A})$. The products in Θ_m and $\boldsymbol{U}$ are no longer products in $\boldsymbol{A}$ or in $\boldsymbol{B}$; instead, we have only the associative product functor $\oplus : \boldsymbol{B} \times \boldsymbol{B} \to \boldsymbol{B}$, with the properties that $\oplus(m, n) = m + n$ on objects, and any morphism in $\boldsymbol{B}(m, n)$ has uniquely the form $\beta_1 \oplus \beta_2 \oplus \cdots \oplus \beta_n$, with $\beta_i \in \boldsymbol{B}(m_i, 1)$ and $m = \sum_i m_i$. Conversely, from such a category $\boldsymbol{B}$ we may reconstruct canonically the split cotheory $\boldsymbol{U}$. Thus a $\boldsymbol{U}$-space is determined by a functor from $\boldsymbol{B}$ that preserves the product functor $\oplus$.

Similarly for Θ_{cm}, except that because $\lambda_n \circ \pi^* = \lambda_n$ for any permutation $\pi \in S_n$, it is desirable to retain all the permutations. Let $\boldsymbol{S}$ be the same category of sets as in §3; that is, $\boldsymbol{S}(m, n)$ is the set of all maps $\{1, 2, \ldots, m\} \to \{1, 2, \ldots, n\}$. We extend $\boldsymbol{A} \subset \Theta_{cm}$ to a functor $\boldsymbol{S} \subset \Theta_{cm}$ (not to be confused with the set operations, which give rise to $\boldsymbol{S}^{\mathrm{op}} \subset \Theta_{cm}$) by regarding $\boldsymbol{A}(m, n)$ as the subset of $\boldsymbol{S}(m, n)$ consisting of the order-preserving maps. We make a map $\alpha \in \boldsymbol{S}(m, n)$ correspond to the element $(y_1, y_2, \ldots, y_n) \in (Fm)^n \cong \Theta_{cm}(m, n)$, where the elements $y_i \in Fm$ are determined by $y_1 y_2 \cdots y_n = \lambda_m = z_1 z_2 \cdots z_m$ and z_i appears in the factor $y_{\alpha i}$. In particular, for a permutation π we observe that $\pi^{-1} = \pi^*$, the set operation induced by π. We find that every element of $\Theta_{cm}(m, n)$ has the form $\alpha \circ \sigma^*$, where $\alpha \in \boldsymbol{S}$ and σ is a set operation, uniquely up to the obvious relations

$$(\alpha \circ \pi^*) \circ \sigma^* = \alpha \circ (\sigma \circ \pi)^*.$$

DEFINITION 5.2. We call $\boldsymbol{U}$ a *split cotheory over* Θ_{cm} if we are given a cotheory functor $P: \boldsymbol{U} \to \Theta_{cm}$ such that every morphism in $\boldsymbol{U}$ has the form $\alpha \circ \sigma^*$, where σ is a set operation and $P\alpha \in \boldsymbol{S}$, uniquely up to permutations, and has the appropriate topology. We call $P^{-1}(\boldsymbol{S})$ a *cotheory over* $\boldsymbol{S}$.

In the category $\boldsymbol{B} = P^{-1}(\boldsymbol{S})$ we again have the functor $\oplus : \boldsymbol{B} \times \boldsymbol{B} \to \boldsymbol{B}$, and the product property in $\boldsymbol{U}$ is reflected in $\boldsymbol{B}$ by the fact that every morphism in $\boldsymbol{B}(m, n)$ has the form $(\beta_1 \oplus \beta_2 \oplus \cdots \oplus \beta_n) \circ \pi^*$, where $\beta_i \in \boldsymbol{B}(m_i, 1)$ and $\pi \in S_m$, uniquely up to tl e obvious relations

$$((\beta_1 \circ \pi_1^*) \oplus \cdots \oplus (\beta_n \circ \pi_n^*)) \circ \pi^* = (\beta_1 \oplus \cdots \oplus \beta_n) \circ (\pi \circ (\pi_1 \oplus \cdots \oplus \pi_n))^*.$$

We also have inclusions $S_n \subset \boldsymbol{B}(n, n)$, satisfying the relations detailed in §2 of [BV]. Conversely, from any category $\boldsymbol{B}$ with appropriate functor $\oplus$ and inclusions $S_n \subset \boldsymbol{B}(n, n)$ we can reconstruct the split cotheory $\boldsymbol{U}$ over Θ_{cm}. Then a $\boldsymbol{U}$-space is determined by a functor from $\boldsymbol{B}$ that preserves the product functor $\oplus$ and the permutations $S_n \subset \boldsymbol{B}(n, n)$ for all n.

This is a more general concept because Θ_m is itself a split cotheory over Θ_{cm}, and hence any split cotheory over Θ_m also splits over Θ_{cm}.

Our principal concern will be E-spaces.

DEFINITION 5.3. We call a cotheory $\boldsymbol{B}$ over $\boldsymbol{S}$ an *E-cotheory* if each space $\boldsymbol{B}(n, 1)$ is contractible, in other words, if $P: \boldsymbol{B} \to \boldsymbol{S}$ is topologically a homotopy equivalence. An *E-space* (or *homotopy-everything H-space* in [BV]) is a space with a $\boldsymbol{B}$-action, for some E-cotheory $\boldsymbol{B}$. Similarly, an *E-map* from the E-space X to the E-space Y is an action on the pair (X, Y) of some cotheory $\boldsymbol{C}$ over $\boldsymbol{S} \otimes \boldsymbol{L}_1$, such that $P: \boldsymbol{C} \to \boldsymbol{S} \otimes \boldsymbol{L}_1$ is a homotopy equivalence on the morphism spaces.

We shall also need split cotheories in color. These are defined in the obvious way as split cotheories over $\Theta_m \otimes \boldsymbol{L}_n$ or $\Theta_{cm} \otimes \boldsymbol{L}_n$.

[Our whole theory developed in part from the theory of PROP's and PACT's propounded by Adams and Mac Lane. A topological PROP is what we call a cotheory over $\boldsymbol{S}$, and a PACT is the analogue for chain complexes. A Steenrod PACT is the analogue of an E-space.]

We also need to know about split cotheories that interchange. Since

$$\Theta_m \otimes \Theta_m \cong \Theta_m \otimes \Theta_{cm} \cong \Theta_{cm} \otimes \Theta_{cm} \cong \Theta_{cm},$$

it follows that, given two cotheories $\boldsymbol{U}$ and $\boldsymbol{V}$, each over $\boldsymbol{A}$ or $\boldsymbol{S}$, we obtain a cotheory $\boldsymbol{U} \otimes \boldsymbol{V}$ over $\boldsymbol{S}$.

6. **The bar construction.** Given a cotheory Θ, we define a new cotheory $W\Theta$ that is homotopically like Θ, but categorically and topologically free in some sense.

As motivation, let us consider monoids, or $\boldsymbol{A}$-spaces. Suppose X is an $\boldsymbol{A}$-space, and that Y is homotopy-equivalent to X. Corresponding to each operation

$\lambda_n : X^n \to X$ we can choose (up to homotopy) a corresponding operation $[\lambda_n] : Y^n \to Y$. Hence the action on Y of a cotheory W^0A over A, that is essentially A with its relations removed. We write $[\alpha \mid \beta]$ for $[\alpha] \circ [\beta]$ and $[\alpha \oplus \beta]$ for $[\alpha] \oplus [\beta]$. Thus any morphism in W^0A has the form $[\alpha_0 \mid \alpha_1 \mid \cdots \mid \alpha_n]$. However, this is misleading; by §2 the morphisms in W^0A are really the tree composites of the operations $[\lambda_n]$.

In Y we have $[\alpha \mid \beta] \simeq [\alpha \circ \beta]$, instead of equality. Choose a homotopy $H : [\lambda_2 \mid \lambda_2 \oplus 1] \simeq [\lambda_2 \mid 1 \oplus \lambda_2] : Y^3 \to Y$, so that $H(a, b, c)$ is a path in Y joining $(ab)c$ and $a(bc)$, where ab denotes $\lambda_2(a, b)$. The following diagram is obligatory in any work on this topic. From λ_2 and H we form the Stasheff pentagon in Y, for any points $a, b, c, d \in Y$,

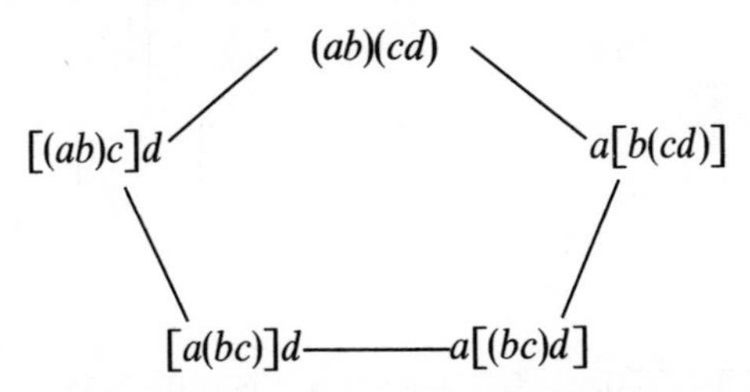

where the five edges, in clockwise order from the top, are $H(a, b, cd)$, $aH(b, c, d)$, $H(a, bc, d)$, $H(a, b, c)d$, and $H(ab, c, d)$. We know that if H is chosen intelligently, this loop can be filled in to form a map $D^2 \to Y$ of the 2-disk (really $D^2 \times Y^4 \to Y$), but usually not if it is not. Once we have chosen such a map, we can go on to consider 5 points, construct a 2-sphere, try to fill that in, etc. This is the classical approach, which led to Stasheff's theory of A_∞-structures. And we have not yet mentioned how the basepoint is to behave.

We need to be more explicit. Instead of comparing $[\lambda_2 \mid \lambda_2 \oplus 1]$ and $[\lambda_2 \mid 1 \oplus \lambda_2]$ directly, we compare each to $[\lambda_3]$; that is, we remove the bars. In terms of trees, this means shrinking the edges away. This suggests our general construction. We shall bisect each edge of the pentagon, place $[\lambda_4]$ in the center, and divide the pentagon into 5 squares. However, we wish to retain the essential properties of the set operations, especially the identity operation.

DEFINITION 6.1. Given a cotheory Θ, we define a new cotheory $W\Theta$. We consider the space of all trees with structure as in §2, with vertex labels in Θ, and in addition for each internal edge a real number in $[0, 1]$, called its *length*. We impose three kinds of relation:

(a) *We may remove any edge of length* 0: we unite the vertices at the two ends to form a new vertex, whose label is the tree composite in Θ of the old vertex labels.

(b) *We may remove any vertex labeled by* $1 \in \Theta(1, 1)$: we give the resulting edge length $t_1 * t_2 = t_1 + t_2 - t_1 t_2$, where t_1 and t_2 are the lengths of the edges above and below this vertex.

(c) *We may replace any vertex label* $\alpha \circ \sigma^*$ *by* α, *where* $\alpha \in \Theta(m, 1)$ *and* $\sigma \in \mathbf{S}(m, n)$, *by changing the part of the tree above this vertex*: if $C_1, C_2, \ldots, C_n$ are the subtrees above the inputs to $\alpha \circ \sigma^*$, we take $C_{\sigma 1}, C_{\sigma 2}, \ldots, C_{\sigma m}$ as the subtrees over α.

We give $W\Theta$ the obvious quotient topology; a tree is to be a continuous function of its edge lengths and vertex labels as long as its shape remains the same.

When we compose trees or copses by grafting, we give the new edges length 1. (By convention, all roots and twigs have length 1.)

The *augmentation* functor $\varepsilon: W\Theta \to \Theta$ is defined on a tree by ignoring the lengths and taking the tree composite in Θ of the vertex labels as in Definition 2.1.

Pictorially, the relations are:

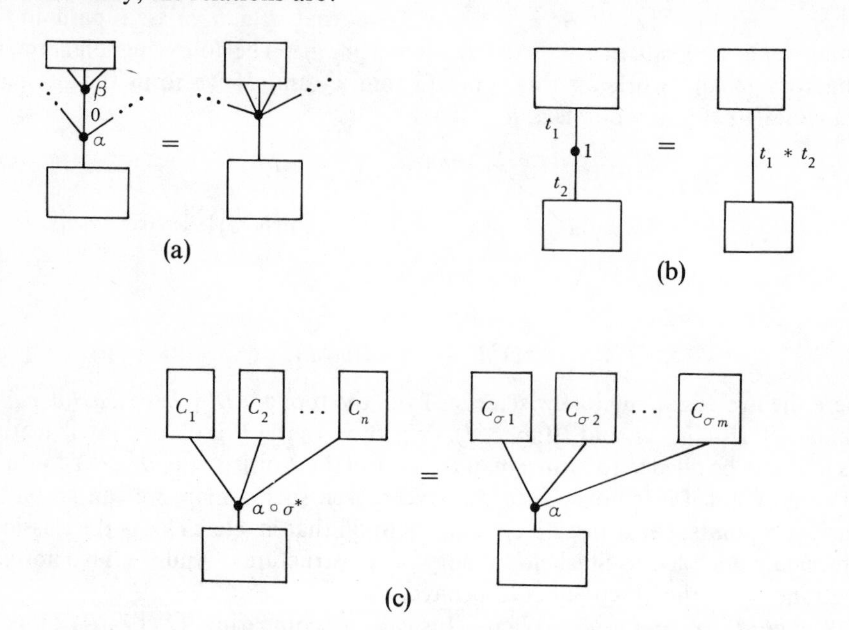

REMARK. Instead of relation (b), we may use relation (b′), for which we replace $t_1 * t_2$ by the simpler formula $\max(t_1, t_2)$. It matters little which we use, except in §9, where we must use (b).

This all works equally well for a cotheory in color. In the useful case $W(\boldsymbol{U} \otimes \boldsymbol{L}_n)$, by Lemma 4.2 we may label the vertices simply by operations in $\boldsymbol{U}$, with some care when applying relation (b): a vertex labeled in this way by 1 is not really an identity unless the incoming and outgoing edges have the same color.

In case $\boldsymbol{U}$ is a split cotheory over Θ_{cm}, with $\boldsymbol{B} = P^{-1}(\boldsymbol{S})$, relation (c) shows that we need only consider trees whose vertex labels lie in $\boldsymbol{B}$. We deduce that $W\boldsymbol{U}$ is again a split cotheory over Θ_{cm}, and we write $W\boldsymbol{B}$ for the part over $\boldsymbol{S}$. Thus in $W\boldsymbol{B}$ we need relation (c) only for permutations, and a tree with n twigs lies in $W\boldsymbol{B}(n, 1)$ if and only if its twig labels are $1, 2, \ldots, n$ in some order. In view of relation (b), *we assume from now on that the inclusion* $\{1\} \subset \boldsymbol{B}(1, 1)$ *has the* HEP.

In the more special case when the cotheory $\boldsymbol{U}$ is split over Θ_m, and $\boldsymbol{B} = P^{-1}(\boldsymbol{A})$, relation (c) again shows that $W\boldsymbol{U}$ consists of trees whose vertex labels lie in $\boldsymbol{B}$. Hence $W\boldsymbol{U}$ is a split theory over Θ_m, and we write $W\boldsymbol{B}$ for the part over $\boldsymbol{A}$; the subspace $W\boldsymbol{B}(n, 1)$ of $W\boldsymbol{U}(n, 1)$ consists of those trees whose twig labels are

$1, 2, \ldots, n$ in order. Thus in $\boldsymbol{WB}$ twig labels are unnecessary, and relation (c) becomes vacuous.

Let us take a cotheory $\boldsymbol{B}$ over $\boldsymbol{S}$ and consider the spaces $\boldsymbol{WB}(n, 1)$ in more detail. We define the *k-skeleton* subcategory $W^k\boldsymbol{B}$ of $\boldsymbol{WB}$ as generated under composition by trees with at most k internal edges. Consider the space P_λ of trees in $\boldsymbol{WB}(n, 1)$ with a given shape λ; a tree in P_λ is specified by its edge lengths t_i, its vertex labels α_j, and its twig labels, which are some permutation of $\{1, 2, \ldots, n\}$. Thus P_λ has the form

$$P_\lambda = I^k \times \boldsymbol{B}(1, 1)^m \times \prod_{r \neq 1} \boldsymbol{B}(r, 1)^{m_r} \times S_n,$$

which we call a "generalized cube". Let G be the symmetry group of the tree shape λ, so that G acts on P_λ by relation (c). In particular, G permutes the n twigs, and we obtain a homomorphism $\theta: G \to S_n$. We note for future use that G acts freely on P_λ if each space $\boldsymbol{B}(n, 1)$ is S_n-free, where $S_n \subset \boldsymbol{B}(n, n)$ acts on the right by composition.

The relations apply to P_λ in various ways: (i) if some $t_i = 0$, relation (a) applies, to simplify the tree, (ii) if some $t_i = 1$, the tree decomposes, and lies in $W^{k-1}\boldsymbol{B}(n, 1)$, and (iii) if some $\alpha_j = 1$, relation (b) applies and the tree simplifies. Let Q_λ be the subspace of P_λ consisting of points satisfying at least one of these, and R_λ the subspace of points satisfying (i) or (iii). Then we have the G-equivariant characteristic map

$$u_\lambda: (P_\lambda, Q_\lambda) \to (W^k\boldsymbol{B}(n, 1), W^{k-1}\boldsymbol{B}(n, 1)),$$

where G acts on $W^k\boldsymbol{B}(n, 1)$ by means of $\theta: G \to S_n$ and composition. Call the set of shapes obtained from λ by relation (c) a shape orbit; as λ runs through one representative from each orbit we account for all the points of $\boldsymbol{WB}$. The tree differential calculus assures us that the identifications are consistent. If each $\boldsymbol{B}(n, 1)$ is a CW-complex, so are the spaces $\boldsymbol{WB}(n, 1)$. We have here a counterexample to the dictum that all good up-to-date spaces are constructed from simplexes rather than cubes.

LEMMA 6.2. (a) *Suppose given a functor $G_{k-1}: W^{k-1}\boldsymbol{B} \to \boldsymbol{C}$ of cotheories over $\boldsymbol{S}$. Suppose given a collection of G-equivariant maps $g_\lambda: P_\lambda \to \boldsymbol{C}(n, 1)$ extending $G_{k-1} \circ (u_\lambda \mid Q_\lambda)$, one for each shape orbit of trees with k internal edges. Then there is a unique functor $G_k: W^k\boldsymbol{B} \to \boldsymbol{C}$ that extends G_{k-1} and satisfies $G_k \circ u_\lambda = g_\lambda$ for all λ.*

(b) *Suppose given for each k a functor $G_k: W^k\boldsymbol{B} \to \boldsymbol{C}$ such that $G_{k-1} = G_k \mid W^{k-1}\boldsymbol{B}$. Then we obtain a unique functor $G: \boldsymbol{WB} \to \boldsymbol{C}$ by setting $G \mid W^k\boldsymbol{B} = G_k$ for all k.*

Compare the construction of maps from a CW-complex.

Hence a similar result, with no group actions, for a cotheory over $\boldsymbol{A}$.

We next show that $\varepsilon: \boldsymbol{WB} \to \boldsymbol{B}$ is a homotopy equivalence.

LEMMA 6.3. *For each n, $\varepsilon: \boldsymbol{WB}(n, 1) \to \boldsymbol{B}(n, 1)$ is a S_n-equivariant homotopy equivalence.*

PROOF. We filter $WB(n, 1)$ by the subspaces F_k of trees with at most k internal edges. Then F_k is obtained from F_{k-1} by attaching spaces by the maps

$$u_\lambda : (P_\lambda, R_\lambda) \to (F_k, F_{k-1}),$$

one for each λ. We know from Lemma 1.1 that each R_λ is equivariantly a deformation retract of P_λ, and hence that F_{k-1} is a deformation retract of F_k. It follows by induction that $WB(n, 1)$ has the copy F_0 of $B(n, 1)$ as deformation retract. (If we use relation (b′) instead of (b), the proof becomes trivial: an explicit homotopy replaces each edge length t_i by tt_i, where t runs from 1 to 0.)

A more pictorial description of a WB-action on a space X is sometimes useful. Rather than give maps from $WB(n, 1)$ to the space of maps from X^n to X, we consider the maps $WB(n, 1) \times X^n \to X$.

DEFINITION 6.4. A *cherry tree* on X is simply a tree in WB, except that to each twig is assigned a point of X, which we call a *cherry*, instead of a twig label, that is, it is a point of the space $WB(n, 1) \times_{S_n} X^n$.

EXAMPLES.

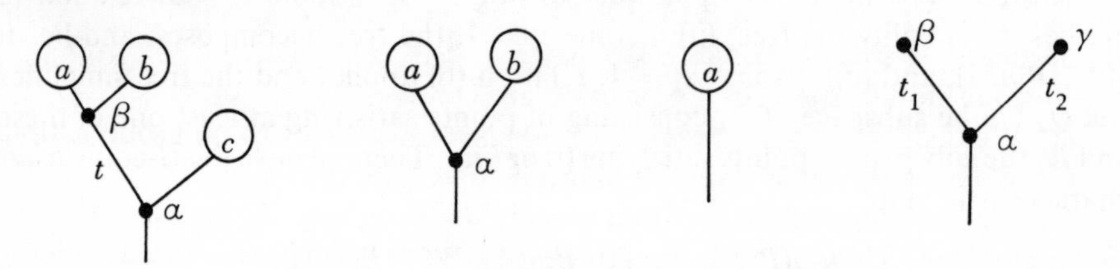

LEMMA 6.5. *A WB-action on X is equivalent to a continuous function from the space of cherry trees on X to X, that factors through the following relations*:

(a) *as before, for an edge of length* 0;
(b) *as before, for a vertex labeled with* 1;
(c) *as before, for permutation relations*;
(d) *we can cut down a fully-grown cherry tree* (*after G. Washington*);
(e) *the value of the trivial cherry tree with cherry x is x.*

Relation (d) demands some explanation. We say a cherry tree is *fully grown* if some internal edge has length 1; to *cut it down* we replace the subtree sitting on that edge by its value, regarded as a cherry in X, and the cut branch becomes a twig.

EXAMPLE.

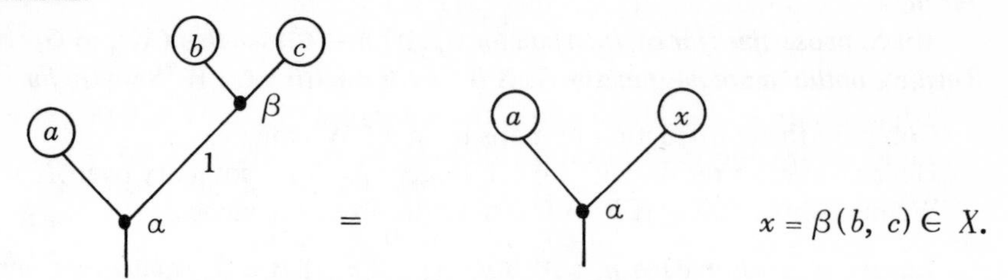

Of course, relations (d) and (e) say that the WB-action is functorial.

The lemma is really a special case of Theorem 3.2. The space of all cherry trees on X is just the free $\boldsymbol{WB}$-space FX.

7. **The Lifting Theorem.** We use the following theorem to replace a naturally occurring cotheory by an artificial bar construction cotheory introduced in §6.

THEOREM 7.1. *Suppose $\boldsymbol{B}$, $\boldsymbol{C}$, and $\boldsymbol{D}$ are cotheories over $\boldsymbol{S}$, with functors $F:\boldsymbol{B}\to\boldsymbol{C}$ and $G:\boldsymbol{D}\to\boldsymbol{C}$. Then there exists a functor $H:W\boldsymbol{B}\to\boldsymbol{D}$ and a homotopy of functors $K(t):W\boldsymbol{B}\to\boldsymbol{C}$ $(t\in I)$ from $F\varepsilon$ to GH, provided that for all $n\geqq 0$:*

(a) *$G:\boldsymbol{D}(n,1)\to\boldsymbol{C}(n,1)$ is a S_n-equivariant homotopy equivalence, or*

(b) *$G:\boldsymbol{D}(n,1)\to\boldsymbol{C}(n,1)$ is an ordinary homotopy equivalence, and $\boldsymbol{B}(n,1)$ is S_n-free and paracompact.*

Moreover, the functor H is unique up to homotopy of functors.

PROOF. We construct H and $K(t)$ by induction on the skeleton subcategories of $W\boldsymbol{B}$ by Lemma 6.2. Suppose we already have H_{k-1} and $K(t)_{k-1}$, defined on $W^{k-1}\boldsymbol{B}$. To continue, we need, for each λ, maps $h_\lambda:P_\lambda\to\boldsymbol{D}(n,1)$ and $k_\lambda(t):P_\lambda\to\boldsymbol{C}(n,1)$, already given on Q_λ and satisfying $k_\lambda(0)=F\circ\varepsilon\circ u_\lambda$. These maps are provided by the equivariant HELP in §1.

Then any $\boldsymbol{D}$-space becomes a $W\boldsymbol{B}$-space.

Various modifications of the theorem are useful:

(1) We deduce a similar result for cotheories over $\boldsymbol{A}$, with no group actions.

(2) We have the same result for colored cotheories over $\boldsymbol{S}\otimes\boldsymbol{L}_n$.

(3) We may suppose H and $K(t)$ already given on a subcotheory of $W\boldsymbol{B}$ that is topologically closed and is the union of certain of the generalized cubes P_λ (the analogue of a subcomplex of a CW-complex).

REMARK. One would obviously like to have $GH=F\varepsilon$. To achieve this, further assumptions are necessary. The homotopy equivalence $\boldsymbol{D}(n,1)\to\boldsymbol{C}(n,1)$ must be fibrewise as well as equivariant, and we need the identity $1\in\boldsymbol{B}(1,1)$ to live in a good neighborhood that is mapped to 1 under F (which is obviously true if 1 is an isolated point).

As an application, the cotheory $\boldsymbol{Q}_1$ over $\boldsymbol{S}$ acts naturally on the loop space ΩY (see [BV]). This cotheory lifts in the obvious way to a cotheory $\boldsymbol{Q}$ over $\boldsymbol{A}$, for which $\boldsymbol{Q}(n,1)$ is contractible for all n. The lifting theorem applies to the functor $\boldsymbol{Q}\to\boldsymbol{A}$, with $F=1$, to yield a functor $W\boldsymbol{A}\to\boldsymbol{Q}$. Thus

LEMMA 7.2. *The loop space ΩY can be made a $W\boldsymbol{A}$-space, naturally in Y.*

REMARKS. The usual loop space ΩY is not a monoid, but the functor Ω does preserve products. Moore's loop space $\Omega_M Y$, which contains ΩY as deformation retract, is the space of pairs $(\omega,a)\in Y^R\times R$ such that $a\geqq 0$ and $\omega:R\to Y$ is a "path of length a", such that $\omega(t)$ is the basepoint of Y for $t\leqq 0$ or $t\geqq a$. It is a monoid under the multiplication defined by $(\omega_1,a_1)\cdot(\omega_2,a_2)=(\omega,a_1+a_2)$, where

$$\omega(t)=\begin{cases}\omega_1(t) & \text{if } 0\leqq t\leqq a_1,\\ \omega_2(t-a_1) & \text{if } a_1\leqq t\leqq a_1+a_2,\end{cases}$$

but the functor Ω_M fails to preserve products. *It is easy to see that no loop space functor L can preserve products and be monoid-valued.* For then LLY would admit an action of $\Theta_m \otimes \Theta_m \cong \Theta_{cm}$, and the Dold-Thom theorem asserts that any commutative monoid has the homotopy type of a product of Eilenberg-Mac Lane spaces; which is obviously not the case in general for $\Omega^2 Y$.

8. **Homotopy invariance.** The property of having the homotopy type of some $\boldsymbol{B}$-space is obviously a homotopy invariant. We show that this property is the same as admitting a $\boldsymbol{WB}$-structure, which implies Theorem E of [BV].

First, let $p: Y \to Z$ be a map, where Y is a $\boldsymbol{WB}$-space, and Z is a $\boldsymbol{B}$-space and therefore a $\boldsymbol{WB}$-space by means of $\varepsilon: \boldsymbol{WB} \to \boldsymbol{B}$. The general concept of morphism of $\boldsymbol{WB}$-spaces is simply a $W(\boldsymbol{B} \otimes \boldsymbol{L}_1)$-action on (Y, Z); that is, we consider trees with vertex labels in $\boldsymbol{B}$ and with each edge colored Y or Z. In our situation, the edge lengths are irrelevant in the given induced action of $\boldsymbol{WB}$ on the $\boldsymbol{B}$-space Z. This suggests considering the quotient cotheory $\boldsymbol{C}$, say, of $W(\boldsymbol{B} \otimes \boldsymbol{L}_1)$ in which the action of any tree is independent of the lengths of its Z-edges, which might as well be 0; let us say the Z-edges are *ignorable*. We now describe $\boldsymbol{C}$. Any morphism is a composite of four kinds of morphism:

(i) $A: Y^n \to Y$, where $A \in \boldsymbol{WB}(n, 1)$ and is indecomposable;

(ii) $(A, t): Y^n \to Z$, where $A \in \boldsymbol{WB}(n, 1)$ is indecomposable and $t \in I$, given pictorially by

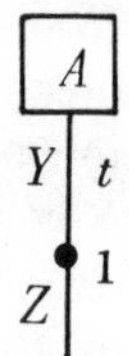

(iii) $p: Y \to Z$, given by

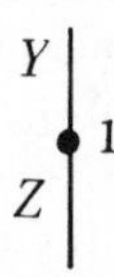

(iv) the morphisms $\beta: Z^n \to Z$, where $\beta \in \boldsymbol{B}(n, 1)$.

As sufficient relations, we have the relations in $\boldsymbol{B}$ for the morphisms (iv), $(A, 1) = p \circ A$, and

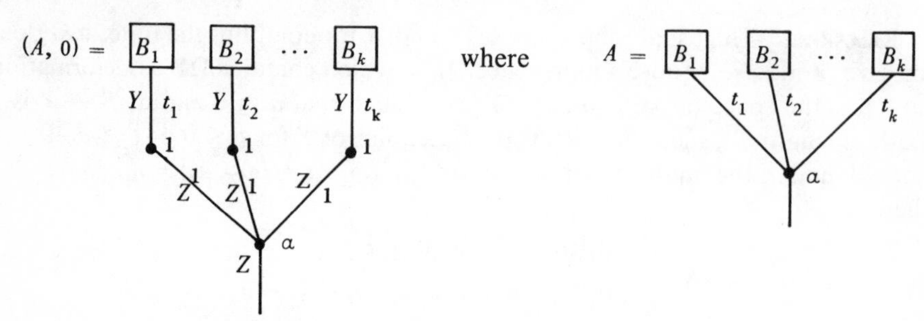

THEOREM 8.1. *Suppose given a homotopy equivalence $p: Y \to Z$, where Z is a $\boldsymbol{B}$-space. Then we can give Y a $W\boldsymbol{B}$-structure, and $p: Y \to Z$ a $W(\boldsymbol{B} \otimes \boldsymbol{L}_1)$-structure with ignorable Z-edges.*

PROOF. We need actions $F(A): Y^n \to Y$ and $G(A, t): Y^n \to Z$ satisfying $G(A, 1) = p \circ F(A)$, for $A \in W\boldsymbol{B}$. Suppose these are already defined by induction for $A \in W^{k-1}\boldsymbol{B}$. By Lemma 6.2, to extend over $W^k\boldsymbol{B}$ we need G-equivariant maps $f_\lambda: P_\lambda \times Y^n \to Y$ and $g_\lambda(t): P_\lambda \times Y^n \to Z$, already given on $Q_\lambda \times Y^n$, where G acts on Y^n by means of $\theta: G \to S_n$ and permutation of the factors of Y^n, and G acts trivially on Y and Z. Also, $g_\lambda(0)$ is already known. Thus $p: Y \to Z$ is trivially an equivariant homotopy equivalence, and the equivariant HELP applies.

The converse is even simpler. Whereas it was known by circuitous means that any $W\boldsymbol{A}$-space is homotopy-equivalent to a monoid, the Moore loop space on a classifying space, a direct proof is due to Adams (unpublished). Moreover, Adams' proof is completely general and holds for any cotheory.

THEOREM 8.2 (ADAMS; Compare Theorem H of [BV]). *We can embed any $W\boldsymbol{B}$-space X naturally as a strong deformation retract of a $\boldsymbol{B}$-space UX, and give the inclusion $i: X \subset UX$ a $W(\boldsymbol{B} \otimes \boldsymbol{L}_1)$-action with ignorable UX-edges. Moreover, UX is universal for these actions: given any $\boldsymbol{B}$-space Y, any such action from X to Y factors through a unique $\boldsymbol{B}$-homomorphism $UX \to Y$.*

PROOF. We construct UX explicitly, as the space of all cherry trees with cherries in X, root colored UX, all internal edges and twigs colored X, and vertex labels in $\boldsymbol{B}$, subject to the usual relations (a), (b), (c), and the cutting-down relation (d) applied to any proper subtree using the given $W\boldsymbol{B}$-action on X, as interpreted by Lemma 6.5. We define $i: X \to UX$ by taking $i(x)$ as the cherry tree

The $W(\boldsymbol{B} \otimes \boldsymbol{L}_1)$-action on (X, UX) and the universal property are immediate.

To see that X is a deformation retract of UX, consider the homotopy $h_t: UX \to UX$ defined by

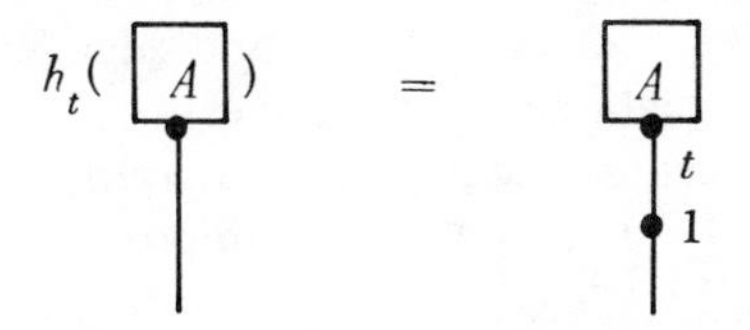

Then $h_0 = 1$, h_t leaves iX fixed, and the image cherry tree under h_1 can be cut down to iX.

We shall in §11 strengthen the result to say that X and UX are homotopy-equivalent $W\boldsymbol{B}$-spaces.

REMARK. The functor U does not preserve products. In fact we saw in §7 that there is no product-preserving functor that takes WA-spaces to homotopy-equivalent monoids, not even for loop spaces.

[We have not yet mentioned basepoints. We suggest four possibilities for basepoint-watchers, in the case of the theory $\boldsymbol{A}$:

(1) Have *no* basepoints! The operations λ_0 etc. are well able to take care of themselves, up to all the right homotopies.

(2) Have a basepoint e, which is preserved by all the operations. In tree language this means that all trees with no twigs have the same value e. This is the most usual and useful hypothesis. It allows trees to have stumps, if they have twigs as well.

(3) Have a basepoint e which is a strict "identity" for all the operations. In tree language, the natural (somewhat stronger) demand is that any stumps on a tree may be pruned away without affecting the operation. Unfortunately, such a condition is totally unacceptable when dealing with interchanges. Suppose we have an operation $\alpha:2 \to 1$ in $\boldsymbol{B}$ with $e:0 \to 1$ as two-sided strict identity, and similarly $\beta:2 \to 1$ and $f:0 \to 1$ in $\boldsymbol{C}$. Then in $\boldsymbol{B} \otimes \boldsymbol{C}$, $e = f$ by Lemma 4.3, and α and β must coincide and be commutative, since

$$\alpha(x, y) = \alpha(\beta(x, e), \beta(e, y)) = \beta(\alpha(x, e), \alpha(e, y)) = \beta(x, y),$$

and

$$\alpha(x, y) = \alpha(\beta(e, x), \beta(y, e)) = \beta(\alpha(e, y), \alpha(x, e)) = \beta(y, x) = \alpha(y, x).$$

(4) Forget all about basepoints, identity operations, and λ_0; that is, replace $\boldsymbol{A}$ by its subcategory generated by λ_2. Then the theorem replaces a strongly homotopy-associative multiplication on X by a strictly associative multiplication on a homotopy-equivalent space UX, an observation due to Adams.]

9. **Universal G-spaces.** Let G be a topological group, or monoid. We consider G not as an $\boldsymbol{A}$-space, but as a category to act on other spaces. Define the cotheory $\boldsymbol{B}$ over $\boldsymbol{A}$ or $\boldsymbol{S}$ by $\boldsymbol{B}(1, 1) = G$ and $\boldsymbol{B}(n, 1) = \varnothing$ if $n \neq 1$, with composition inherited from G, so that a $\boldsymbol{B}$-space is a G-space as usually understood. Given a $W\boldsymbol{B}$-space X, Theorem 8.2 constructs the universal $\boldsymbol{B}$-space UX. In particular, when X is a single point P with its unique $W\boldsymbol{B}$-action, the space UP is a contractible G-free G-space. This is the universal space for the group G.

In this case, a cherry tree is linear and vertical, and may be specified by giving in order, going up the tree, the vertex labels and edge lengths, as $(g_0, t_1, g_1, t_2, g_2, \ldots, g_k)$. We have the relations (a), (b), and (d).

Milgram takes in effect all sequences $(g_0, u_1, g_1, \ldots, g_k)$, where $k \geqq 0$, $g_i \in G$, and $0 \leqq u_1 \leqq u_2 \leqq \cdots \leqq u_k \leqq 1$, and imposes relations:

(a) replace "$g_i, 0, g_{i+1}$" by "$g_i g_{i+1}$";

(b) omit "$u_i, 1,$";

(d) replace "$g_{k-1}, 1, g_k)$" by "$g_{k-1})$".

Call this space EG, with appropriate identification topology. We map UP to EG by

$$(g_0, t_1, g_1, t_2, g_2, \ldots, t_k, g_k) \mapsto (g_0, u_1, g_1, u_2, g_2, \ldots, u_k, g_k),$$

where $u_1 = t_1$, $u_2 = t_1 * t_2$, $u_3 = t_1 * t_2 * t_3, \ldots, u_k = t_1 * t_2 * \cdots * t_k$. Then our relations (a), (b), and (d) on cherry trees in UP correspond in order to the relations in EG. Moreover, this is a homeomorphism of G-spaces (but not if we use relation (b′) instead of (b)). The important property of Milgram's construction is that, unlike Milnor's, it *does* preserve products: $E(G \times H) \cong EG \times EH$.

The *classifying space* BG is the orbit space EG/G, so that the functor B also preserves products. (We are not concerned with whatever BG classifies.) So for BG, we take the same cherry trees as for UP, but ignore the label on the root vertex.

For any monoid G, there is a canonical map $jG: G \to \Omega BG$ defined by $((jG)g)(t) = (1 - t, g) \in BG$. For any space Y, there is a canonical map $eY: B\Omega_M Y \to Y$, where $\Omega_M Y$ is the Moore loop space, given by

$$e(u_1, x_1, u_2, x_2, \ldots, u_k, x_k) = \omega \sum_i (1 - u_i)a_i\,,$$

where $x_i = (\omega_i, a_i)$ in $\Omega_M Y$ and $x_1 x_2 \cdots x_k = (\omega, \sum_i a_i)$. It is clear that the composite

$$\Omega Y \xrightarrow[j\Omega Y]{} \Omega B\Omega Y \subset \Omega B\Omega_M Y \xrightarrow[\Omega e Y]{} \Omega Y$$

is the identity.

THEOREM 9.1. *The canonical map $jG: G \to \Omega BG$ is a homotopy equivalence under suitable conditions, e.g. G a* CW*-complex with $\pi_0(G)$ a group. It admits a $W(A \otimes L_1)$-structure, natural in G. The canonical map $eY: B\Omega_M Y \to Y$ is a homotopy equivalence, if Y is a connected* CW*-complex.*

10. **The noncategory of WB-spaces.** Although WA-spaces are well established, maps between them have generally been neglected as being too complicated. This does not deter us. It is already apparent that the appropriate kind of morphism from one WB-space X to another WB-space Y is a $W(B \otimes L_1)$-structure on the pair (X, Y) that extends the given WB-actions on X and Y. (Smaller less extravagant categories are available, but are technically harder to work with.) Such morphisms occur in §§8 and 9. Unfortunately we cannot take these as morphisms in a category, for lack of a definition of the composite of two morphisms, unless one of them happens to be a WB-homomorphism.

The phenomenon is seen at its simplest in the case of homotopy homomorphisms $f: X \to Y$ and $g: Y \to Z$ between monoids, with given homotopies $f \circ \lambda_2 \simeq \lambda_2 \circ (f \oplus f): X^2 \to Y$ and $g \circ \lambda_2 \simeq \lambda_2 \circ (g \oplus g): Y^2 \to Z$. We deduce that $h \circ \lambda_2 \simeq \lambda_2 \circ (h \oplus h)$, where $h = g \circ f$, but not by any homotopy that is going to make composition associative.

For the present, therefore, we just have to learn to live with this situation. Categories are not everything. Instead we look again and see—a simplicial set (or class).

DEFINITION 10.1. We define the simplicial set K by taking as an n-simplex any action of $W(B \otimes L_n)$ on any $(n + 1)$-tuple of spaces $(X_0, X_1, \ldots, X_n)$, with face and degeneracy operations in K induced by the corresponding functors between the various categories L_r.

In particular, the vertices of K are $\boldsymbol{WB}$-spaces, and an edge of K is a morphism between two $\boldsymbol{WB}$-spaces which are the two ends of this edge.

THEOREM 10.2. *The simplicial set K satisfies the restricted Kan extension condition, which is the usual Kan extension condition, except that the omitted face in the data is not allowed to be the first or the last.*

PROOF. The face $d_i W(\boldsymbol{B} \otimes \boldsymbol{L}_n)$ is the subcategory of trees containing no edge colored i. Suppose the omitted face is d_k. The data yield an action on some $(n + 1)$-tuple $(X_0, X_1, \ldots, X_n)$ of the subcategory $\boldsymbol{C}$ of $W(\boldsymbol{B} \otimes \boldsymbol{L}_n)$ generated by the subcategories $d_i W(\boldsymbol{B} \otimes \boldsymbol{L}_n)$ for $i \neq k$. We show there is a retraction functor $W(\boldsymbol{B} \otimes \boldsymbol{L}_n) \to \boldsymbol{C}$.

We consider the pairs of spaces $(W(\boldsymbol{B} \otimes \boldsymbol{L}_n)(b, 1), \boldsymbol{C}(b, 1))$. Our first step is to shrink all internal edges colored 0 by the homotopy used in Lemma 6.3. We next replace each twig colored 0 by

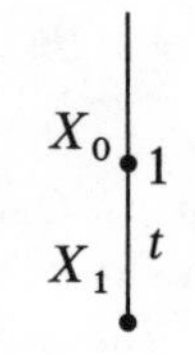

where t runs from 0 to 1. This homotopy deforms $W(\boldsymbol{B} \otimes \boldsymbol{L}_n)$ into $\boldsymbol{C}$, keeping $\boldsymbol{C}$ inside $\boldsymbol{C}$, which shows by Lemma 6.3 that $\boldsymbol{C} \to \boldsymbol{B} \otimes \boldsymbol{L}_n$ is an equivariant homotopy equivalence. The relative form of the Lifting Theorem 7.1 applies, to extend $1 : \boldsymbol{C} \to \boldsymbol{C}$ to a functor $W(\boldsymbol{B} \otimes \boldsymbol{L}_n) \to \boldsymbol{C}$.

Further, if the given trees have ignorable X_n-edges (see §8), we can arrange this also for the action of $W(\boldsymbol{B} \otimes \boldsymbol{L}_n)$.

Given morphisms $f : X \to Y$ and $g : Y \to Z$ of $\boldsymbol{WB}$-spaces, this theorem yields a 2-simplex having f and g as two of its edges. The third edge, which is some morphism $h : X \to Z$, we call *a composite* of f and g.

DEFINITION 10.3. Given any simplicial set K satisfying the restricted Kan extension condition, we call two edges in K *homotopic* if they are two edges of a 2-simplex whose remaining edge is degenerate. The *fundamental category* of K has as objects the vertices of K and as morphisms the homotopy classes of edges of K.

It follows from Theorem 10.2 that we have defined a category. (Compare the theory of the fundamental group, or better, groupoid, of a Kan complex. We obtain a category rather than a groupoid because we do not have the full Kan extension condition.)

11. **Homotopy equivalences.** By §10 we have the perfectly respectable category of $\boldsymbol{WB}$-spaces and homotopy classes of morphisms. In particular, we have the concept of homotopy equivalence of $\boldsymbol{WB}$-spaces. Even the trivial case

is interesting. Let $\boldsymbol{I}$ be the category with objects 0 and 1 and inverse isomorphisms $0 \cong 1$ and $1 \cong 0$, so that an $\boldsymbol{I}$-space is a homeomorphism $X_0 \cong X_1$.

LEMMA 11.1. *Suppose $p: Y \to Z$ is a homotopy equivalence. Then p admits a $W\boldsymbol{I}$-structure.*

PROOF. The trees in question, after simplification by relation (b), are vertical linear trees, with edges colored alternately Y and Z. Instead of skeletons, we use induction on a slightly different filtration: we give an indecomposable tree with k internal edges filtration k if it has a Z-root, or filtration $k + 1$ if it has a Y-root, and let $\boldsymbol{C}_k$ be the subcategory of $W\boldsymbol{I}$ generated by the trees of filtration $\leqq k$. We are given the action of $\boldsymbol{C}_0$.

Suppose given the action of $\boldsymbol{C}_{2n}$. To extend over $\boldsymbol{C}_{2n+1}$ we need the actions of the following trees

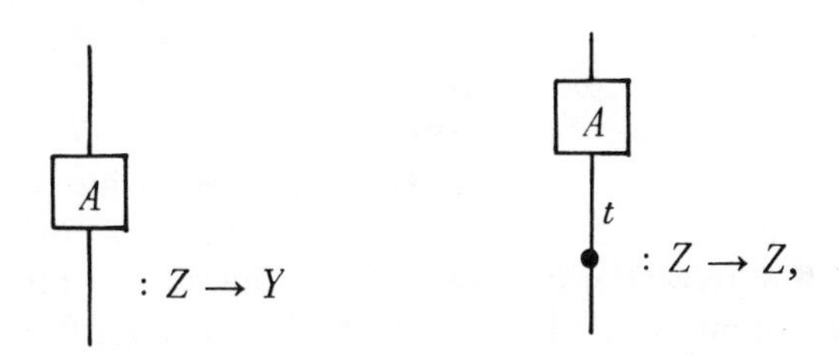

where A stands for a tree with $2n$ edges, which we represent by its edge lengths as a point in I^{2n}. That is, we require maps

$$f: I^{2n} \times Z \to Y, \qquad h_t: I^{2n} \times Z \to Z,$$

which are already given on $\partial I^{2n} \times Z$. Now h_0 is known in terms of $\boldsymbol{C}_{2n}$, and we require $h_1 = p \circ f$. The maps f and h_t are provided by HELP.

Similarly to extend from $\boldsymbol{C}_{2n+1}$ to $\boldsymbol{C}_{2n+2}$.

With this categorical description of a homotopy equivalence, we pass to the general theorem.

THEOREM 11.2 (hence Theorem F of [BV]). *Suppose $p: Y \to Z$ is a homotopy equivalence of spaces, and that Z is a $W\boldsymbol{B}$-space. Then*

(a) *we can introduce a $W\boldsymbol{B}$-structure on Y and give p the structure of a morphism of $W\boldsymbol{B}$-spaces;*

(b) *any action as in* (a) *is a homotopy equivalence of $W\boldsymbol{B}$-spaces.*

PROOF. By Lemma 11.1 we give p a $W\boldsymbol{I}$-action. In either case we seek to give p a $W(\boldsymbol{B} \otimes \boldsymbol{I})$-action, where we are in effect given the action of a certain subcategory $\boldsymbol{C}$ of $W(\boldsymbol{B} \otimes \boldsymbol{I})$. We extend by applying the relative case of the Lifting Theorem 7.1 to obtain a retraction functor $W(\boldsymbol{B} \otimes \boldsymbol{I}) \to \boldsymbol{C}$, for which we need only show that $\boldsymbol{C} \to \boldsymbol{B} \otimes \boldsymbol{I}$ is a homotopy equivalence.

In case (a), $\boldsymbol{C}$ is generated by $W\boldsymbol{I}$ and $W\boldsymbol{B}$. We take the spaces $\boldsymbol{C}(b, 1)$ and contract all the internal Y-edges as in Lemma 6.3. This reduces each morphism to one of the following forms:

(i) $q \circ \alpha \circ (u_1 \oplus u_2 \oplus \cdots \oplus u_n): Y_1 \times Y_2 \times \cdots \times Y_n \to Y$,
(ii) $\alpha \circ (u_1 \oplus u_2 \oplus \cdots \oplus u_n): Y_1 \times Y_2 \times \cdots \times Y_n \to Z$,
(iii)

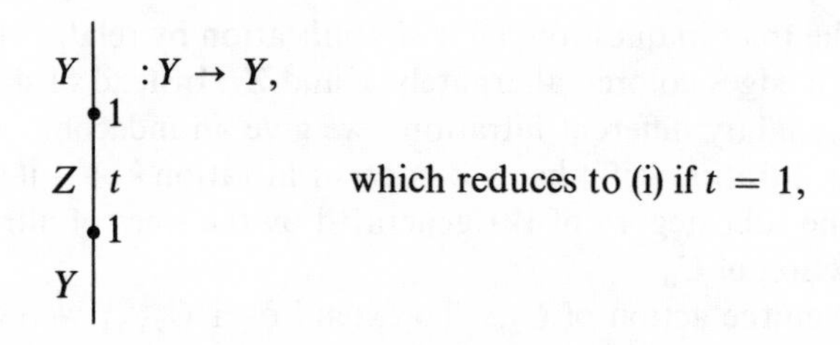

where q is the homotopy inverse to p given by the $\boldsymbol{WI}$-structure, $\alpha \in \boldsymbol{WB}(n, 1)$, and for each i, $u_i = p$ and $Y_i = Y$, or $u_i = 1$ and $Y_i = Z$. The homotopy equivalence now follows.

In case (b), $\boldsymbol{C}$ is generated by $\boldsymbol{WI}$ and $W(\boldsymbol{B} \otimes \boldsymbol{L}_1)$. This is considerably more complicated. We filter the spaces $\boldsymbol{C}(b, 1)$ by the subspaces F_k of trees with at most k vertex labels (in $\boldsymbol{B}$) different from 1, and show by tree surgery that F_{k-1} is a deformation retract of F_k, and that F_1 is homotopy equivalent to $\boldsymbol{B} \otimes \boldsymbol{I}$.

By this theorem, X and UX in Adams' Theorem 8.2 are homotopy-equivalent $\boldsymbol{WB}$-spaces, in our strong sense.

We also need a relative version, where we are already given the action of $W(\boldsymbol{D} \otimes \boldsymbol{I})$, where $\boldsymbol{D}$ is a subcategory of $\boldsymbol{B}$ such that $\boldsymbol{D}(n, 1) \subset \boldsymbol{B}(n, 1)$ is closed and has the S_n-equivariant HEP for all n.

12. **Infinite loop spaces and classifying spaces.** We defined E-spaces in §5. It is not difficult to construct or find an E-cotheory $\boldsymbol{E}$ in which each space $\boldsymbol{E}(n, 1)$ is a CW-complex on which S_n acts freely, for example $\boldsymbol{Q}_\infty$ (see below). Then $\boldsymbol{WE}$ is another E-cotheory, and by Theorem 7.1 we can replace any E-structure on a space X by a $\boldsymbol{WE}$-structure, and any E-map between $\boldsymbol{WE}$-spaces by a $W(\boldsymbol{E} \otimes \boldsymbol{L}_1)$-action. Unlike the case of monoids, however, we have no canonical $\boldsymbol{E}$ in sight.

THEOREM 12.1 (Theorem A in [BV]). *A CW-complex X is an infinite loop space if and only if it admits an E-structure that makes $\pi_0(X)$ a group. Any such E-space has a "classifying space" BX, which satisfies $X \simeq \Omega BX$ and is again an E-space, uniquely up to homotopy equivalence of E-spaces.*

PROOF. If X is an infinite loop space, J. P. May has noted that we may replace X by a homotopy-equivalent space Y_0, for which there is a sequence of homeomorphisms $Y_0 \cong \Omega Y_1$, $Y_1 \cong \Omega Y_2$, $Y_2 \cong \Omega Y_3, \ldots$. Since $Y_0 \cong \Omega^n Y_n$, the nth little-cube category $\boldsymbol{Q}_n$ (see §2 of [BV]) acts on Y_0. As n varies, we have canonical inclusions $\boldsymbol{Q}_n \subset \boldsymbol{Q}_{n+1}$, and an action of the E-cotheory $\boldsymbol{Q}_\infty = \bigcup_n \boldsymbol{Q}_n$ on Y_0. By Theorem 8.1, the E-cotheory $\boldsymbol{WQ}_\infty$ acts on X.

Conversely, if X is an E-space, we can by the Lifting Theorem 7.1 replace the given E-structure by an action of $W(\boldsymbol{A} \otimes W\boldsymbol{E})$. (We need to verify that S_n acts freely on $(\boldsymbol{A} \otimes W\boldsymbol{E})(n, 1)$.) By Adams' Theorem 8.2, X is homotopy-equivalent to some $(\boldsymbol{A} \otimes W\boldsymbol{E})$-space Y. In particular Y is a monoid and a CW-complex, and the E-category $W\boldsymbol{E}$ acts on its classifying space BY. By Theorem 9.1, Y is homotopy-equivalent to ΩBY. We put $BX = BY$ and continue the induction.

It is desirable to have more precise information.

LEMMA 12.2. *Suppose $\boldsymbol{B}$ is a cotheory over $\boldsymbol{S}$ such that each diagonal inclusion $\Delta: \boldsymbol{B}(n, 1) \subset \boldsymbol{B}(n, 1) \times \boldsymbol{B}(n, 1)$ has* HEP. *Then the augmentation*

$$\varepsilon \otimes 1 : (W\boldsymbol{A} \otimes W\boldsymbol{B})(n, 1) \to (\boldsymbol{A} \otimes W\boldsymbol{B})(n, 1)$$

is a S_n-equivariant homotopy equivalence for all n.

PROOF. This is not trivial. We find a canonical representative for each morphism in $W\boldsymbol{A} \otimes W\boldsymbol{B}$ and $\boldsymbol{A} \otimes W\boldsymbol{B}$, and show by tree surgery that each fibre is contractible. Although $\varepsilon \otimes 1$ is not a fibration, we are able to use this fact to prove that $\varepsilon \otimes 1$ is a homotopy equivalence.

With this lemma we can determine when the classifying space BX of a monoid X admits a given structure. If BX is a $W\boldsymbol{B}$-space, then ΩBX is a $(W\boldsymbol{A} \otimes W\boldsymbol{B})$-space by Lemma 7.2, and since $X \simeq \Omega BX$ by Theorem 9.1, Theorem 8.1 gives X a $W(W\boldsymbol{A} \otimes W\boldsymbol{B})$-action. By Lemma 12.2 and the Lifting Theorem 7.1, we can replace this by a $W(\boldsymbol{A} \otimes W\boldsymbol{B})$-action. Conversely, if X is a $W(\boldsymbol{A} \otimes W\boldsymbol{B})$-space, we find $W\boldsymbol{B}$ acts on BX as in the proof of Theorem 12.1. With more care we can prove the following result:

THEOREM 12.3. *Suppose BX is a classifying space for a $W\boldsymbol{A}$-space X. Then BX admits a $W\boldsymbol{B}$-action if and only if the given $W\boldsymbol{A}$-action on X extends to a $W(\boldsymbol{A} \otimes W\boldsymbol{B})$-action, provided X is a* CW-*complex and $\pi_0(X)$ is a group.*

As an application we have:

THEOREM 12.4. *A* CW-*complex X is an nth loop space if and only if it admits a $W\boldsymbol{Q}_n$-action that makes $\pi_0(X)$ a group.*

PROOF. This makes use of the obvious functors

$$W\boldsymbol{A} \otimes \boldsymbol{Q}_{n-1} \to \boldsymbol{Q}_1 \otimes \boldsymbol{Q}_{n-1} \to \boldsymbol{Q}_n,$$

and induction.

Envoi. We have seen that if $\boldsymbol{B}$ is any cotheory over $\boldsymbol{A}$ such that $\boldsymbol{B}(n, 1)$ is contractible for all n, we can replace the $\boldsymbol{B}$-action on any space X by a $W\boldsymbol{A}$-action, and then by Adams' theorem replace X by a homotopy-equivalent monoid. This becomes false if we replace $\boldsymbol{A}$ by $\boldsymbol{S}$ throughout; we cannot hope to replace an E-space by an $\boldsymbol{S}$-space, or commutative monoid, because we know that any $\boldsymbol{S}$-space has the homotopy type of a product of Eilenberg-Mac Lane spaces. The essential difference between these two situations becomes clear if we return to cotheories as defined in general. In the first case, for the cotheory Θ obtained from $\boldsymbol{B}$, the maps $\Theta(n, 1) \to \Theta_m(n, 1)$ are homotopy equivalences. In the second case, the maps $\Theta(n, 1) \to \Theta_{cm}(n, 1)$ are not usually homotopy equivalences, because over the point $z_1^n \in F1 \cong \Theta_{cm}(1, 1)$ we find the orbit space $\boldsymbol{E}(n, 1)/S_n$ where $\boldsymbol{E}$ is the given E-cotheory. Finally, one can show that if Θ is a cotheory over Θ_{cm} such that each map $\Theta(n, 1) \to \Theta_{cm}(n, 1)$ is a homotopy equivalence, then any Θ-space is indeed homotopy-equivalent to a commutative monoid.

JOHNS HOPKINS UNIVERSITY

HOMOTOPY WITH RESPECT TO A RING

A. K. BOUSFIELD AND D. M. KAN

1. **Introduction.** This is a report on a construction which assigns to a *space* (i.e. simplicial set with base point) X and a *ring* (with unit) R, a tower of principal fibrations

$$\cdots \longrightarrow R_nX \xrightarrow{p_n} R_{n-1}X \longrightarrow \cdots \longrightarrow R_0X \longrightarrow R_{-1}X = *,$$

where the fibre of each p_n is a simplicial (left) R-module, together with maps,

$$X \xrightarrow{i_n} R_nX$$

such that $p_n i_n = i_{n-1}$ for all n. *This tower contains, in some sense, all the information concerning the homotopy type of X that can be expressed in terms of the reduced homology $\tilde{H}_*(X; R)$ and operations thereon of all orders* and *the inverse limit $R_\infty X$ of this tower is a kind of completion of X with respect to R.*

The tower is constructed by first "resolving the space X with respect to the ring R" into a cosimplicial space (i.e. a cosimplicial object over the category of spaces) $\mathbf{R}X$ and then associating with this cosimplicial space $\mathbf{R}X$ the desired tower of fibrations. This will be done in §§2 and 3.

The homotopy spectral sequence $\{E_r(X; R)\}$ of the tower $\{R_nX\}$ is a spectral sequence with Whitehead products as well as smash and composition pairings. We list some of its additional properties for $R = Z_p$ (the integers modulo a prime p) in §§4 and 7, for $R = Q$ (the rationals) in §5 and for $R = Z$ (the integers) in §6. A discussion of the completion for $R \subset Q$ and for $R = Z_p$ will be given in §§6 and 7 respectively.

We end this introduction with the definition of a class of spaces which turns out to be very useful.

AMS 1970 *subject classifications.* Primary 55D15; Secondary 55H15.

1.1. *Nilpotent spaces.* A space X will be called *nilpotent* if
(i) X is connected,
(ii) $\pi_1 X$ is nilpotent,
(iii) $\pi_1 X$ acts nilpotently on each of the higher homotopy groups.
To explain this last statement, let $\Gamma_1 \pi_n X = \pi_n X$ and let $\Gamma_j \pi_n X \subset \pi_n X$ denote the subgroup generated by the elements of the form $a(b) - b$, where $a \in \pi_1 X$ and $b \in \Gamma_{j-1}\pi_n X$. Then we mean that for each n there is an integer i_n such that $\Gamma_{i_n}\pi_n X = 0$.

This is equivalent with saying that *the Postnikov tower of X can be refined to a tower of principal fibrations with abelian fibres.*

Clearly *every simply connected space and every connected H-space is nilpotent.*

2. **Resolving a space X with respect to a ring R.** This construction is based on

2.1. *The functor R.* Let RX denote the simplicial (left) R-module which is freely generated by the simplices of X, except that the base point of X (and its degeneracies) are put equal to 0, let $R^n X = R \cdots RX$ for $n > 1$ and let

$$X \xrightarrow{\varphi} RX \quad \text{and} \quad R^2 X \xrightarrow{\psi} RX$$

be the map given by $\varphi x = 1 \cdot x$ for all $x \in X$, respectively the R-module homomorphism given by $\psi(1 \cdot y) = y$ for all $y \in RX$. *The functor R and the natural transformations*

$$\mathrm{Id} \xrightarrow{\varphi} R \quad \text{and} \quad R^2 \xrightarrow{\psi} R$$

so defined have the following important properties:

(i) *there is a natural isomorphism $\pi_* RX \approx \tilde{H}_*(X; R)$ such that*

$$\pi_* X \xrightarrow{\varphi_*} \pi_* RX \approx \tilde{H}_*(X; R)$$

is the Hurewicz homomorphism, and

(ii) *$\{R, \varphi, \psi\}$ is a triple in the sense of Eilenberg-Moore, i.e.*

$$(\varphi R)\varphi = (R\varphi)\varphi, \qquad \psi(\psi R) = \psi(R\psi), \qquad \psi(\varphi R) = \psi(R\varphi) = \mathrm{Id}.$$

Now we are ready for

2.2. *The resolution of X with respect to R.* This is the cosimplicial space $\mathbf{R}X$ with

$$(\mathbf{R}X)^k = R^{k+1} X$$

for its "space in codimension k" and

$$(\mathbf{R}X)^{k-1} \xrightarrow{d^i} (\mathbf{R}X)^k = R^k X \xrightarrow{R^i \varphi R^{k-i}} R^{k+1} X, \qquad 0 \leqq i \leqq k,$$

$$(\mathbf{R}X)^{k+1} \xrightarrow{s^i} (\mathbf{R}X)^k = R^{k+2} X \xrightarrow{R^i \psi R^{k-i}} R^{k+1} X, \qquad 0 \leqq i \leqq k,$$

for its coface and codegeneracy maps; it is an easy consequence of property 2.1(ii) that all cosimplicial identities indeed hold.

3. **The tower of a cosimplicial space.** To build the tower we need *the cosimplicial standard simplex* $\mathbf{\Delta}$, i.e. the cosimplicial simplicial set which in codimension k is the standard k-simplex $\Delta[k]$ and which has as coface and

codegeneracy maps the standard maps between the standard simplices, as well as *its simplicial n-skeleton* $\mathbf{\Delta}^{[n]}$ (which in codimension k is the n-skeleton of $\Delta[k]$).

3.1. *The general case.* For a cosimplicial space $\mathbf{Y}$ one can form the function complexes

$$\hom(\mathbf{\Delta}^{[n]}, \mathbf{Y}), \qquad -1 \leqq n \leqq \infty,$$

which are the set complexes with as q-simplices the cosimplicial maps

$$\mathbf{\Delta}^{[n]} \times \Delta[q] \to \mathbf{Y}$$

and the *tower* of $\mathbf{Y}$ then consists of the maps

$$\hom(\mathbf{\Delta}^{[n]}, \mathbf{Y}) \xrightarrow{p_n} \hom(\mathbf{\Delta}^{[n-1]}, \mathbf{Y})$$

induced by the inclusions $\mathbf{\Delta}^{[n-1]} \subset \mathbf{\Delta}^{[n]}$. Clearly

(i) $\hom(\mathbf{\Delta}^{[-1]}, \mathbf{Y}) = *$,

(ii) $\hom(\mathbf{\Delta}^{[0]}, \mathbf{Y})$ *can (and will) be identified with the space* Y^0, and

(iii) $\hom(\mathbf{\Delta}, \mathbf{Y})$ *is exactly the inverse limit of the tower.*

3.2. *The case* $\mathbf{Y} = \mathbf{R}X$. Applying the above to $\mathbf{R}X$ one gets the desired tower with

$$R_nX = \hom(\mathbf{\Delta}^{[n]}, \mathbf{R}X), \qquad -1 \leqq n \leqq \infty,$$

and readily verifies that

(i) *each* p_n *is indeed a principal fibration with a simplicial R-module as fibre,*

(ii) *the homotopy type of the fibre of each* p_n *depends only on the homotopy type of* RX *and hence* (2.1(i)) *on* $\tilde{H}_*(X; R)$,

(iii) *if* $f: X \to Y$ *induces an isomorphism* $\tilde{H}_*(X; R) \approx \tilde{H}_*(Y; R)$, *then* $R_nf: R_nX \to R_nY$ *is a homotopy equivalence for all* $n \leqq \infty$.

It remains to construct

3.3. *The maps* $i_n: X \to R_nX$. Let $\mathbf{X}$ denote the constant cosimplicial space over X (i.e. $\mathbf{X}^k = X$ for all k and all the d^i and s^i are the identity) and let $\boldsymbol{\varphi}: \mathbf{X} \to \mathbf{R}X$ be the (unique) cosimplicial map which in codimension 0 coincides with the map $\varphi: X \to RX$. The tower of $\mathbf{X}$ is rather trivial; all its maps are isomorphisms and hence all its spaces can (and will) be identified with X. It thus makes sense to define the maps $i_n: X \to R_nX$ $(-1 \leqq n \leqq \infty)$ as those induced by $\boldsymbol{\varphi}$

$$X = \hom(\mathbf{\Delta}^{[n]}, \mathbf{X}) \xrightarrow{\hom(\Delta^{[n]}, \varphi)} \hom(\mathbf{\Delta}^{[n]}, \mathbf{R}X) = R_nX.$$

Clearly $p_ni_n = i_{n-1}$. Also *the map* $i_0: X \to R_0X$ *coincides with* $\varphi: X \to RX$.

4. **The associated spectral sequence for** $R = Z_p$. This seems to be a good candidate for the title of *the* unstable Adams spectral sequence as it has the following properties:

4.1. *Convergence.* If X is *connected* and $\tilde{H}_n(X; Z_p)$ *is finite for each n*, then the spectral sequence $\{E_r(X; Z_p)\}$ converges to $\pi_*R_\infty X$, i.e. there is a filtration $\{F^s\pi_*R_\infty X\}$ of $\pi_*R_\infty X$ such that

$$F^s\pi_*R_\infty X/F^{s+1}\pi_*R_\infty X \approx E_\infty^{s,*}(X; Z_p),$$

$$\pi_*R_\infty X = \operatorname{proj\,lim}\,(\pi_*R_\infty X/F^s\pi_*R_\infty X).$$

If X is *nilpotent* and $\pi_n X$ *is finitely generated for each* n, then (see §7) $\pi_* R_\infty X$ is the p-completion of $\pi_* X$ and hence in this case $\{E_r(X; Z_p)\}$ converges to

$$\pi_* X/(\text{elements divisible by all powers of } p),$$

i.e. there is a filtration $\{F^s \pi_* X\}$ of $\pi_* X$ such that

$$F^s\pi_* X/F^{s+1}\pi_* X \approx E_\infty^{s,*}(X; Z_p),$$

$$\bigcap_s F^s \pi_* X = \pi_* X \cap (\text{elements divisible by all powers of } p).$$

4.2. *The E_2-term.* The E_1-term depends functorially on $\tilde{H}_*(X; Z_p)$ and the differential d_1 (and hence the E_2-term) depends on only primary operations thereon. In fact, the E_2-term is an *unstable* Ext depending only on the structure of $\tilde{H}_*(X; Z_p)$ as a coalgebra over the Steenrod algebra. One can also describe $E_2(X; Z_p)$ more "geometrically" in terms of $\pi_* \mathbf{R}X$, as the cohomology of the cochain complex with coboundary $\delta = \sum (-1)^i d^i_*$,

$$\pi_*(\mathbf{R}X)^0 \longrightarrow \cdots \longrightarrow \pi_*(\mathbf{R}X)^k \xrightarrow{\delta} \pi_*(\mathbf{R}X)^{k+1} \longrightarrow \cdots.$$

4.3. *The spectral sequence is an unstable Adams spectral sequence*, i.e. it coincides, in the stable range, with the Adams spectral sequence.

4.4. *Comparison with other unstable Adams spectral sequences.* These are

(i) *The Massey-Peterson spectral sequence* has the "right" E_2-term (i.e. the "unstable Ext" mentioned above), but is only defined for "very nice" spaces,

(ii) *The accelerated p-lower central series spectral sequence of Bousfield-Curtis-Rector*, of which the E_2-term depends in general on higher order operations, but is "right" for "nice" spaces (these include loop spaces and spheres, but not wedges of spheres),

(iii) *The p-derived series spectral sequence* of which also the E_2-term depends in general on higher order operations, but is "right" for "nice" spaces.

As mentioned above our spectral sequence always has the "right" E_2-term. Moreover for "nice" spaces it coincides, from E_2 on, with spectral sequence (iii), while for "very nice" spaces it coincides, from E_2 on, with all three spectral sequences (i), (ii), and (iii).

5. **The associated spectral sequence for $R = Q$.** For connected X this spectral sequence is closely related to another spectral sequence of Adams, the (rational) *cobar spectral sequence* which abuts to $\tilde{H}^*(\Omega X; Q)$. In fact, if the latter is denoted by $E_r X$, then one has, for $2 \leqq r \leqq \infty$,

(i) *$E_r X$ is a primitively generated Hopf algebra,*

(ii) *$E_r(X; Q)$ is a Lie algebra (under the Whitehead product), and*

(iii) *$E_r(X; Q)$ is naturally isomorphic with the Lie algebra of the primitive elements of $E_r X$.*

6. **The completion for $R \subset Q$.** One has

6.1. *Homology behavior.* If X is nilpotent, then the map $i_\infty : X \to R_\infty X$ induces an isomorphism

$$\tilde{H}_*(X; R) \approx \tilde{H}_*(R_\infty X; R).$$

In view of 3.2(iii) this implies that *for nilpotent X the inverse limit $R_\infty X$ determines, up to homotopy, the tower $\{R_n X\}$.*

6.2. *Homotopy behavior.* If X is nilpotent, then the map $i_\infty : X \to RX$ induces an isomorphism

$$\pi_* X \otimes R \to \pi_* R_\infty X,$$

where, for π a nonabelian group, $\pi \otimes R$ should be interpreted as $\pi_1 R_\infty K(\pi, 1)$.

This means that if R consists of the rationals with denominators prime to p, then $R_\infty X$ is *the localization of X at the prime p.*

For $R = Z$ one has as an immediate consequence the following result of E. Dror, which suggested to us the usefulness of nilpotent spaces.

6.3. *A generalization of the Whitehead theorem.* Let X and Y be nilpotent spaces. A map $f: X \to Y$ then is a (weak) homotopy equivalence if and only if

$$\tilde{H}_*(X; Z) \overset{f_*}{\approx} \tilde{H}_*(Y; Z).$$

One proves 6.1 and 6.2 by first verifying them for $K(\pi, n)$'s (π abelian) and then combining the nilpotency with

6.4. *A fibre lemma.* Let $p: E \to B$ be a fibre map with fibre F and let $\pi_0 F = \pi_0 E = \pi_1 B = 0$. Then $R_\infty p: R_\infty E \to R_\infty B$ is a fibre map and the inclusion of $R_\infty F$ in its fibre is a homotopy equivalence.

Using a similar argument and the Whitehead product structure of $\{E_r(X; Z)\}$ one can prove

6.5. *A characterization of nilpotent spaces.* A connected space is nilpotent if and only if the spectral sequence $\{E_r(X; Z)\}$ converges strongly to $\pi_* X$ (i.e. there is a finite filtration on each $\pi_n X$ for which the associated graded group is isomorphic to $E_\infty^{*,n}(X; Z)$).

Of course one would like to know what happens to the above results if X were not nilpotent. That this may be rather nontrivial, although interesting, can be seen from the following

6.6. *Example.* Let Σ_∞ denote the infinite symmetric group. Then

$$\pi_i R_\infty K(\Sigma_\infty, 1) \approx \pi_i \Omega^\infty S^\infty \otimes R, \qquad i \geqq 1.$$

This follows immediately from an observation of Barratt-D. S. Kahn-Priddy that there exists a map from $K(\Sigma_\infty, 1)$ to the constant component of $\Omega^\infty S^\infty$ which induces an isomorphism of integral homology.

7. **The completion for $R = Z_p$.** In this case things are more complicated. We still have

7.1. *Homology behavior.* This is the same as in 6.1.

But 6.2 becomes

7.2. *Homotopy behavior.* If X is nilpotent and $Z_{p\infty}$ denotes the p-primary component of Q/Z, then there is for every $n \geqq 1$ a short exact sequence

$$0 \to \operatorname{Ext}(Z_{p\infty}, \pi_n X) \to \pi_n R_\infty X \to \operatorname{Hom}(Z_{p\infty}, \pi_{n-1} X) \to 0,$$

where for nonabelian π one should interpret $\mathrm{Ext}(Z_{p\infty}, \pi)$ and $\mathrm{Hom}(Z_{p\infty}, \pi)$ as

$$\mathrm{Ext}(Z_{p\infty}, \pi) = \pi_1 R_\infty K(\pi, 1), \qquad \mathrm{Hom}(Z_{p\infty}, \pi) = \pi_2 R_\infty K(\pi, 1).$$

Note that *for π nilpotent and finitely generated one has that* $\mathrm{Hom}(Z_{p\infty}, \pi) = 0$ *while* $\mathrm{Ext}(Z_{p\infty}, \pi)$ *is the p-completion of* π.

We end with the observation that, although the associated spectral sequence converges to $\pi_* R_\infty X$ whenever X is connected with $\tilde{H}_n(X; Z_p)$ finite for each n (4.1), this need not be the case without this assumption, even for X a $K(\pi, n)$. Still, for a nilpotent X the tower $\{R_n X\}$ and the completion $R_\infty X$ determine each other up to homotopy (6.1) and hence it should be possible to find out what homotopy information about $R_\infty X$ is contained in $E_\infty(X; Z_p)$.

BRANDEIS UNIVERSITY AND
MASSACHUSETTS INSTITUTE OF TECHNOLOGY

THE KERVAIRE INVARIANT OF A MANIFOLD

EDGAR H. BROWN, JR.

1. **Introduction.** Surgery is one of the main tools of differential topology. Typically, one wishes to construct a manifold with certain properties. By some means, usually a transverse regularity argument, one constructs a manifold enjoying some of the desired properties. Then, by surgery, one attempts to modify this manifold to meet all the required conditions. Usually, when the dimension of the manifold $\equiv 2 \bmod 4$ one meets Kervaire or Arf invariant obstruction. As an example of this we describe surgery on a map.

Suppose X is a finite CW-complex.

PROBLEM. When does X have the same homotopy type as a smooth, closed, compact orientable manifold?

Necessary conditions are that the homology and cohomology of X satisfy Poincaré duality (see [**1**] for details) and that there is a vector bundle η over X such that the top homology class of the Thom space $T(\eta)$ is spherical. Assuming these conditions hold, one can construct an m-manifold M and maps

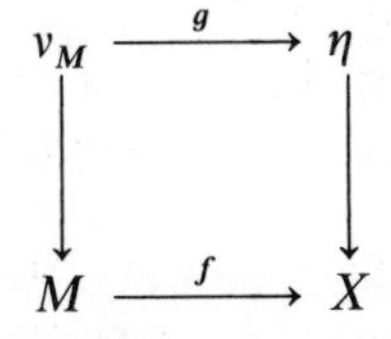

where ν_M is the normal bundle of M in R^{m+k} (k large), such that $f_*: H_m(M) \approx H_m(X)$. Surgery in this situation proceeds as follows: Let $\alpha \in \ker f_* \pi_l(M) \to \pi_l(X)$. Represent α by an embedding $i: S^l \subset M$ (if possible). Extend i to an embedding $j: S^l \times D^{m-l} \subset M$ (if possible). Let $N = M \times I \cup D^{l+1} \times D^{m-l}$ with $(j(x, y), 1)$ and $(x, y) \in S^l \times D^{m-l}$ identified (smooth corners). Choose j so that g can be extended to $G: \nu_N \to \eta$ (again if possible). $\partial N = M \times \{0\} \cup M'$. M' and $g' = G \mid \nu_{M'}$ are said to be obtained from (M, g) by surgery on α. One can apply

AMS 1970 *subject classifications*. Primary 57D65, 57D90.

this technique, by induction on i to produce an (M, g) such that f_* is a monomorphism on $\pi_i(M)$ for $i < [m/2]$. Furthermore, if one can carry through this construction to produce a monomorphism for $i = [m/2]$, f will be a homotopy equivalence. Wall has defined a group $L_m(\pi_1(X))$, depending only on the group $\pi_1(X)$, and an element $\sigma(M, g)$ in this group such that the surgery can be performed if and only if $\sigma(M, g) = 0$.

We consider the case $\pi_1(X) = 0$ and $m = 2n > 4$. Suppose $f_*:\pi_i(M) \to \pi_i(X)$ is a monomorphism for $i < n$ (it will be an isomorphism by Poincaré duality). Let $K = \ker(H_n(M) \to H_n(X))$. If we can kill K by surgery, f will be a homotopy equivalence. Each element of K can be represented by an embedding $i:S^n \subset M$ but in general, the normal bundle v_i of $i(S^n)$ in M will not be trivial. v_i is stably trivial because $i(S^n)$ represents an element of K. If n is even, v_i is characterized by its Euler number, and when n is odd v_i is either trivial or isomorphic to the tangent bundle of S^n. Let

$$\varphi:K \to \begin{cases} Z, & n \text{ even,} \\ Z_2, & n \text{ odd,} \end{cases}$$

be defined as follows: Let $u \in K$ and let $i:S^n \subset M$ represent u. Let $\varphi(u)$ be the Euler number of v_i if n is even and 0 or 1 as v_i is trivial or not when n is odd, $n \neq 1, 3, 7$. When n is even, $\varphi(u) = u \cap u$, where "$\cap$" denotes the intersection pairing, but for n odd φ cannot be expressed in terms of the intersection pairing. One does have the relation:

$$\varphi(u + v) = \varphi(u) + \varphi(v) + u \cap v \bmod 2 \tag{1.1}$$

for n odd. To perform the surgeries making $f: M \to X$ into a homotopy equivalence it is necessary and sufficient that there is a symplectic basis λ_i, μ_i, $i = 1, 2, \ldots, l$ $(\lambda_i \cap \mu_j = \delta_{ij}$, $\lambda_i \cap \lambda_j = \mu_i \cap \mu_j = 0)$ such that $\varphi(\lambda_i) = \varphi(\mu_i) = 0$. By Poincaré duality, $\cap$ is nonsingular on K. Suppose n is odd, $n \neq 1, 3, 7$. In this case $\cap$ is skew so there is a symplectic basis for K. The Arf invariant of φ is an algebraic invariant given by $A(\varphi) = \sum \varphi(\lambda_i)\varphi(\mu_i) \in Z_2$ for any symplectic basis λ_i, μ_i. An algebraic result about A is that $A(\varphi) = 0$ if and only if λ_i, μ_i can be chosen so that $\varphi(\lambda_i) = \varphi(\mu_i) = 0$. In this case the Wall group $L_{4k+2}(0)$ is Z_2 and $\sigma(M, g) = A(\varphi)$.

Suppose n is even. $\cap$ on K is nonsingular and $u \cap u =$ Euler number v_i is even. It follows from results on quadratic forms over Z, that there is a symplectic basis if and only if the signature of φ is zero. Thus $L_{4k}(0) = Z$ and $\sigma(M, g) =$ signature φ. Furthermore, we can give a formula for $\sigma(M, g)$ in terms of X and η. $H_n(M)$ splits, with respect to the intersection pairing, as $H_n(M) \approx K \oplus H_n(X)$, where the pairing on $H_n(X)$ comes from the fact that it satisfies Poincaré duality. Thus

THEOREM 1.2. *If n is even,*

$$\sigma(M, g) = I(M) - I(X),$$

where I denotes the index.

Using the Hirzebruch index theorem we have

$$I(M) = \bar{L}_{2n}(p(\nu))(M) = \bar{L}_{2n}(p(\eta))(X),$$

where $\bar{L}_{2n}$ is the L-polynomial and p is the Pontrjagin class. Hence

THEOREM 1.3. $\sigma(M,g) = \bar{L}_{2n}(p(\eta))(X) - I(X)$.

Theorems 1.2 and 1.3 provide a model for what we would like to do when n is odd. In subsequent lectures we describe a version of Theorem 1.2 for n odd.

An important special case of the above is $X = S^{2n}$, η trivial and n odd. Then g is a framing of the normal bundle of M and $\sigma(M, g)$ is the Kervaire invariant **[2]**. This defines a homomorphism

$$K:\Omega_{2n}(\text{framed}) \to Z_2.$$

Kervaire and Milnor conjectured that $K = 0$, $n \neq 1, 3, 7$. (K is defined for $n = 1, 3, 7$ as $\sigma(M, g)$ but in these cases $\sigma(M, g)$ is the obstruction to finding G; see page 65. $K \neq 0, n = 1, 3, 7$.) The results on this conjecture are:

Kervaire: $K = 0$ for $n = 5, 7$ **[2]**,

Brown-Peterson: $K = 0$ for $n \equiv 1 \bmod 4$ **[3]**,

Browder: $K = 0$ for $n \neq 2^i - 1$ **[4]**, and $K \neq 0$ for $n = 2^i - 1$, if and only if, h_i^2 lives to E_∞ in the Adams spectral sequence for homotopy groups of spheres. ($K \neq 0$, if $n = 30$.)

We will indicate some of the methods used by Browder to prove his results.

2. **Algebra of the Arf invariant.** Let V be a finite-dimensional vector space over Z_2. The Arf invariant is defined on quadratic functions $\varphi: V \to Z_2$. Both for algebraic and geometric reasons it is useful to consider functions into Z_4.

DEFINITION 2.1. $\varphi: V \to Z_4$ is (nonsingular) quadratic if

$$\varphi(u + v) = \varphi(u) + \varphi(v) + j\mu(u \otimes v),$$

where $\mu: V \otimes V \to Z_2$ is a nonsingular pairing and $j: Z_2 \to Z_4$ is the nontrivial homomorphism.

REMARK. $2\varphi(u) = j\mu(u \otimes u)$. Thus considering φ with values in Z_4 instead of Z_2 allows us to deal with the case in which $\mu(u \otimes u) \neq 0$. This allows us to deal with manifolds in which cup product to the top dimension is nonzero.

If $\varphi_1: V_1 \to Z_4$ and $\varphi_2: V_2 \to Z_4$ are quadratic, we define $\varphi_1 \approx \varphi_2$ if there is a linear isomorphism $\lambda: V_1 \to V_2$ such that $\varphi_2\lambda = \varphi_1$. We define

$$\varphi_1 + \varphi_2: V_1 \oplus V_2 \to Z_4$$

by

$$(\varphi_1 + \varphi_2)(u, v) = \varphi_1(u) + \varphi_2(v)$$

and we define $\varphi_1\varphi_2: V_1 \otimes V_2 \to Z_4$ by

$$\varphi_1\varphi_2(u \otimes v) = \varphi_1(u)\varphi_2(v).$$

(Use the quadratic property to extend this to all of $V_1 \otimes V_2$.) Let $(-\varphi)(u) = -\varphi(u)$.

We wish to show that the Grothendieck group of these functions is Z_8. We state this in the following form:

THEOREM 2.2. *There is a unique function σ from quadratic functions to Z_8 such that*

(i) *If $\varphi_1 \approx \varphi_2$, $\sigma(\varphi_1) = \sigma(\varphi_2)$.*

(ii) $\sigma(\varphi_1 + \varphi_2) = \sigma(\varphi_1) + \sigma(\varphi_2)$.

(iii) $\sigma(-\varphi) = -\sigma(\varphi)$.

(iv) $\sigma(\gamma) = 1$, *where* $\gamma: Z_2 \to Z_4$ *by* $\gamma(0) = 0$, $\gamma(1) = 1$.

Furthermore

(v) $\sigma(\varphi_1\varphi_2) = \sigma(\varphi_1)\sigma(\varphi_2)$.

(vi) *If $\psi: V \to Z_2$ is nonsingular quadratic, $\sigma(j\psi) = k(\operatorname{Arf}\psi)$ where $k: Z_2 \to Z_8$ by $k(1) = 4$ and $k(0) = 0$.*

(vii) *If $\psi: U \to Z$ is a unimodular quadratic form over Z, $\tilde{\psi}: U/2U \to Z_4$ is well defined and quadratic and $\sigma(\tilde{\psi}) =$ signature ψ mod 8.*

(viii) *For any φ there is a $\bar{\varphi}$ such that $\varphi + (\bar{\varphi} + (-\bar{\varphi})) \approx n\gamma + m(-\gamma)$ and $\sigma(\varphi) = n - m$.*

(ix) $\sigma(\varphi) = \dim V \bmod 2$.

PROOF. We describe a trick due to Paul Monsky for defining σ. Let $i = (-1)^{1/2}$ and consider

$$\alpha(\varphi) = \sum_{u \in V} i^{\varphi(u)} \in \mathbf{C}.$$

It is trivial to check that $\alpha(\varphi_1 + \varphi_2) = \alpha(\varphi_1)\alpha(\varphi_2)$, $\alpha(-\varphi) = \overline{\alpha(\varphi)}$, if $L: V \to Z_4$ is linear, $\alpha(L) = 0$ if $L \neq 0$ and $\alpha(L) = 2^{\dim V}$ if $L = 0$. From the quadratic property one then sees that

$$\alpha(\varphi)\overline{\alpha(\varphi)} = 2^{\dim V} \quad \text{and} \quad \alpha(2\varphi) = \pm i 2^{\dim V}.$$

Hence $\alpha(8\varphi)$ is real and

$$\alpha(\varphi) = \sqrt{2}^{\dim V} \cdot \text{8th root of } 1 = \sqrt{2}^{\dim V}\left(\frac{1+i}{\sqrt{2}}\right)^{\sigma(\varphi)}.$$

Continuing in this vein one can prove (i) – (ix).

Suppose $\mu: V \otimes V \to Z_2$ is a nonsingular symmetric pairing. Let $Q(V, \mu) = V \times Z_2$ with the abelian group structure given by

$$(u, n) + (v, m) = (u + v, \mu(u \otimes v) + n + m).$$

It is trivial to check that quadratic functions $\varphi: V \to Z_4$ whose associated bilinear form is μ are in one-to-one correspondence with homomorphisms $\psi: Q(V, \mu) \to Z_4$ such that $\psi(0, 1) = 2$, under the correspondence $\varphi(u) = \psi(u, 0)$.

3. **The Kervaire invariant of a manifold.** Suppose M is a closed $2n$-manifold (or a Poincaré space). Let $K_n = K(Z_2, n)$, $H^n(M) = [M^+, K_n]$, and

$$\{M^+, K_n\} = \lim_{k \to \infty} [S^k M^+, S^k K_n].$$

Let $\theta:[M^+, K_n] \to \{M^+, K_n\}$ by $\theta[f] = \{f\}$. Let $d:M \to S^{2n}$ be a map of degree 1. $\{S^{2n}, K_n\} \approx Z_2$.

PROPOSITION 3.1 $\theta \times d^*:Q(H^n(M), \cup) \approx \{M^+, K_n\}$, *where* $\cup$ *denotes cup product.*

PROOF. One shows that $\theta(u + v) = \theta(u) + \theta(v) + (u \cup v)(M)\alpha$, where $\alpha = d^*1$, by using $S(K_n \times K_n) = S(K_n) \vee SK_n \vee S(K_n \wedge K_n)$. The methods for proving Proposition 3.3 then yield Proposition 3.1.

Let ν_M be the normal bundle of M in R^{2n+k}. Recall M^+ is the S-dual of $T(\nu_M)$. Hence $\{M^+, K_n\} \approx \{S^{2n+k}, T(\nu_M) \wedge K_n\}$. α corresponds to $\bar{\alpha}$ = image of the generator of $\{S^{2n+k}, S^k \wedge K_n\} \approx Z_2$ under the inclusion S^k in $T(\nu_M)$ as a fibre. Combining the results of §2 and Proposition 3.1 we have:

PROPOSITION 3.2. *The quadratic functions on* $H^n(M)$ *associated to cup product are in one-to-one correspondence with homomorphisms*

$$\{S^{2n+k}, T(\nu_M) \wedge K_n\} \to Z_4$$

taking $\bar{\alpha}$ *into* 2.

Let Y be a 0-connected spectrum such that $H^0(Y; Z_2) \approx Z_2$ and let $U:Y \to K(Z_2)$ represent the generator. A *Y orientation* for M is a map $V:T(\nu_M) \to Y_k$ such that UV is the Thom class of $T(\nu_M)$. Hence, a Y orientation of M gives a map

$$\{S^{2n+k}, T(\nu_M) \wedge K_n\} \to \{S^{2n+k}, Y_k \wedge K_n\}$$

and $\bar{\alpha}$ maps into an obvious canonical element $\bar{\bar{\alpha}}$.

PROPOSITION 3.3. $\bar{\bar{\alpha}}$ *is at most divisible by* 2 *and* $\bar{\bar{\alpha}} \neq 0$ *if and only if*

$$\chi(Sq^{n+1})U = 0.$$

PROOF.

$$\{S^{2n+k}, Y_k \wedge K_n\} \approx \{S^{2n+k+1}, Y_k \wedge SK_n\}.$$

For the dimensions under consideration, the two stage Postnikov system of SK_n, namely (K_{n+1}, Sq^{n+1}), suffices to compute this group. This gives an exact sequence

$$H_{k+n+1}(Y_k) \xrightarrow{Sq^{n+1}} H_k(Y_k) \longrightarrow \{S^{2n+k}, Y_k \wedge SK_n\} \longrightarrow H_{n+k}(Y_k) \longrightarrow 0.$$

Q.E.D.

Suppose $\chi(Sq^{n+1})U = 0$. Choose a homomorphism

$$\lambda:\{S^{2n+k}, Y_k \wedge K_n\} \to Z_4$$

such that $\lambda(\bar{\bar{\alpha}}) = 2$. Suppose V is a Y orientation of M. We then have a Kervaire invariant $K(M, V) \in Z_8$ given by $\sigma(\varphi)$, where $\varphi:H^n(M) \to Z_4$ assigns to u, λ on:

$$S^{2n+k} \xrightarrow{t} T(\nu_M) \xrightarrow{\Delta} T(\nu_M) \wedge M^+ \xrightarrow{V \wedge U} Y_k \wedge K_n,$$

where t is the Thom construction and Δ is the diagonal map.

4. **Kervaire invariant and cobordism.** Suppose $\{MG_k\}$ are the Thom spaces for some cobordism theory and $\chi(Sq^{n+1})U = 0$, where U is the Thom class of MG_k. Taking $Y = \{MG_k\}$ (and choosing λ as above) we obtain a Kervaire invariant for each G manifold of dimension $2n$.

THEOREM 4.1. *K defines a homomorphism*

$$K:\Omega_{2n}(G) \to Z_8.$$

PROOF. The proof of this is somewhat tedious but straightforward.

EXAMPLE 1. $MG_k = S^k$. λ is unique.

THEOREM 4.2. *$K:\Omega_{2n}(framed) \to Z_8$ has its image in $\{0, 4\}$ and is the Kervaire invariant.*

EXAMPLE 2. $MG_k = M\,\mathrm{Spin}_k$, $n \equiv 1 \bmod 4$. For certain choices of λ, K is the Kervaire invariant defined by Brown-Peterson.

EXAMPLE 3. $MG_k = MSU_k$, $n \equiv 1 \bmod 4$. λ is unique.

EXAMPLE 4. $MG_k = MSO_k$, n even.

CONJECTURE. For a certain choice of λ, $\varphi:H^n(M) \to Z_4$ is the Pontrjagin square and K is the index mod 8.

EXAMPLE 5. $MG_k = S^{k-1}RP_\infty$. λ is unique. $\Omega_2(G)$ is the cobordism group of surfaces immersed in R^3 and K is an isomorphism. $\varphi(u)$ may be obtained as follows: Suppose $i:S \to R^3$ is an immersion of a surface S. Represent the Poincaré dual of u by an embedded circle (or disjoint circles). Let $\varphi(u)$ = number of half twists (in R^3) of a tubular neighborhood of this circle (Mobius band has one half twist).

EXAMPLE 6. Let $v_{n+1} \in H^{n+1}(BO_k)$ be the Wu class given by $v_{n+1}U = \chi(Sq^{n+1})U$ in $H^*(MO_k)$. Let $B_k^{(n)} \xrightarrow{p} BO_k$ be the fibration with k-invariant v_{n+1}. Let $MB_k^n = T(p^*\zeta_k)$, where ζ_k is the canonical k-plane bundle. This is the cobordism theory utilized by Browder to deal with the Kervaire-Milnor conjecture. One has a commutative diagram

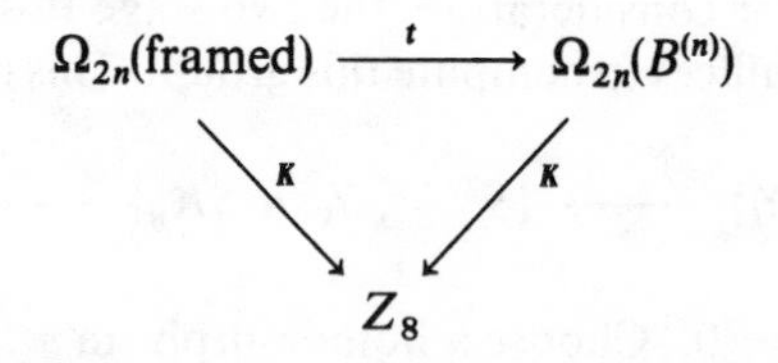

Browder shows that $t = 0$ if $n \neq 2^i - 1$ by constructing a Postnikov system for MB_k^n up to dimension $2n$.

Suppose Y is a spectrum as in §3, $\chi(Sq^{n+1})U = 0$ and suppose λ has been chosen. Let X be a 1-connected Poincaré space of dimension $2n$, n odd, ξ its Spivak normal bundle, V a Y orientation of ξ and $\alpha \in \pi_{2n+k}(T(\xi))$ an element representing the top homology class of $T(\xi)$. The methods of §3 give an invariant

$K(X, \xi, \alpha, V) \in Z_8$. Suppose

$$\begin{array}{ccc} \nu_M & \xrightarrow{g} & \xi \\ \downarrow & & \downarrow \\ M & \xrightarrow{f} & X \end{array}$$

is as in §1. Let $\beta \in \pi_{2n+k}(T(\nu_M))$ be the element obtained from the Thom construction.

THEOREM 4.3. $\sigma(M, g) = K(M, \nu_M, \beta, g^*V) - K(X, \xi, T(g)_*(\beta), V)$.

REFERENCES

1. C. T. C. Wall, *Surgery on non-simply connected manifolds*, Ann. of Math. (2) **84** (1966), 217–276. MR **35** #3692.

2. M. Kervaire, *A manifold which does not admit any differentiable structure*, Comment Math. Helv. **34** (1960), 257–270. MR **25** #2608.

3. E. H. Brown and F. S. Peterson, *Kervaire invariant of* (8k + 2) *manifolds*, Bull. Amer. Math. Soc. **71** (1965), 190–193. MR **30** #584.

4. W. Browder, *Kervaire invariant and its generalizations*, Ann. of Math. (2) **90** (1969), 157–186. MR **40** #4963.

BRANDEIS UNIVERSITY

HOMOTOPY EQUIVALENCES OF ALMOST SMOOTH MANIFOLDS

GREGORY W. BRUMFIEL

1. **Introduction.** Let $M_0^k, k \geqq 8$, be an oriented, differentiable manifold whose boundary is homeomorphic to a sphere. In general, $\partial M_0^k \in \Gamma_{k-1}$ is not a homotopy invariant of M_0^k. (Γ_{k-1} denotes the group of oriented differentiable structures on S^{k-1}.) In this paper we will study this noninvariance.

Specifically, let $B_h(M_0^k) \subset \Gamma_{k-1}$ denote the set of boundaries of homotopy smoothings of M_0^k. That is, $\Sigma^{k-1} \in B_h(M_0^k)$ if and only if there is a smooth manifold M'_0, with $\partial M'_0 = \Sigma^{k-1}$, and an orientation preserving homotopy equivalence of pairs $h: M'_0, \partial M'_0 \to M_0, \partial M_0$. We will give a homotopy theoretic description of the set of differences $\Delta_h(M_0^k) = \{B_h(M_0^k) - \partial M_0^k\} \subset \Gamma_{k-1}$, for certain classes of manifolds.

Following Sullivan, two homotopy smoothings $h: M'_0, \partial M'_0 \to M_0, \partial M_0$ and $g: M''_0, \partial M''_0 \to M_0, \partial M_0$ are called equivalent if there is a diffeomorphism $f: M'_0 \xrightarrow{\sim} M''_0$ such that h is homotopic to gf. The set of equivalence classes is denoted $hS(M_0^k)$. Sullivan has shown that if M_0^k is simply connected there is a bijection $\theta: hS(M_0^k) \xrightarrow{\sim} [M_0^k, F/O]$, where F/O is the fibre of the map $BSO \to BSF$. Thus, if $h: M'_0 \to M_0$ represents an element of $hS(M_0^k)$, the formula

$$d\theta(M'_0, h) = \partial M'_0 - \partial M_0 \in \Gamma_{k-1}$$

defines a map $d: [M_0^k, F/O] \to \Gamma_{k-1}$ and $\Delta_h(M_0^k) = \text{image}(d)$. The map d is not necessarily linear, nor is it functorial.

Let $B_{\text{th}}(M_0^k)$ (resp. $B_{\text{c}}(M_0^k)$) denote the set of homotopy spheres that bound manifolds M'_0 which are tangentially homotopy equivalent to M_0 (resp. combinatorially equivalent to M_0). Let $B_{\text{tc}}(M_0^k) \subset B_{\text{th}}(M_0^k) \cap B_{\text{c}}(M_0^k)$ denote the set of boundaries of those M'_0 such that some combinatorial equivalence $M'_0 \to M_0$ is also a tangential homotopy equivalence. Set $\Delta_{\text{th}}(M_0^k) = \{B_{\text{th}}(M_0^k) - \partial M_0^k\}$, $\Delta_{\text{c}}(M_0^k) = \{B_{\text{c}}(M_0^k) - \partial M_0^k\}$, and $\Delta_{\text{tc}}(M_0^k) = \{B_{\text{tc}}(M_0^k) - \partial M_0^k\}$.

AMS 1970 *subject classifications*. Primary 57D55, 57D60; Secondary 55D10.

The space F/O fits into a commutative diagram of fibrations

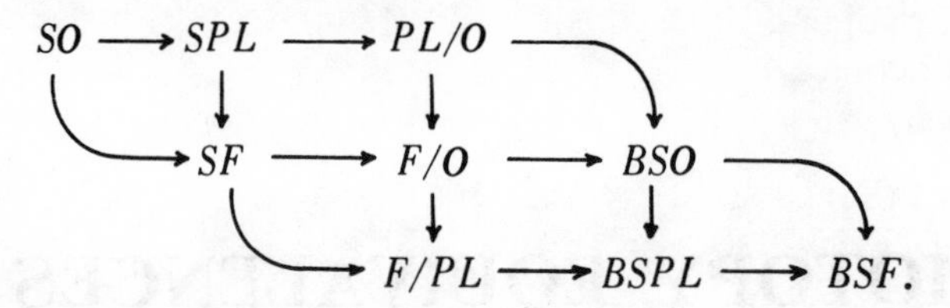

If we denote also by d the compositions $[M_0^k, X] \to [M_0^k, F/O] \xrightarrow{d} \Gamma_{k-1}$, where $X = SF$, PL/O, or SPL, then $\Delta_{\text{th}}(M_0^k) = d([M_0^k, SF])$, $\Delta_c(M_0^k) = d([M_0^k, PL/O])$, and $\Delta_{\text{tc}}(M_0^k) = d([M_0^k, SPL])$.

2. **The structure of** Γ_{k-1}. The group Γ_{k-1} can be described as follows. If $k \neq 2^j - 1$ or $2^j - 2$ then $\Gamma_{k-1} \cong bP_k \oplus (\pi_{k-1}^s/\text{im}(J))$, where $bP_k \subset \Gamma_{k-1}$ is the cyclic subgroup of homotopy spheres that bound π-manifolds. Projections onto the summands are given by the Kervaire-Milnor map $\rho:\Gamma_{k-1} \to \pi_{k-1}^s/\text{im}(J)$ and our invariants $f_R:\Gamma_{4n+1} \to Z_2 = bP_{4n+2}$ and $f_R:\Gamma_{4n-1} \to Z_{\theta_n} = bP_{4n}$, where $\theta_n = a_n \cdot 2^{2n-2} \cdot (2^{2n-1} - 1) \cdot \text{num}(B_n/4n)$. (If k is odd, $bP_k = 0$.) $\Gamma_{2^j-2} \cong \text{kernel}(\pi_{2^j-2}^s \xrightarrow{\psi} Z_2)$, where ψ is the Arf invariant. If the element $h_{j-1}^2 \in \text{Ext}_A(Z_2, Z_2)$ survives to E_∞ in the Adams spectral sequence then $\psi \neq 0$. Moreover, in this case $\Gamma_{2^j-3} \cong \pi_{2^j-3}^s/\text{im}(J)$. Thus an element $\Sigma^{k-1} \in \Gamma_{k-1}$ is determined by $\rho(\Sigma^{k-1}) \in \pi_{k-1}^s/\text{im}(J)$ and $f_R(\Sigma^{k-1}) \in bP_k$.

The invariants $f_R:\Gamma_{4n-1} \to Z_{\theta_n}$ and $f_R:b\,\text{spin}_{8n+2} \to Z_2$ are natural, and can be computed, where $b\,\text{spin}_{8n+2} \subset \Gamma_{8n+1}$ is the subgroup (of index 2) of homotopy spheres that bound spin manifolds. However, $f_R:\Gamma_{8n+5} \to Z_2$ and the extension $f_R:\Gamma_{8n+1} \to Z_2$ depend on choices, and cannot be effectively computed.

3. **The composition** $\rho d:[M_0^k, F/O] \to \pi_{k-1}^s/\text{im}(J)$. Let $\partial: S^{k-1} \to M_0^k$ represent the homotopy class of the inclusion of the boundary $\partial M_0^k \to M_0^k$. If $u \in [M_0^k, F/O]$ then $\partial^*(u) \in [S^{k-1}, F/O] = \pi_{k-1}(F/O)$ is a torsion element. Moreover, the torsion subgroup of $\pi_{k-1}(F/O)$ is isomorphic to $\pi_{k-1}^s/\text{im}(J)$, and we have the following well-known result:

THEOREM 3.1. *If $u \in [M_0^k, F/O]$ then $\rho(du) = \partial^*(u) \in \pi_{k-1}^s/\text{im}(J) \subset \pi_{k-1}(F/O)$.*

COROLLARY 3.2. *The composition $\rho d:[M_0^k, F/O] \to \pi_{k-1}^s/\text{im}(J)$ is a group homomorphism. Thus, if $u, v \in [M_0^k, F/O]$ then $du + dv - d(u + v) \in bP_k \subset \Gamma_{k-1}$.*

COROLLARY 3.3. *If $g:N_0^k, \partial N_0^k \to M_0^k, \partial M_0^k$ is a map of degree one and $u \in [M_0^k, F/O]$ then $\rho(dg^*(u)) = \rho(du)$. Thus $dg^*(u) - du \in bP_k \subset \Gamma_{k-1}$.*

REMARK 3.4. As a set $[M_0^k, SF] \cong [S^q \Lambda M_0^k, S^q]$, q large. Thus

$$\partial^*:[M_0^k, SF] \to [S^{k-1}, SF] = \pi_{k-1}^s$$

can be computed as $(\text{Id}\,\Lambda\partial)^*:[S^q\Lambda M_0^k, S^q] \to [S^q \Lambda S^{k-1}, S^q] = \pi_{q+k-1}(S^q)$.

REMARK 3.5. We have $\pi_{k-1}(PL/O) \cong \Gamma_{k-1}$ and $d = \partial^*:[M_0^k, PL/O] \to [S^{k-1}, PL/O] \cong \Gamma_{k-1}$.

4. **The surgery obstruction.** Let $M^k = M_0^k \cup_{\partial M_0} D^k$ be the closed PL manifold obtained by adjoining a disc along the boundary of M_0^k. Sullivan has defined a surgery obstruction $s:[M^k, F/O] \to P_k$, where $P_k = 0, \mathbf{Z}_2, 0, \mathbf{Z}$ if $k \equiv 1, 2, 3, 4$ (mod 4), respectively.

THEOREM 4.1 (SULLIVAN). *If $u \in [M^k, F/O]$ and $u_0 = u|_{M_0} \in [M_0^k, F/O]$ then $du_0 \in bP_k$. In fact, $du_0 = s(u) \in \mathbf{Z}_2$ if $k = 4n + 2 \neq 2^j - 2$ and $du_0 = b(su) \in \mathbf{Z}_{\theta_n}$ if $k = 4n$, where $b:\mathbf{Z} \to \mathbf{Z}_{\theta_n}$ is the projection.*

If $u \in [M^k, F/O]$, let $\xi(u) = i_*(u) \in KO^0(M^k)$, where $i:F/O \to BSO$ is the inclusion. If $k = 4n$ there is a formula expressing $s(u)$ in terms of Pontrjagin classes.

THEOREM 4.2 (SULLIVAN). *If $u \in [M^{4n}, F/O]$ then*

$$s(u) = (\tfrac{1}{8})\langle L(M)(1 - L(\xi(u))), [M^{4n}]\rangle \in \mathbf{Z},$$

where L is the Hirzebruch polynomial.

If $k = 8n + 2$ and M_0^k is a spin manifold, there is also a simple formula for s.

THEOREM 4.3. *If $u \in [M^{8n+2}, F/O]$ and $w_2(M^{8n+2}) = 0$ then*

$$s(u) = \langle V_{4n}^2(M) \cdot u^*(k_2), [M^{8n+2}]_2\rangle \in \mathbf{Z}_2,$$

where $V_{4n} \in H^{4n}(M, \mathbf{Z}_2)$ is the Wu class and $k_2 \in H^2(F/O, \mathbf{Z}_2) = \mathbf{Z}_2$ is the generator.

Theorem 4.3 is essentially a consequence of the work of Anderson, Brown, and Peterson on spin cobordism and the Arf invariant.

5. **Some consequences of the Adams conjecture.** Consider the exact sequence $[X, SF] \to [X, F/O] \xrightarrow{i_*} KO^0(X) \xrightarrow{J} J(X) \to 0$ induced by the fibrations $SF \to F/O \to BSO \to BSF$. Adams has conjectured the following:

5.1. Kernel(J) = image(i_*) coincides with the subgroup of $KO^0(X)$ generated by the elements $k^{e(k)}(\psi^k - 1)(\xi)$, where $e(k)$ is a sufficiently large integer, ψ^k is the Adams operation, and $\xi \in KO^0(X)$.

This has been proved recently by Quillen and Sullivan. As a consequence, we have

COROLLARY 5.2. *If $k \not\equiv 2$ (mod 8), or if M_0^k is a spin manifold, then each element $w \in [M_0^k, F/O]$ can be written as a sum $w = u + v$, where*

$$u \in \text{image}([M^k, F/O] \to [M_0^k, F/O]) \quad \text{and} \quad v \in \text{image}([M_0^k, SF] \to [M_0^k, F/O]).$$

We have mentioned that in general $d:[M_0^k, F/O] \to \Gamma_{k-1}$ is not linear. However, we can prove

THEOREM 5.3. *If $u, v \in [M_0^k, F/O]$ and $v \in \text{image}([M_0^k, SF] \to [M_0^k, F/O])$ then $d(u + v) = du + dv$.*

COROLLARY 5.4. *If $k \not\equiv 2$ (mod 8), or if M^k is a spin manifold, then*

$$\Delta_h(M_0^k) = (\Delta_h(M_0^k) \cap bP_k) + \Delta_{th}(M_0^k) \subset \Gamma_{k-1}.$$

REMARK 5.5. If $k = 4n$, it follows from Theorems 4.1 and 4.2, and 5.1 that $\Delta_h(M_0^{4n}) \cap bP_{4n}$ can be computed in terms of $ph(KO^0(M^{4n})) \subset H^{**}(M^{4n}, Q)$. If $k = 8n + 2$ and M_0^{8n+2} is a spin manifold, it is a consequence of Theorem 4.3 and 5.1 that $\Delta_h(M_0^{8n+2}) \cap bP_{8n+2} = 0$ if and only if $V_{4n}^2(M) \cdot w_2(\gamma) = 0$ for all $\gamma \in KO^0(M^{8n+2})$.

6. **The map** $d:[M_0^k, SF] \to \Gamma_{k-1}$. In this section we complete our description of $\Delta_h(M_0^k)$ if $k \equiv 0, 1, 3, 4, 5$, or $7 \pmod 8$ or if $k \equiv 2 \pmod 8$ and M_0^k is a spin manifold. After the results of §§2, 3, 4, and 5 it suffices to compute the composition $f_R d:[M_0^k, SF] \to bP_k$.

First, recall the invariant of Adams $e_R:\pi_{4n-1}^s \to Q/Z$, which splits off $\mathrm{im}(J) \subset \pi_{4n-1}^s$ as a direct summand. Image(e_R) consists of integral multiples of $1/j_n$, where $j_n = \mathrm{denom}(B_n/4n)$. Thus there is a unique homomorphism $\tilde{e}_R:\pi_{4n-1}^s \to Q/Z$ defined by $\mathrm{num}(B_n/4n)\tilde{e}_R(\alpha) = e_R(\alpha) \in Q/Z$. If α is the image of a generator of $\pi_{4n-1}(SO) = \mathbf{Z}$ then $\tilde{e}_R(\alpha) = 1/j_n$, so $\tilde{e}_R$ can be thought of as a "normalization" of e_R. It will be convenient to regard our invariant $f_R:\Gamma_{4n-1} \to \mathbf{Z}_{\theta_n}$ as a homomorphism $f_R:\Gamma_{4n-1} \to Q/\mathbf{Z}$, by embedding $\mathbf{Z}_{\theta_n} \subset Q/\mathbf{Z}$ in the obvious way.

THEOREM 6.1. *If* $u \in [M_0^{4n}, SF]$ *then* $f_R(du) = \tilde{e}_R(\partial^*(u)) \in Q/\mathbf{Z}$, *where* $\partial^*(u) \in [S^{4n-1}, SF] = \pi_{4n-1}^s$.

THEOREM 6.2. *If* $u \in [M_0^{8n+2}, SF]$ *and* M_0^{8n+2} *is a spin manifold, then* $f_R(du) = \langle V_{4n}^2(M) \cdot u^*j^*(k_2), [M^{8n+2}]_2\rangle \in \mathbf{Z}_2$, *where* $j:SF \to F/O$ *is the natural map.*

7. **Some consequences of the main results.** In this section we list a number of results which follow from the work above. First, we consider $4n$-manifolds.

COROLLARY 7.1. *The maps* $d:[M_0^{4n}, SF] \to \Gamma_{4n-1}$ *and* $d:[M_0^{4n}, PL/O] \to \Gamma_{4n-1}$ *are group homomorphisms which depend only on the homotopy type of* M_0^{4n}.

COROLLARY 7.2. *If* $u \in [M_0^{4n}, SF]$ *or* $u \in [M_0^{4n}, PL/O]$ *then* $f_R(du) \in \mathbf{Z}_{\theta_n}$ *has order a power of* 2. *If* M_0^{4n} *is a spin manifold,* $f_R(du) = 0$. *If* M_0^{4n} *is a weakly complex manifold,* $a_n f_R(du) = 0$.

COROLLARY 7.3. *If* $u, v \in [M_0^{4n}, F/O]$ *then*

$$du + dv - d(u + v) = (\tfrac{1}{8})\langle L(M)(L(\xi(u)) - 1)(L(\xi(v)) - 1), [M^{4n}]\rangle \in \mathbf{Z}/\theta_n \cdot \mathbf{Z} = bP_{4n}.$$

COROLLARY 7.4. *If* $g:N_0^{4n} \to M_0^{4n}$ *is a map of degree one and* $u \in [M_0^{4n}, F/O]$, *then*

$$dg^*(u) - du = (\tfrac{1}{8})\langle (g^*(L(M)) - L(N))(g^*(L(\xi(u)) - 1)), [N^{4n}]\rangle \in \mathbf{Z}/\theta_n \cdot \mathbf{Z} = bP_{4n}.$$

Thus if g *is a homotopy equivalence and* $g^*(L(M)) = L(N) \in H^{**}(N^{4n}, Q)$, *then* $\Delta_h(M_0^{4n}) = \Delta_h(N_0^{4n})$.

Next we consider $(8n + 2)$-spin manifolds.

COROLLARY 7.5. *If* $u \in [M_0^{8n+2}, F/O]$ *and* M_0^{8n+2} *is a spin manifold, then* $f_R(du) = \langle V_{4n}^2(M) \cdot u^*(k_2), [M^{8n+2}]_2 \rangle \in \mathbf{Z}_2$.

COROLLARY 7.6. *If* $u \in [M_0^{8n+2}, PL/O]$ *then* $f_R(du) = 0$.

COROLLARY 7.7. *The map* $d:[M_0^{8n+2}, F/O] \to \Gamma_{8n+1}$ *is a group homomorphism, which depends only on the homotopy type of* M_0^{8n+2}.

Finally, we consider manifolds M_0^k with a fibre homotopically trivial stable normal bundle (fht-manifolds).

COROLLARY 7.8. *If* M_0^k *is an fht-manifold then* $\Delta_c(M_0^k) = \Delta_{th}(M_0^k) = 0$. *Thus* $\Delta_h(M_0^k) \subset bP_k$. *If* $k = 8n + 2$ *then* $\Delta_h(M_0^k) = 0$.

8. S^1 **actions on homotopy spheres.** It is known that the equivariant diffeomorphism classes of free S^1 actions on homotopy $(2n - 1)$-spheres, $n \geqq 4$, correspond bijectively with the elements of $hS(\mathbf{C}P(n - 1))$. (A principal fibration $S^1 \to \Sigma^{2n-1} \to P^{2n-2} = \Sigma^{2n-1}/S^1$ corresponds to a classifying map $f: P^{2n-2} \to \mathbf{C}P(n - 1)$ and a spectral sequence argument shows that f is a homotopy equivalence.)

There are homotopy equivalences $\mathbf{C}P(n - 1) \xrightarrow{i} \mathbf{C}P(n)_0 \xrightarrow{\pi} \mathbf{C}P(n - 1)$, since $\mathbf{C}P(n)_0$ is the total space, $E(H)$, of a D^2-bundle over $\mathbf{C}P(n - 1)$. Thus, there is a commutative diagram

$$\begin{array}{ccccc} hS(\mathbf{C}P(n-1)) & \xrightarrow{\theta} & [\mathbf{C}P(n-1), F/O] & \xrightarrow{s} & P_{2n-2} \\ \Big\downarrow{\scriptstyle i_*} & & \Big\uparrow{\scriptstyle \wr i^*} & & \\ hS(\mathbf{C}P(n)_0) & \xrightarrow[\approx]{\theta} & [\mathbf{C}P(n)_0, F/O] & \xrightarrow{d} & \Gamma_{2n-1} \end{array} \tag{8.1}$$

where if $(P^{2n-2}, f) \in hS(\mathbf{C}P(n - 1))$ then $i_*(P^{2n-2}, f)$ is the homotopy smoothing $\hat{f}: E(f^*H) \to E(H) = \mathbf{C}P(n)_0$. Let

$$\tilde{B}_h(\mathbf{C}P(n)_0) = d(\theta i_*(hS(\mathbf{C}P(n - 1)))) \subset d([\mathbf{C}P(n)_0, F/O]) = B_h(\mathbf{C}P(n)_0).$$

THEOREM 8.2. $\tilde{B}_h(\mathbf{C}P(n)_0)$ *is the set of homotopy spheres that admit free* S^1 *actions.*

From (8.1), $\tilde{B}_h(\mathbf{C}P(n)_0) = d(\text{kernel}(si^*))$. Now

$$[\mathbf{C}P(n)_0, F/O] \cong Z^{[n-1/2]} \oplus [\mathbf{C}P(n)_0, SF].$$

We can apply Theorems 4.2 and 4.3 to compute $s:[\mathbf{C}P(n - 1), F/O] \to P_{2n-2}$ if $n \equiv 1, 2$, or $3 \pmod 4$. We find

LEMMA 8.3. $si^*([\mathbf{C}P(n)_0, SF]) = 0$ *if* $n \equiv 1, 2$, *or* $3 \pmod 4$. *Generators of the summand* $\mathbf{Z}^{[n-1/2]} \subset [\mathbf{C}P(n)_0, F/O]$ *can be chosen such that*

$$si^*(m_1 \cdots m_{[n-1/2]}) = m_1 \pmod 2 \in Z_2 \qquad \text{if } n \equiv 2 \pmod 4,$$
$$si^*(m_1, m_2) = -4m_1^2 + 10m_1 + 28m_2 \in Z \qquad \text{if } n = 5$$

and

$si^*(m_1, m_2, m_3)$

$$= (-m_1(32m_1^2 + 301)/3) + 84m_1^2 + 224m_1m_2 - 384m_2 - 496m_3 \in Z \quad \text{if } n = 7.$$

REMARK 8.4. $[\mathbf{C}P(4)_0, SF] = \mathbf{Z}_2$ and $si^*:[\mathbf{C}P(4)_0, SF] \to \mathbf{Z}_2$ is nonzero. In general, if $a_j \in [\mathbf{C}P(2^j)_0, SF]$ is the element $\mathbf{C}P(2^j)_0 \xrightarrow{\pi} \mathbf{C}P(2^j - 1) \xrightarrow{p} S^{2^{j+1}-2} \xrightarrow{h_j^2} SF$ then $si^*(a_j) \neq 0$. One expects that $si^*(\mathbf{Z}^{(2^{j-1}-1)}) = 0$, where $\mathbf{Z}^{(2^{j-1}-1)} \subset [\mathbf{C}P(2^j)_0, F/O]$ and that $si^*([\mathbf{C}P(n)_0, F/O]) = 0$ if $n \equiv 0 \pmod 4$, $n \neq 2^j$.

The summand $\mathbf{Z}^{[n-1/2]} \subset [\mathbf{C}P(n)_0, F/O]$ is contained in

$$\operatorname{image}([\mathbf{C}P(n), F/O] \to [\mathbf{C}P(n)_0, F/O]).$$

Thus, combining Lemma 8.3 and Remark 8.4 with Remark 3.4, Theorems 4.1, 5.3, 6.1, and 6.2 we can compute $\tilde{B}_h(\mathbf{C}P(n)_0)$ for $n = 4, 5, 6$, and 7.

THEOREM 8.5. (1) $\Gamma_7 = bP_8 = \mathbf{Z}/28\mathbf{Z}$ *and*

$$\tilde{B}_h(\mathbf{C}P(4)_0) = \{10m - 4m^2 \mid m \in \mathbf{Z}\} = \{0, 4, \pm 6, \pm 8, -10, 14\} \subset \mathbf{Z}/28 \cdot \mathbf{Z}.$$

(2) $\Gamma_9 = bP_{10} \oplus (\pi_9^s/\mathrm{im}(J)) = \mathbf{Z}_2 \oplus \mathbf{Z}_2^2$ *and*

$$\tilde{B}_h(\mathbf{C}P(5)_0) = \{\nu^3\} = \mathbf{Z}_2 \subset \pi_9^s/\mathrm{im}(J).$$

(3) $\Gamma_{11} = bP_{12} = \mathbf{Z}/992\mathbf{Z}$ *and*

$$\tilde{B}_h(\mathbf{C}P(6)_0) = \{(-m_1(32m_1^2 + 301)/3 + 84m_1^2 + 224m_1m_2 - 384m_2 \mid m_1, m_2 \in \mathbf{Z}, m_1 \equiv 0 \pmod 2\} \subset \mathbf{Z}/992 \cdot \mathbf{Z}.$$

(4) $\Gamma_{13} = \pi_{13}^s = \mathbf{Z}_3$ *and* $\tilde{B}_h(\mathbf{C}P(7)_0) = \Gamma_{13}$.

9. **Inertia groups.** Let N^k be a closed differentiable manifold. The inertia group of N^k, $I(N^k) \subset \Gamma_k$, is the subgroup consisting of those homotopy spheres Σ^k such that the manifolds N^k and $N^k \# \Sigma^k$ are diffeomorphic. Equivalently, if we identify Γ_k with the group of pseudo-isotopy classes of orientation preserving diffeomorphisms of S^{k-1}, then $I(N^k)$ is the subgroup of diffeomorphisms $\sigma: s^{k-1} \xrightarrow{\sim} S^{k-1}$ such that there is a diffeomorphism $h_0: N_0^k \xrightarrow{\sim} N_0^k$ with $h_0|_{\partial N_0} = \sigma$. Let $h: N^k \to N^k$ be the PL isomorphism defined by coning $h_0|_{\partial N_0}$ over $D^k \subset N^k$. It is not difficult to see that the mapping torus of h, $T_h = N^k \times I/(x, 0) \equiv (h(x), 1)$, is an almost smooth manifold with $\partial(T_h)_0 = \Sigma^k$, where $\Sigma^k \in \Gamma_k$ corresponds to $\sigma: s^{k-1} \xrightarrow{\sim} S^{k-1}$.

Define the homotopy inertia group $I_h(N^k) \subset I(N^k)$ (resp. concordance inertia group $I_c(N^k) \subset I(N^k)$) to be the subgroup consisting of those diffeomorphisms $\sigma: s^{k-1} \xrightarrow{\sim} S^{k-1}$ with the property that $h_0: N_0^k \xrightarrow{\sim} N_0^k$ can be chosen such that $h: N^k \to N^k$ is homotopic to the identity (resp. PL pseudo-isotopic to the identity). Then there is a homotopy equivalence (resp. PL isomorphism) $H:(T_h)_0 \to (N^k \times S^1)_0$, with $H|_{N^k \times 0} = \mathrm{Id}$.

Now $N^k \times S^1$ is not simply connected. However, the map

$$\theta: hS((N^k \times S^1)_0) \to [(N^k \times S^1)_0, F/O]$$

is still defined. Moreover, there is a natural decomposition

$$[(N^k \times S^1)_0, F/O] \cong [N^k, F/O] \oplus [N_0^k \wedge S^1, F/O]$$

and, if N^k is nice enough, the homotopy smoothings $H:(T_h)_0 \to (N^k \times S^1)_0$ above correspond bijectively under θ to the second summand. Thus,

$$I_h(N^k) = d([N_0^k \wedge S^1, F/O]) \subset \Gamma_k$$

(similarly, $I_c(N^k) = d([N_0^k \wedge S^1, PL/O])$), and the results of §3 through §7 can be applied to compute $I_h(N^k)$.

If N^k is a manifold such that each self-homotopy equivalence of N^k is homotopic to a diffeomorphism, then $I_h(N^k) = I(N^k)$. For example, $S^1 \times \mathbf{C}P(n)$ has this property. Using Theorems 4.1 and 4.2, we can show that $I_h(S^1 \times \mathbf{C}P(3)) = I(S^1 \times \mathbf{C}P(3)) = \mathbf{Z}_7 \subset \mathbf{Z}_{28} = \Gamma_7$. On the other hand, for each integer j there is a manifold P_j^6, homotopy equivalent to $\mathbf{C}P(3)$, with $p_1(P_j^6) = (4 + 24j)z^2$, $z \in H^2(P_j^6, Z)$. It turns out that $I_h(S^1 \times P_j^6) = 0$ if $j \equiv 1 \pmod 7$ and $I_h(S^1 \times P_j^6) = Z_7$ if $j \not\equiv 1 \pmod 7$. Thus $I_h(N^k)$ is not a homotopy invariant in general. ($I_h(N^k)$ is a homotopy invariant if k is even, $k = 2^j - 3$, or if $k \equiv 1 \pmod 8$ and N^k is a spin manifold.)

We can also prove from Corollary 7.8 and Theorem 4.2 the following

THEOREM 9.1. *If N^k is an fht-manifold then $I_c(N^k) = 0$ and $I_h(N^k) \subset bP_{k+1}$. If $k \equiv 1 \pmod 8$ then $I_h(N^k) = 0$. If N^k is a π-manifold and $k \not\equiv 5 \pmod 8$ then $I_h(N^k) = 0$.*

REFERENCES

1. G. Brumfiel, *On the homotopy groups of* BPL *and* PL/O. I, Ann. of Math. (2) **88** (1968), 291–311. MR **38** #2775.

———, II, Topology **8** (1969), 305–311. MR **40** #2080.

———, III, Michigan Math. J. **17** (1970), 217–224.

2. M. Kervaire and J. Milnor, *Groups of homotopy spheres*. I, Ann. of Math. (2) **77** (1963), 504–537. MR **26** #5584.

3. D. Sullivan, *Smoothing homotopy equivalences*, University of Warwick, 1966 (mimeographed).

UNIVERSITY OF CALIFORNIA, BERKELEY

A FIBERING THEOREM FOR INJECTIVE TORAL ACTIONS

PIERRE CONNER AND FRANK RAYMOND

This is a report by Frank Raymond on part of our joint work entitled *Injective operations of the toral groups* [4].

The presence of a compact connected (Lie) subgroup G in the group of homeomorphisms of a space X often forces the space to display large amounts of detectable symmetry. In what follows, X will always be, at least, paracompact, pathwise connected, locally pathwise connected, and either locally compact and semi 1-connected or the homotopy type of a CW complex.

EXAMPLE 1. (G, X^n), where X^n is a closed aspherical n-manifold ($\pi_i(X) = 0$, $i > 1$). Then $G = T^k$, a k-dimensional torus for some $k \geqq 1$, and the universal covering is Euclidean n-space ($(k, n) \neq 1, 4$) [3].

EXAMPLE 2. Weighted homogeneous affine varieties in $\mathbf{C}^{n+1}$ [5]. Let us look at the special case of Brieskorn varieties. The more general case is exactly the same.

Let V be the set of zeroes of the complex polynomial $p(z) = z_0^{a_0} + \cdots + z_n^{a_n}$. Let $c = \operatorname{lcm}\{a_0, \ldots, a_n\}$ and put $c_i = c/a_i$. Define

$$S^1 \times \mathbf{C}^{n+1} \to \mathbf{C}^{n+1} \quad \text{by} \quad (t, z_0, \ldots, z_n) \to (t^{c_0}z_0, \ldots, t^{c_n}z_n).$$

It leaves S^{2n+1} and V invariant.

There is a map

$$\gamma : S^{2n+1} - V \to S^1 \quad \text{defined by} \quad \gamma(z) = p(z)/\|p(z)\|.$$

Observe that

$$\gamma(tz) = t^c\gamma(z), \quad \text{for all } t \in S^1, z \in S^{2n+1} - V.$$

Now recall a few facts that follow immediately from the existence of slices.

If $G \to G/H$ is a principal H-fibering then we can define $(G, G/H)$ by just left multiplication on the cosets. If $\gamma : (G, X) \to (G, G/H)$ is an *equivariant* map, then $W = \gamma^{-1}(\{H\})$ is an H-slice in X. That is, $(X = G(W), G/H, W, H)$ is a fiber

AMS 1970 *subject classifications*. Primary 57E10; Secondary 55F10.

bundle with fiber W and structure group H associated with the principal fibering $G \to G/H$, $G(W)$ denotes the G image of the slice W.

Define (S^1, S^1) by $(t, \tau) \to t^c\tau$. Now apply the paragraph above. *This exhibits the complement of the variety as fibered over the circle* with fiber W and structure group Z_c:

$$\begin{array}{ccc} (S^1, S^1 \times W) & \longrightarrow & (S^1, S^1) \\ \downarrow /Z_c & & \downarrow /Z_c \\ (S^1, S^1 \times_{Z_c} W) & \xrightarrow{/W} & (S^1, S^1/Z_c) \end{array}$$

In this example it is clear that:

$$(S^1, 1) \xrightarrow{f^x} (X, x) \xrightarrow{g} (S^1/Z_c, 1)$$

is a map of degree c, where $f^x:(S^1, 1) \to (X, x)$ is defined by $f^x(t) = tx$.

Let (T^k, X) be an action of a k-dimensional torus. We say that the action (T^k, X) is *injective* if $f^x_*:\pi_1(T^k, 1) \to \pi_1(X, x)$ is a *monomorphism* $(f^x(t) = tx)$. All actions in Example 1 are injective. Let us call the action (T^k, X) *algebraic* if $f^x_\# : H_1(T^k, 1) \to H_1(X, x)$ is a *monomorphism*. (If $H^2(X/T^k; Z)$ is a torsion group, and (T^k, X) is in Example 1, then (T^k, X) is algebraic. The examples in 2 and 3 are also algebraic.)

EXAMPLE 3. Let X be a *flat* Riemannian manifold (i.e. a Riemannian manifold whose universal covering with induced Riemannian structure is Euclidean space with the ordinary metric). Then it can be seen (Calabi [1]) that if rank $H_1(X; Z) = k$ then T^k acts on X and fibers over T^k with $(Z_n)^k$ as structure group for some integer n.

FIBERING THEOREM. *Let (T^k, X) be an action and $H_1(X; Z)$ be finitely generated. Then (T^k, X) fibers (equivariantly) over T^k, if and only if, $f^x_\# : H_1(T^k, 1) \to H_1(X, x)$ is a monomorphism.*

Clearly, if $f^x_\#$ is to be a monomorphism then f^x_* must be a monomorphism. It behooves us to look at the influence of the fundamental group, here.

Suppose $\mathrm{im}(f^x_*) \subseteqq H \subseteqq \pi_1(X, x)$. Let B_H be the covering space associated with the group H and $b_0 \in B_H$ be a base point corresponding to the constant path at x. In [3] we described how one can lift (T^k, X) to (T^k, B_H). The action (T^k, B_H) covers the action (T^k, X). Furthermore, if H is normal then the covering transformations $\pi_1(X, x)/H$ commutes with the action of T^k.

(Roughly the description is as follows: Choose $b \in B_H$ and a path α from b_0 to b. If we wish to define $g(b)$ we take a path β in T^k from 1 to g. We project α to the path α' in X. We form the composition $\alpha' * \beta(t)(\alpha'(1))$; that is, first α' then the β image of the endpoint of α'. Lift this path to b_0 and endpoint is $g(b)$. One can also form $\beta(t)(\alpha'(0)) * \beta(1)\alpha'$ or $\beta(t)(\alpha'(t))$ and their endpoints. They are homotopic with fixed endpoints.)

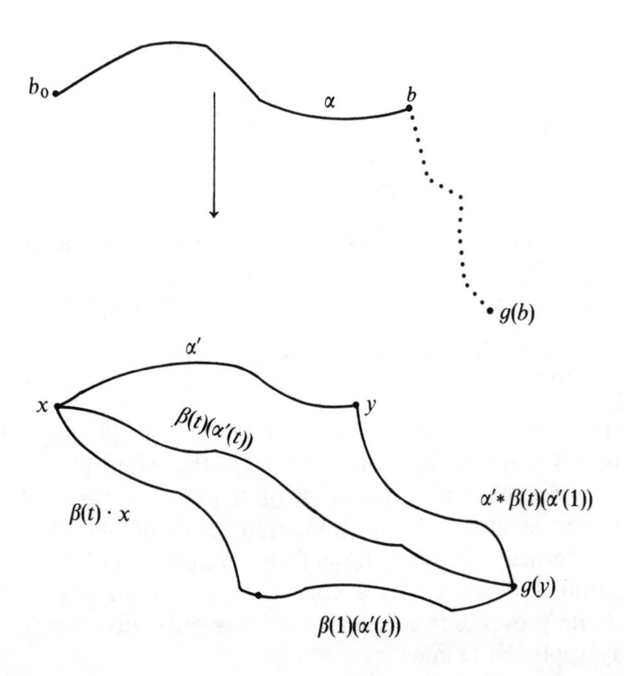

Let $\phi: \pi_1(X, x) \to L \to 1$ be an epimorphism so that $\operatorname{im}(f_*^x) \cap \operatorname{kernel} \phi = \{1\}$, where (T^k, X) is injective. Define $N = L/\phi(\operatorname{im}(f_*^x))$ and let B_H be the covering space associated with $H = \operatorname{kernel}(\pi_1(X, x) \to N)$. The most important geometric fact concerning injective actions is the

SPLITTING THEOREM. *The covering action (T^k, B_H) is equivariantly homeomorphic to $(T^k, T^k \times Y)$, where the T^k action is just left translation on the first factor.*

An important *special case* to keep in mind is the epimorphism

$$\pi_1(X, x) \xrightarrow{\text{identity}} \pi_1(X, x).$$

Consequently, $(T^k, B_{\operatorname{im}(f_*^x)}) = (T^k, T^k \times Y)$, where Y is simply connected. Notice that this fact implies the last part of Example 1.

From the splitting action $(T^k, T^k \times Y) = (T^k, B_H)$ we obtain, by projecting onto $Y = B_H/T^k$, an $N = \pi_1(X, x)/H$ action induced from the covering transformations on B_H. This N-action (Y, N) is *properly discontinuous*, i.e. all stability groups are finite and the slice theorem holds.

We may now begin with (Y, N) a properly discontinuous action and ask how may we produce an N-action on $T^k \times Y$ so that it is equivariant with (Y, N) under the projection map and commutes with the usual action of T^k on $T^k \times Y$.

A calculation shows that such $(T^k - N)$-actions on $T^k \times Y$ are in 1:1 correspondence with elements of $H^1(N; \operatorname{Maps}(Y, T^k))$ with the N-module structure on $\operatorname{Maps}(Y, T^k)$ given by $(\alpha f)(y) = f(y\alpha)$, for $f \in \operatorname{Maps}(Y, T^k)$, $\alpha \in N$.

If $m \in Z^1(N; \mathrm{Maps}(Y, T^k))$ then this cocycle represents an action described as follows:

$$\begin{array}{ccccc} (t', (t, y), \alpha) & \xrightarrow{t'} & ((t't, y), \alpha) & \xrightarrow{\alpha} & (t'tm(y, \alpha), y\alpha) \\ \cap & & \cap & & \cap \\ (T^k, T^k \times Y, n) & & (T^k \times Y, N) & & T^k \times Y \end{array}$$

If m has *finite order* n, then there exists $g: Y \to T^k$ so that $g(y)(g(y\alpha))^{-1} \equiv m(y, \alpha)^n$ for all $y \in T^k$, $\alpha \in N$.

The main ingredient in the proof of the Fibering Theorem is:

LEMMA. *If $m \in H^1(N; \mathrm{Maps}(Y, T^k))$ has finite order n, then the resulting T^k action on $(T^k \times Y)/N$ fibers over the torus with structure group $(Z_n)^k$.*

To *prove* the *Fibering Theorem* observe that if $f_{\#}^x: H_1(T^k, 1) \to H_k(X, x)$ is a monomorphism, then $\mathrm{im}(f_*^x)$ is not in the kernel of the epimorphism $\phi: \pi_1(X, x) \to H_1(X, x)$. We can choose a summand L of $H_1(X, x)$ so that $\mathrm{im}(f_{\#}^x) \subseteqq L$ and $L/\mathrm{im}(f_{\#}^x)$ is a finite group N. We apply the splitting theorem to the covering space B_H, where $H = \mathrm{kernel}(\pi_1(X, x) \to L/\mathrm{im}(f_{\#}^x))$. We obtain $(T^k, T^k \times Y = B_H, N)$ and the N-action is described by a cohomology class in $H^1(N; \mathrm{Maps}(Y, T^k))$. Since N is a finite group each element in H^1 has order dividing the order of N. Hence we may apply the Lemma.

On the other hand, if (T^k, X) fibers equivariantly over the torus $(T^k, T^k/F)$, where F is a finite subgroup of T^k, then the map

$$H_1(T^k; Z) \xrightarrow{f_{\#}^x} H_1(X; Z) \longrightarrow H_1(T^k/F; Z)$$

is a monomorphism and consequently $f_{\#}^x$ is a monomorphism. This completes the proof of the Fibering Theorem.

A SKETCH OF THE PROOF OF THE SPLITTING THEOREM. First one shows that $H \cong Z^k \times \mathrm{kernel}\, \phi$ in a natural way. Then one proceeds by induction on k.

Sketch for case of *circle*: Let ω be the generator of $\pi_1(S^1, 1)$ and represented by $\exp(2\pi it)$, $0 \leqq t \leqq 1$. $f_*^{b_0}(\omega) = \exp(2\pi it)b_0$ represents the generator of the Z-factor in $\pi_1(B_H)$, where b_0 is the base point. Let $n =$ largest integer so that $\exp(2\pi it/n)b_0$, $0 \leqq t \leqq 1$, is a closed loop. This makes $f_*^{b_0}(\omega)$ divisible by n which is only possible if $n = 1$ by the naturality of the splitting.

Next one shows that the orbit through any $b \in B_H$ is also a principal orbit by a similar argument. So therefore (S^1, B_H) is free. Now one applies induction and has (T^k, B_H) *is free.*

To see that this principal T^k-fibering is trivial we use the Leray-Hirsch theorem and the fact that the splitting $\pi_1(T^k, 1) \overset{f_{\#}^x}{\rightleftarrows} Z^k \times \mathrm{kernel}\, \phi$ implies that $H^1(B_H; Z) \to H^1(T^k; Z)$ is surjective and hence the action is translation, and the theorem is proved.

SKETCH OF THE PROOF OF THE LEMMA. First we take $C = \{(\tau, y) \mid \tau^n g(y) = 1\}$. The groups $(Z_n)^k$ act freely on $T^k \times C$ with quotient space $T^k \times C/(Z_n)^k = T^k \times Y$ and commutes with the action of T^k and N. We now draw a diagram that illustrates what is to be done:

$$\begin{array}{ccccc}
(T^k, T^k \times C, (Z_n)^k \times N) & \xrightarrow[\text{free}]{/(Z_n)^k \text{ on } T^k \times C \text{ factor}} & (T^k, T^k \times Y, N) & \xrightarrow{/T^k} & (Y, N) \\
\downarrow {\scriptstyle /N \text{ on } C\text{-factor only}} & & \downarrow {\scriptstyle /N \text{ on } T^k \times Y} & & \downarrow {\scriptstyle /N} \\
(T^k, T^k \times (C/N), (Z_n)^k) & \xrightarrow[\text{free}]{/(Z_n)^k \text{ on } T^k \times (C/N)} & (T^k, \underbrace{(T^k \times Y)/N}_{X}) & \xrightarrow{/T^k} & Y/N = X/T^k
\end{array}$$

One checks that this diagram all makes sense and that $(T^k, (T^k \times C/N)/(Z_n^k))$ is equivariantly homeomorphic to $(T^k, (T^k \times Y)/N) = (T^k, X)$. Furthermore, the $(Z_n)^k$ action on T^k factor of $T^k \times C/N$ is by translation.

Let us look at the bottom line. This is:

$$\begin{array}{ccccc}
T^k & \longleftarrow & T^k \times (C/N) & \longrightarrow & C/N \\
\downarrow {\scriptstyle /(Z_n)^k \text{ free}} & & \downarrow {\scriptstyle /(Z_n)^k \text{ free}} & & \downarrow {\scriptstyle /(Z_n)^k} \\
T^k/(Z_n)^k & \xleftarrow{C/N} & T^k \times_{(Z_n)^k} (C/N) = X & \xrightarrow{/T^k} & (C/N)/(Z_n)^k = Y/N = X/T^k
\end{array}$$

It clearly exhibits (T^k, X) as an (equivariant) fibering over $(T^k, T^k/(Z_n)^k)$ with fiber C/N.

An alternative way of looking at the Fibering Theorem. If we choose $B_H = B_{\operatorname{im}(f_*^x)}$, then Y is simply connected and $N = \pi_1(X, x)/\operatorname{im}(f_*^x)$. It can be seen that $H^1(N; \operatorname{Maps}(Y, T^k)) \xrightarrow[\cong]{\delta} H^2(N; Z^k)$ which can be identified with the set of central extensions of Z^k by N. In fact, if (T^k, X) is an injective action and $(T^k, T^k \times Y, N)$ is the splitting action corresponding to $N = \pi_1(X, x)/\operatorname{im}(f_*^x)$, then the cocycle $m \in H^1(N; \operatorname{Maps}(Y, T^k))$ yielding the $(T^k \times Y, N)$ has the property that $\delta[m] \in H^2(N; Z^k)$ is the central extension:

$$0 \longrightarrow \pi_1(T^k, 1) \xrightarrow{f_*^x} \pi_1(X, x) \longrightarrow N \longrightarrow 1.$$

THEOREM. *If (T^k, X) is an injective action the element $\delta[m]$ representing the central extension*

$$0 \longrightarrow \pi_1(T^k, 1) \xrightarrow{f_*^x} \pi_1(X, x) \longrightarrow N \longrightarrow 1$$

has finite order if and only if (T^k, X) fibers equivariantly over $(T^k, T^k/(Z_n)^k)$ for suitable n.

Some other examples and applications. One can find conditions which guarantee that every element of $H^2(N; Z^k)$ will be of finite order. For example, if (Y, N) is properly discontinuous, Y is acyclic and Y/N is compact and $H^2(Y/N; Z)$ has only elements of finite order, then $H^2(N; Z^k)$ is a torsion group. This has the direct implication that:

Every action of the circle on a compact 3-manifold without fixed points and whose orbit space is not S^2 or $P_2(\mathbf{R})$ is injective. Furthermore, if the orbit space is not a closed orientable surface, the 3-manifold fibers (equivariantly) over the circle.

In particular, if the 3-manifold is nonorientable or has boundary, this always holds. (A form of this result first proved by P. Orlik, E. Vogt, and H. Zieschang.)

It can be seen from the work of E. Calabi that all closed flat Riemannian manifolds M (Euclidean space forms) have injective T^k actions which fiber over the torus provided that $k \leqq \text{rank } H_1(M; Z)$ (Example 3).

A topological form of Calabi's theorem can be stated as follows:

TOPOLOGICAL CALABI THEOREM. *Let (T^k, X) be an injective action and suppose there exists a normal finitely generated abelian subgroup $A \subset \pi_1(X, x)$ for which π_1/A is a finite quotient group. Then, X fibers over T^k with a finite (abelian) structure group.*

(In the flat case the holonomy group plays the role of the finite quotient group.) The idea of the proof consists in embedding $\text{im}(f_*^x)$ in A and utilizing the splitting action on B_A. Since $\pi_1(X, x)/A$ is finite this $(\pi_1(X, x)/A)$-action on B_A yields a cohomology class of finite order, whence the fibering.

We may also deduce an interesting application to the well-known question of whether or not every flat manifold is a boundary. (The theorem appears to be related to recent results of A. Vasquez.)

THEOREM. *Let Γ be the holonomy group of a flat (orientable) closed Riemannian manifold M. Suppose $(Z_2)^s$ is the largest product of (Z_2)'s embeddable in Γ. Then,* dimension $H_1(M; Q) > s$ *implies that M fibers over a torus of rank greater than s, and M bounds a closed (orientable) smooth manifold.*

This generalizes to injective actions as follows:

THEOREM. *Let (T^k, X) be an injective smooth action on a smooth closed (orientable) manifold. Let $\phi: \pi_1(X, x) \to L$ be an epimorphism so that* kernel $\phi \cap \text{im}(f_*^x) = \{e\}$. *If $(Z_2)^{s+1}$ is not contained as a subgroup of $L/\text{im}(f_*^x)$ and $k > s$, then X bounds a smooth (orientable) manifold.*

The point here is that $L/\text{im}(f_*^x)$ contains the stability groups of (T^k, X). Since (T^k, X) has no fixed points, the rational Pontrjagin numbers must vanish [2] and because $(Z_2)^k$ has no fixed points, the Stiefel-Whitney numbers must vanish [2].

REFERENCES

1. E. Calabi, *Closed locally Euclidean, 4-dimensional manifolds*, Bull. Amer. Math. Soc. **63** (1957), 135.

2. P. Conner and E. Floyd, *Differentiable periodic maps*, Ergebnisse der Mathematik und ihrer Grenzgebiete, Band 23, Academic Press, New York; Springer-Verlag, Berlin, 1964. MR **31** #750.

3. P. Conner and F. Raymond, *Actions of compact Lie groups on aspherical manifolds*, Proc. Conf. on Manifolds, University of Georgia, Athens, Ga., 1969.

4. ———, *Injective operations of the toral groups*, Topology (to appear).

5. J. Milnor, *Singular points of complex hypersurfaces*, Ann. of Math. Studies, no. 61, Princeton Univ. Press, Princeton, N.J.; Tokyo Press,Tokyo, 1968. MR **39** #969.

UNIVERSITY OF VIRGINIA,
UNIVERSITY OF MICHIGAN

IMMERSION AND EMBEDDING OF MANIFOLDS

S. GITLER

1. **Introduction.** We will only consider C^∞-manifolds, and mappings between them will be C^∞-mappings unless otherwise stated. We do not consider at all the *PL* and topological cases. Let M, N be manifolds and $f: M \to N$ a mapping. Recall that if $T(M)$, $T(N)$ are their corresponding tangent bundles, f induces a bundle homomorphism $df: T(M) \to T(N)$. The mapping f is called an *immersion* if df is a monomorphism and f is called an *embedding* if f is 1-1 and an immersion. The following natural problem arises:

1.1. CLASSIFICATION PROBLEM. *Given manifolds M and N, classify immersions (embeddings), up to homotopy of M in N.*

This problem, in particular contains the

1.2. EXISTENCE PROBLEM. *Given manifolds M and N, does there exist an immersion (embedding) of M in N, in symbols $M \subseteqq N$ ($M \subset N$)?*

The first general answers to (1.1) and (1.2) were given by H. Whitney in [**57**] and [**58**], when N is Euclidean space, and by W. T. Wu in [**59**] for (1.1).

1.3. THEOREM (WHITNEY). *Every n-manifold immerses in R^{2n-1} and embeds in R^{2n}.*

1.4. THEOREM (WU). *Any two embeddings of M^n in R^{2n+1} are isotopic.*

2. **Immersions of manifolds.** The immersion problems (1.1) and (1.2) were reduced to homotopy problems by the fundamental work of S. Smale [**53**] and M. Hirsch [**28**]. One has:

AMS 1970 *subject classifications.* Primary 57D40, 57-02, 55B15, 55G40.

2.1. THEOREM (HIRSCH). *Let* $m < n$, *then the correspondence* $f \to df$ *establishes a one-to-one correspondence between regular homotopy classes of* M^m *in* N^n *and homotopy classes of bundle monomorphisms* $T(M^m) \to T(N^n)$.

There is the following:

2.2. COROLLARY. M^m *immerses in* R^{m+k} *if and only if there exists a bundle* ξ *of dimension* k *such that* $T(M) \oplus \xi$ *is a trivial bundle.*

There is another formulation of the existence problem of immersions in Euclidean space,[1] at least in the stable range that stems from Haefliger [**26**] and Wu [**60**], and was considered by Ging-tzung Yo [**61**]. Let $P(M^m)$ be the associated projective space bundle of $T(M^m)$. With respect to some Riemannian metric, let $S(M^m)$ denote the sphere tangent bundle, then $S(M^m) \to P(M^m)$ is a twofold covering and let ζ be the associated line bundle. Let $2k \geqq m$, then

2.3. THEOREM. $M^m \subseteqq R^{m+k}$ *if and only if there exists a continuous mapping* $f: P(M^m) \to RP^{m+k-1}$ *such that* $f * \xi = \zeta$, *where* ξ *is the canonical Hopf bundle over* RP^{m+k-1}.

Most of the results that have been obtained, deal with (1.2). With respect to immersions, there is the following:

2.4. CONJECTURE. *Every* n*-manifold* M^n *immerses in* $R^{2n-\alpha(n)}$, *where* $\alpha(n) =$ *number of nonzero terms in the dyadic expansion of* n.

With respect to this conjecture, if we let $X^n = \prod_{i=1}^k RP^{2^{t_i}}$, where $n = \sum_{i=1}^k 2^{t_i}$ is the dyadic expansion, it is easily seen using Stiefel-Whitney classes that $X^n \nsubseteqq R^{2n-\alpha(n)-1}$. Thus if (2.4) is true, it is best possible. With some vague evidence, and optimistically, let me set:

2.5. CONJECTURE. *Every* n*-manifold* M^n *embeds in* $R^{2n-\alpha(n)+1}$.

Again if (2.5) be true, it would be best possible using the manifolds X^n as above.

Given the manifold M^m we can associate with it two numbers, $i(M^m)$, the immersion dimension of M^m in Euclidean space, i.e. the least k such that $M^m \subseteqq R^k$, and similarly $e(M^m)$, the embedding dimension of M^m. We define $d(M^m)$, the deficiency of M^m, to be the difference $e(M^m) - i(M^m)$, so that $d(M^m) \geqq 0$. Hsiang and Sczarba in [**32**], have given examples of parallelizable manifolds M such that $d(M)$ becomes arbitrarily large. By Hirsch's theorem (2.2), $i(M^m) = m + 1$ for M^m a parallelizable manifold. However, let me advance the following courageously: Let $\mathscr{M}_s^m$ be the set of n-manifolds such that $i(M^m) \geqq 3(m+1)/2$, then

2.6. CONJECTURE. (a) *There exists an integer* $q(m)$ *such that for* $M^m \in \mathscr{M}_s^m$, $d(M^m) \leqq q(m)$, *with* $q(m) < \frac{1}{4}m$ *and*

(b) $q(m)$ *is independent of* m.

---

[1] I want to thank S. Feder for pointing this out to me.

The only case that I know that $d(M^m) \geqq 2$ and $M^m \in \mathcal{M}_s^m$ is RP^n, where $n = 2^r + 1$, this being a result of J. Levine [34] and M. Mahowald in [41]. The other cases in which $d(RP^n)$ is known, one has $d(RP^n) = 1$, (see Table (II)).

3. **Embedding of manifolds.** Here too, the embedding problems were reduced to a homotopy problem by Haefliger [25], at least in the stable range, i.e. with $n \geqq 3(m + 1)/2$, which we assume for the first three results.

3.1. THEOREM (HAEFLIGER). *Let $f: M^n \to N^n$ be a mapping which admits an extension to $\varphi: M \times M \to N \times N$ which is equivariant and such that $\varphi^{-1}(\Delta_N) \subset \Delta_M$, then f is homotopic to an embedding $g: M \to N$. If $n > 3(m + 1)/2$, then any two such embeddings g are diffeotopic.*

3.2. COROLLARY. *If M embeds in N topologically, then it does so differentiably.*

3.3. $M^m \subset R^n$ *if and only if there exists an equivariant mapping* $M \times M - \Delta_M \to S^{n-1}$.

Another approach to embeddings was given by the work of Novikov [46], Browder [11] and J. Levine [35], using surgery. The most general result of this type is the following, due to Browder [12]. We require some notation.

By a system $S = (\xi^k, f, Y)$ of codimension k on a manifold M^m, we mean a vector bundle ξ^k of k-planes, and if $S(\xi^k)$ denotes its associated sphere bundle, $f: S(\xi^k) \to Y$ is a continuous mapping, where Y is a fixed space. If $M^m \subset S^{m+k}$ is an embedding with normal bundle ν_k, we have a system called a *normal system* $N = (\nu_k, i, S^{m+k} - M^m)$. Two systems S, S' on M are called equivalent if when $S = (\xi, f, Y)$, $S' = (\eta, g, Z)$, there exists $b: \xi \to \eta$ a bundle isomorphism and $h: Y \to Z$ a homotopy equivalence such that

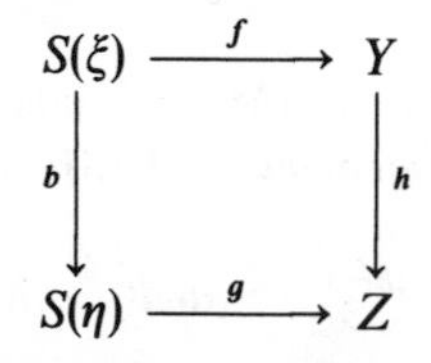

commutes. It is easy to see that for k large any two embeddings give rise to equivalent normal systems. The class of these normal systems is called the *stable normal system* of dim k. Let $S = (\xi, f, Y)$ be a system; we define its suspension $\Sigma S = (\xi \oplus 1, f', \Sigma Y)$, where f' is the suspension of f on each fiber.

3.4. THEOREM (BROWDER). *Let M^m be a manifold, $k \geqq 3$ and $m + k \geqq 5$. Then M^m embeds in S^{m+k} with normal bundle ξ^k if and only if there exists a 1-connected space Y and a system $S = (\xi^k, f, Y)$ such that $\Sigma^r S$ is in the stable normal system of M^m of dimension $k + r$.*

These results of Haefliger and Browder, beautiful as they are, had not given, until recently, strong positive or negative results when applied to specific manifolds in which the connectivity is low. There have been some attempts to use Haefliger's theorem by D. Handel [27] and S. Feder [16]. Recently E. Rees in [49] has used

Browder's theorem to produce embeddings of real projective spaces. See §5.

Browder's theorem seems relevant to the following:

3.5. CONJECTURE. *Every π-manifold M^n embeds in R^m, where $m \geqq 3(n + 1)/2$.*

4. **General nonimmersion and nonembedding results.** A manifold M^m immersed in R^{m+k} has a normal bundle ν, classified by a mapping $f_\nu: M \to BO_k$ and under special circumstances by $f_\nu: M \to BG_k$, where $G_k \subset O_k$. Given a generalized multiplicative theory h^*, f_ν induces $f_\nu^*: h^*(BG_k) \to h^*(M)$ and $\operatorname{Im} f_\nu^*$ is called the h^*-characteristic ring of ν and any class in it is called an h^*-characteristic class. Examples are the Stiefel-Whitney classes and the Pontrjagin classes [**29**], the classes γ^i in KO-theory considered by Atiyah [**6**] and the K-Pontrjagin classes of Anderson-Brown and Peterson [**5**]. The vanishing of some of these characteristic classes is a necessary condition for the immersion. On the other hand these classes are computable from those of the tangent bundle. For example if $W(M) = 1 + W_1(M) \oplus \cdots \oplus W_m(M)$ is the total Stiefel-Whitney class of M, $\overline{W}(M) = 1 + \overline{W}_1(M) \oplus \cdots \oplus \overline{W}_m(M)$, the total normal Stiefel-Whitney class, i.e. $\overline{W}(M) = W(\nu)$, is defined by $W(M) \cup \overline{W}(M) = 1$, and if $M^m \subseteqq R^{m+k}$, $\overline{W}_{k+1}(M) = \cdots = \overline{W}_m(M) = 0$.

Atiyah-Hirzebruch in [**7**] obtained very general and strong results on nonembedding of manifolds. These results were shortly extended to nonimmersions by Sanderson-Shwarzenberger [**51**]. These results together with some significant improvements appeared in K. H. Mayer's work [**42**] as an application of the Atiyah-Singer index theorem [**8**]. They seem to work best for manifolds whose integral cohomology has little or no torsion. We give some of the results of Mayer. Let M be a $2n$-manifold. Given $z \in K(M^{2n})$, we may write $\operatorname{ch}_n(Z) = s(z)/n!$, where $\langle s(z), M\rangle$ is an integer.

4.1. THEOREM (MAYER). *Let $W_3(M) = 0$ and b be a nonnegative integer with 2^b dividing n. If there exists an element $z \in K(M)$ such that $s(z)/2^b$ is an odd integer, then*

$$M^{2n} \not\subseteqq R^{4n-2\alpha(n)-2b-1} \quad \textit{and} \quad M^{2n} \not\subset R^{4n-2\alpha(n)-2b}.$$

4.2. THEOREM (MAYER). *Let M be a $4m$-manifold with $W_2(M) = 0$, b a natural number such that 2^b divides $2m$. Suppose there exists an element $z \in K(X)$ such that $s(z)/2^b$ is an odd integer. We have*:

(i) *If $\alpha(m) + b \equiv 0, 1, 3 \bmod 4$ and $z \in KSp(M)$, or $\alpha(m) + b \equiv 1, 2, 3 \bmod 4$ and $z \in KO(M)$, then $M \not\subseteqq R^{8m-4\alpha(m)-2b}$.*

(ii) *If $\alpha(m) + b \equiv 0, 1 \bmod 4$ and $z \in KSp(M)$, or $\alpha(m) + b \equiv 2, 3 \bmod 4$ and $z \in KO(M)$, then $M \not\subseteqq R^{8m-2\alpha(m)-2b+1}$.*

(iii) *If $\alpha(m) + b \equiv 1 \bmod 4$ and $z \in KSp(M)$, or $\alpha(m) + b \equiv 3 \bmod 4$ and $z \in KO(M)$, then $M \not\subset R^{8m-2\alpha(m)-2b+2}$.*

For special manifolds, results of the same nature have been obtained by S. Feder in [**14**], [**15**], S. Gitler and M. Mahowald in [**22**], and possibly others.

5. **Tables for projective spaces.** In 1963, M. Hirsch wrote up some tables on the immersion and embedding of projective spaces. They proved to be quite

useful to the author. This is the reason we give an up-to-date version of them.

TABLE 5.1. Immersion of the first few real projective spaces

n	$\subseteq$	$\not\subseteq$	n	$\subseteq$	$\not\subseteq$	n	$\subseteq$	$\not\subseteq$
*1	2		*14	22		*27	46	
*2	3		*15	22		28	49	45
*3	4		*16	31		29	50	45
*4	7		*17	31		30	53	45
*5	7		*18	32		*31	53	
*6	7		*19	32		*32	63	
*7	8		*20	34		*33	63	
*8	15		*21	38		*34	64	
*9	15		*22	38		*35	64	
*10	16		*23	38		*36	66	
*11	16		24	39	37	*37	70	
*12	18		*25	46		*38	70	
*13	22		*26	46		*39	70	

* denotes best possible

TABLE 5.2. For real projective spaces

n	$\subseteq$	$\not\subseteq$	$\subset$	$\not\subset$	References
2^r	$2n-1$	$2n-2$	$2n$	$2n-1$	Stiefel-Whitney
2^r+1	$2n-3$	$2n-4$	$2n-1$	$2n-2$	**[34]**, **[41]**
2^r+2	$2n-4$	$2n-5$	$2n-3$	$2n-4$	**[9]**, **[47]**
2^r+3	$2n-6$	$2n-7$			**[52]**, **[3]**
2^r+4	$2n-6$	$2n-7$			**[47]**, **[18]**
2^r+5	$2n-4$	$2n-5$	$2n-3$	$2n-4$	**[49]**, **[4]**
2^r+6	$2n-6$	$2n-7$	$2n-5$	$2n-6$	**[49]**, **[4]**
2^r+7	$2n-8$	$2n-9$	$2n-7$	$2n-8$	**[49]**, **[4]**
$4(2^r+2^s)+1$	$2n-4$	$2n-5$			**[52]**, **[3]**
$4(2^r+2^s)+2$	$2n-6$	$2n-7$			**[52]**, **[3]**
$4(2^r+2^s)+3$	$2n-8$	$2n-9$			**[52]**, **[3]**
$2^r-1, r\equiv 1, 2$ mod 4	$2n-2r+1$	$2n-2r$			**[33]**, **[21]**
$2^r-1, r\equiv 0$ mod 4	$2n-2r$	$2n-2r-1$			**[33]**, **[21]**
$2^r-1, r\equiv 3$	$2n-2r-1$	$2n-2r-2$			**[33]**, **[21]**
$n\equiv 0$ mod 8 $n\neq 2^r$	$2n-9$				**[48]**
$n\equiv 1$ mod 8 $\alpha(n)>3$	$2n-8$				**[48]**
$n\equiv 4$ mod 8 $n\neq 2^r+4$	$2n-7$				**[48]**
$n\equiv 3$ mod 8 $n\neq 2^r+3$			$2n-6$		E. Thomas

The information contained in Tables 5.1 and 5.2 leads to:

5.3. CONJECTURE. *Let $n = 8k + 7, k \neq 0$, then $i(RP^n)$ is given by*:

$$\begin{aligned} i(RP^n) &= 2n - 2\alpha(n) && \text{if } \alpha(n) \equiv 0 \bmod 4, \\ &= 2n - 2\alpha(n) + 1 && \text{if } \alpha(n) \equiv 1, 2 \bmod 4, \\ &= 2n - 2\alpha(n) - 1 && \text{if } \alpha(n) \equiv 3 \bmod 4. \end{aligned}$$

In [**23**], there was an announcement of the negative results contained in (5.3). We found very serious difficulties in evaluating dual operations and we had to abandon the program suggested there.

Two other general results that are interesting are:

5.4. THEOREM (MILGRAM). *For every $N \geqq 0$, there exists a k such that $RP^k \subseteqq R^{2k-N}$.*

5.5. THEOREM (MAHOWALD-MILGRAM). *For every $N \geqq 0$, there exists a k such that $RP^k \subset R^{2k-N}$.*

The result (5.4) appears in [**45**] and (5.5) in [**40**]. Both of these results have extensions to complex and quaternionic projective spaces. It seems that more work is still required, even to make a reasonable guess as to the following:

5.6. PROBLEM. (a) *What should $e(RP^n)$ be?*
(b) *What should $i(RP^n)$ be for $n \not\equiv 7 \bmod 8$?*
(c) *What should $i(RP^{2^r+2^s})$ be for $r > s \geqq 3$?*

6. **Bilinear mappings.** An approach to construct immersions of projective spaces is to use linear algebra [**31**]. In fact Ginsburg in [**17**] observed that Hirsch's theorem implies:

Suppose there exists a nonsingular bilinear mapping

$$\varphi : R^{n+1} \times R^{n+1} \to R^{m+1} \tag{6.1}$$

then RP^n immerses in R^m. Let $i(n) = i(RP^n)$.

Thus if we define $b(n + 1)$ to be the least integer m such that (6.1) exists, we obtain an inequality:

$$b(n + 1) \geqq i(n). \tag{6.2}$$

The equality in (6.2) has been verified for $1 \leqq n \leqq 23$ except possibly for $n = 19$, by K. Y. Lam in [**36**], [**37**] and J. Adem in [**1**].

It was Milgram in [**45**], who gave very general results on $b(n + 1)$. Namely Milgram proved: If n is odd

$$b(n + 1) \leqq 2n - (\alpha(n) + k(n)), \tag{6.3}$$

where

$$\begin{aligned} k(n) &= 0 && \text{if } n \equiv 1, 5 \bmod 8, \\ &= 1 && \text{if } n \equiv 3 \bmod 8, \\ &= 4 && \text{if } n \equiv 7 \bmod 8. \end{aligned}$$

K. Y. Lam in [38] gave another proof of these results together with some new ones.

Now we give two conjectures.

6.4. CONJECTURE. *There exist an infinite number of n's such that* $b(n+1) > i(n)$, *and* $b(n+1) - i(n)$ *is unbounded as a function of n.*

In particular

6.5. CONJECTURE. *For* $n = 19, 24, 28, 29, 30$, $b(n+1) > i(n)$.

We remark that H. Hopf proved that if there exists

$$\varphi : R^{n+1} \times R^{n+1} \to R^{m+1}$$

where φ is nonsingular bilinear and *symmetric*, then φ induces an embedding $RP^n \subset R^m$. See [31].

This approach to embeddings seems much harder.

I would like to stress the importance of solving (6.4) and (6.5). I think they will require essentially new methods.

7. **Obstruction theory.** The most general method for solving problem (1.1) is obstruction theory. As was mentioned before, Hirsch's theorem implies $M^n \subseteqq R^{n+k}$ if and only if in the diagram

$$\begin{array}{ccc} & & BO_k \\ & \overset{g}{\nearrow} & \downarrow \pi \\ M^n & \xrightarrow[f_\nu]{} & BO_N \end{array} \tag{7.1}$$

we can find g making the diagram homotopy commute, where f_ν is the classifying map for the stable normal bundle. In fact Hirsch's theorem says more, namely M^n will immerse with normal bundle classified by g. The lifting problem can be analyzed classically as in Steenrod [55], or else by Postnikov systems as in Spanier [54], or using Mahowald's modified Postnikov towers [39], [20], [56]. The cases treated in [39], [20], and [56] are for orientable bundles, but recently F. Nussbaum [47], J. F. McClendon [43] and C. A. Robinson [50] have extended the theory to cover nonorientable bundles.

In [22], Novikov-Kan-Bousfield-type towers are used and they seem to be quite promising.

Also, recently a spectral sequence has been developed to solve problem (1.1) in the case of immersions. This is a consequence of independent recent work of J. C. Becker [10], J. F. McClendon [43], and J. P. Meyer [44]. Robinson has used this spectral sequence to obtain some results on the classification of immersions of real projective spaces [50]. But certainly much more work needs to be done in this direction to understand the problem better.

With respect to embeddings, one has by Haefliger's theorem (3.3), another formulation of the embedding problem: Let $M^* = M \times M - \Delta/Z_2$, then

$M \subset R^{n+k}$ if and only if in the diagram

(7.2)

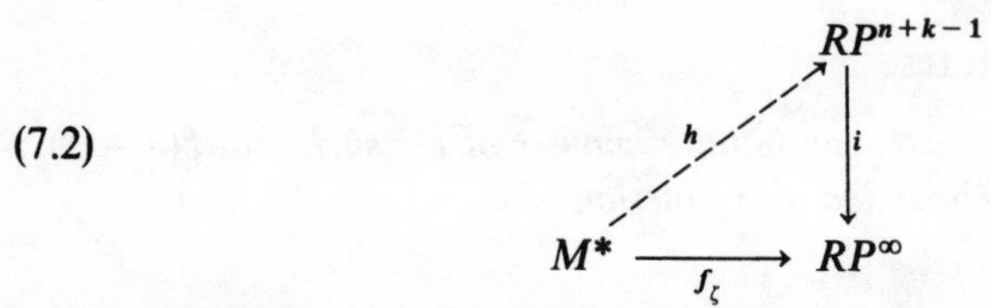

one can find h which makes the diagram commute, and where f_ζ classifies the line bundle associated with $M \times M - \Delta \to M^*$. It seems that Browder's theorem (3.4) should lead to a proof of the following:

7.3. CONJECTURE. *Suppose* $k > 3(n + 1)/2$, *then the mod p obstructions for* $p > 2$ *in* (7.2) *vanish.*

I have omitted considering work on other families of manifolds such as Dold manifolds, lens spaces, Stiefel manifolds, Grassmann manifolds, Lie groups, because of lack of time and knowledge, for which I apologize.

REFERENCES

1. J. Adem, *Some immersions associated with bilinear maps*, Bol. Soc. Mat. Mexicana (2) **13** (1968), 95–104.

2. J. Adem and S. Gitler, *Secondary characteristic classes and the immersion problem*, Bol. Soc. Mat. Mexicana (2) **8** (1963), 53–78. MR **29** #5255.

3. ———, *Non-immersion theorems for real projective spaces*, Bol. Soc. Mat. Mexicana (2) **9** (1964), 37–50. MR **32** #461.

4. J. Adem, S. Gitler and M. Mahowald, *Embedding and immersion of projective spaces*, Bol. Soc. Mat. Mexicana (2) **10** (1965), 84–88. MR **36** #3369.

5. D. W. Anderson, E. H. Brown and F. P. Peterson, *The structure of the spin cobordism ring*, Ann. of Math. (2) **86** (1967), 271–298. MR **36** #2160.

6. M. F. Atiyah, *Immersions and embeddings of manifolds*, Topology **1** (1962), 125–132. MR **26** #3080.

7. M. Atiyah and F. Hirzebruch, *Quelques théorèmes de non-plongement pour les variétés différentiables*, Bull. Soc. Math. France **87** (1959), 383–396. MR **22** #5055.

8. M. Atiyah and I. Singer, *The index of elliptic operators*. I, Ann. of Math. (2) **87** (1968), 484–530. MR **38** #5243.

9. P. Baum and W. Browder, *The cohomology of quotients of classical groups*, Topology **3** (1965), 305–336.

10. J. C. Becker, *Cohomology and the classification of liftings*, Trans. Amer. Math. Soc. **133** (1968), 447–475. MR **38** #5217.

11. W. Browder, *Homotopy type of differentiable manifolds*, Colloquium on Algebraic Topology, 1962, pp. 42–46.

12. ———, *Embedding smooth manifolds*, Proc. Internat. Congress of Math. (Moscow, 1966), "Mir", Moscow, 1968, pp. 712–719. MR **38** #6611.

13. A. Copeland and M. Mahowald, *The odd primary obstructions to finding a section in a V_n-bundle are zero*, Proc. Amer. Math. Soc. **19** (1968), 1270–1272.

14. S. Feder, *Immersions and embeddings in complex projective spaces*, Topology **4** (1965), 143–158. MR **32** #1717.

15. ———, *Non-immersion theorems for complex and quaternionic projective spaces*, Bol. Soc. Mat. Mexicana (2) **11** (1966), 62–67. MR **38** #721.

16. ———, *The reduced symmetric product of a projective space and the embedding problem*, Bol. Soc. Mat. Mexicana (2) **12** (1967), 76–80. MR **38** #5233.

17. M. Ginsburg, *Some immersions of projective spaces in Euclidean space*, Topology **2** (1963), 69–71. MR **26** #6985.

18. S. Gitler, *The projective Stiefel manifolds.* II. *Applications*, Topology **7** (1968), 47–53. MR **36** #3373b.

19. S. Gitler and D. Handel, *The projective Stiefel manifolds.* I, Topology **7** (1968), 39–46. MR **36** #3373a.

20. S. Gitler and M. Mahowald, *The geometric dimension of real stable vector bundles*, Bol. Soc. Mat. Mexicana (2) **11** (1966), 85–107. MR **37** #6922.

21. ———, *Some immersions of real projective spaces*, Bol. Soc. Mat. Mexicana (2) **14** (1969), 9–21.

22. ———, *Obstruction theory and K-theory* (to appear).

23. S. Gitler, M. Mahowald, and J. Milgram, *The non-immersion problem for RP^n and higher order cohomology operations*, Proc. Nat. Acad. Sci. U.S.A. **60** (1968), 432–437. MR **37** #3581.

24. ———, *Secondary cohomology operations and complex vector bundles*, Proc. Amer. Math. Soc. **22** (1969), 223–229. MR **39** #4836.

25. A. Haefliger, *Plongements différentiables des variétés dans variétés*, Comment. Math. Helv. **36** (1961), 47–82. MR **26** #3069.

26. ———, *Plongements différentiables dans le domaine stable*, Comment. Math. Helv. **37** (1962/63), 155–176. MR **28** #625.

27. D. Handel, *An embedding theorem for real projective spaces*, Topology **7** (1968), 125–130. MR **37** #3582.

28. M. W. Hirsch, *Immersions of manifolds*, Trans. Amer. Math. Soc. **93** (1959), 242–276. MR **22** #9980.

29. F. Hirzebruch, *Topological methods in algebraic geometry*, 3rd ed., Die Grundlehren der math. Wissenschaften, Band 131, Springer-Verlag, Berlin, 1966. MR **34** #2573.

30. H. Hopf, *Ein topologischer Beitrag zur reellen Algebra*, Comment. Math. Helv. **13** (1941), 219–239. MR **3**, 61.

31. ———, *Systeme symmetrischer Bilinearformen und euklidische Modelle der projektiven Räume*, Vierteljahrs schrift der Naturforsch. Gesellschaft Zurich **85** (1940) (Festschrift Rudolf Fuster), 165–177. MR **2**, 321.

32. W. C. Hsiang and R. Sczarba, *On the embeddability and non-embeddability of sphere bundles over spheres*, Ann. of Math. (2) **80** (1964), 397–402.

33. I. M. James, *On the immersion problem for real projective spaces*, Bull. Amer. Math. Soc. **69** (1963), 231–238. MR **26** #1900.

34. J. Levine, *Imbedding and immersion of real projective spaces*, Proc. Amer. Math. Soc. **14** (1963), 801–803. MR **27** #5272.

35. ———, *On differentiable imbeddings of simply-connected manifolds*, Bull. Amer. Math. Soc. **69** (1963), 806–809. MR **27** #5270.

36. K. Y. Lam, *Construction of non-singular bilinear maps*, Topology **6** (1967), 423–426. MR **36** #894.

37. ———, *On bilinear and skew linear maps that are non-singular*, Quart. J. Math. Oxford Ser. (2) **19** (1968), 281–288. MR **38** #2791.

38. ———, *Construction of some non-singular bilinear maps*, Bol. Soc. Mat. Mexicana (2) **13** (1968), 88–94.

39. M. E. Mahowald, *On obstruction theory in orientable fiber bundles*, Trans. Amer. Math. Soc. **110** (1964), 315–349. MR **28** #620.

40. M. Mahowald and J. Milgram, *Embedding real projective spaces*, Ann. of Math. (2) **87** (1968), 411–422.

41. M. Mahowald, *On the embeddability of the real projective spaces*, Proc. Amer. Math. Soc. **13** (1962), 763–764. MR **26** #782.

42. K. H. Mayer, *Elliptische Differentialoperatoren und Ganzzahligkeitssätze für characteristiche Zahlen*, Topology **4** (1965), 295–313. MR **33** #6650.

43. J. F. McClendon, *A spectral sequence for classifying liftings in fiber spaces*, Bull. Amer. Math. Soc. **74** (1968), 982–984. MR **38** #2781.

44. J. P. Meyer, *Relative stable homotopy*, Proc. Conf. on Algebraic Topology, Chicago Circle, 1968.

45. J. Milgram, *Immersing projective spaces*, Ann. of Math. (2) **85** (1967), 473–482.

46. S. P. Novikov, *Homotopically equivalent smooth manifolds*. I, Izv. Akad. Nauk SSSR Ser. Mat. **28** (1964), 365–474; English transl., Amer. Math. Soc. Transl. (2) **48** (1965), 271–396. MR **28** #5445.

47. F. Nussbaum, *Obstruction theory of possibly non-orientable fibrations*, Ph.D. Thesis, Northwestern University, Evanston, Ill., 1970.

48. D. Randall, *Some immersion theorems for projective spaces*, Trans. Amer. Math. Soc. **147** (1970), 135–151.

49. E. Rees, *Embeddings of real projective spaces*, Topology (to appear).

50. C. A. Robinson, *Modified Postnikov towers for $BO(n) \to BO$ and immersions of P^n* (to appear).

51. B. J. Sanderson and R. L. E. Schwarzenberger, *Non-immersion theorems for differentiable manifolds*, Proc. Cambridge Philos. Soc. **59** (1963), 319–322. MR **26** #5589.

52. B. J. Sanderson, *Immersions and embeddings of projective spaces*, Proc. London Math. Soc. (3) **14** (1964), 137–153. MR **29** #2814.

53. S. Smale, *The classification of immersions of spheres in Euclidean space*, Ann. of Math. (2) **69** (1959), 327–344. MR **21** #3862.

54. E. H. Spanier, *Algebraic topology*, McGraw Hill, New York, 1966, p. 528. MR **35** #1007.

55. N. E. Steenrod, *The topology of fiber bundles*, Princeton Math. Series, vol. 14, Princeton Univ. Press, Princeton, N.J., 1951. MR **12**, 522.

56. E. Thomas, *Seminar on fiber spaces*, Lecture Notes in Math., no. 13, Springer-Verlag, Berlin and New York, 1966, p. 45. MR **34** #3582.

57. H. Whitney, *The self intersections of a smooth n-manifold in 2n-space*, Ann. of Math. (2) **45** (1944), 220–246. MR **5**, 273.

58. ———, *The singularities of a smooth n-manifold in* $(2n - 1)$*-space*, Ann. of Math. (2) **45** (1944), 247–293. MR **5**, 274.

59. Wen Tsün Wu, *On the isotopy of C^r-manifolds of dimension n in euclidean* $(2n + 1)$*-space*, Sci. Record **2** (1958), 271–275. MR **21** #3027.

60. ———, *On the reduced products and the reduced cyclic products of a space*, Jber. Deutsch. Math. Verein. **61** (1958), 65–75.

61. Ging-Tzung Yo, *Cohomology of the projective bundle of a manifold and its applications to immersions*, Sci. Sinica **14** (1965), 959–963.

Centro de Investigacion del IPN, Mexico City

ON EMBEDDING SURFACES IN FOUR-MANIFOLDS

W. C. HSIANG AND R. H. SZCZARBA[1]

Introduction. There has been, in recent years, much progress in the study of manifolds of dimension greater than four. However, many of the techniques used to bring about this progress break down in dimensions three and four. One such technique, due to Whitney, allows one to realize, for an $(n - 1)$ connected $2n$-manifold M, any element of $H_n(M; Z)$ by a smoothly embedded n-sphere as long as n is greater than two. This is not true for $n = 2$ (see [**5**]) and one of the interesting open questions in four-dimensional topology is to decide, for a simply connected four-manifold M, which elements of $H_2(M; Z)$ can be realized by a smoothly embedded two-sphere.

In this paper, we consider the more general problem of determining the smallest possible genus for a surface representing an element of $H_2(M; Z)$. Our principal result, Theorem 1.1, gives a lower bound for the genus of such a surface in terms of invariants of the homology class and of the manifold.

The idea for the proof of Theorem 1.1 was suggested to us by Massey's solution of the Whitney conjecture [**7**] and goes as follows. Suppose a surface S is embedded in a manifold M realizing an element $u \in H_2(M; Z)$. Then, if u is divisible, $H_1(M - S; Z)$ will be cyclic and we can construct a branched covering $\tilde{M} \to M$ branched along S. Thus we have a periodic diffeomorphism h of $\tilde{M}$ whose fixed point set is S and we apply the G-Signature Theorem of [**1**] to obtain our result.

We note here that our results do not distinguish between manifolds of the same homotopy type.

The paper is organized as follows. The main theorem is stated in §1 with applications to explicit manifolds given in §2. In particular, we recover the results[2]

AMS 1970 *subject classifications*. Primary 57D40, 57D95; Secondary 20E40.

[1] During the preparation of this paper, both authors were partially supported by N.S.F. Grant GP 9452.

[2] We are indebted to C. T. C. Wall for bringing Tristram's paper [**11**] to our attention.

of Tristram [11] on $S^2 \times S^2$ and CP^2. The proof of the main theorem is given in §§3 and 4.

1. **Statement of general results.** Before stating our results, we need a few definitions.

Let M be a simply connected oriented smooth four-manifold. For $u \in H_2(M)$ (integer coefficients understood), define the *divisibility* of u, div(u), by

$$\operatorname{div}(u) = \max\{n \in Z \mid u = nv \text{ for some } v \in H_2(M)\}.$$

We say u is *primitive* if div(u) = 1.

Again for $u \in H_2(M)$, we denote by $u^2 \in Z$ the self-intersection number of u.

Let p be a prime, C the complex numbers and set

$$\begin{aligned} A_p &= \max\{|\operatorname{Re}(z)|, z \in C, z^p = 1, z \neq 1\} \\ &= \max\{|\cos(2m\pi i/p)|, m = 1, \ldots, p-1\}. \end{aligned}$$

We can now state our principal result.

THEOREM 1.1. *Let M be a connected simply connected oriented smooth four-manifold. Let S be a connected oriented surface of genus g and suppose $f: S \to M$ is a smooth embedding with $f_*[S] = u \in H_2(M)$. Set b =* rank $H_2(M)$, $\sigma(M)$ = *index of M, and suppose p is a prime dividing* div(u). *Then*

$$2g \geqq \frac{1}{p(p-1)A_p}\left\{u^2 \operatorname{cosec}^2 \frac{\pi}{p} - p(\sigma(M) + b(p-1)A_p)\right\}. \tag{1.1}$$

ADDENDUM 1. Since $A_p \leqq 1$ and $\sigma(M) \leqq b$, we can replace the inequality (1.1) above by the weaker but often useful

$$2g \geqq \frac{1}{p(p-1)}\left(u^2 \operatorname{cosec}^2 \frac{\pi}{p} - p^2 b\right). \tag{1.2}$$

ADDENDUM 2. According to Wall [10, Theorem 2], any locally flat codimension two piecewise linear submanifold S of M can be isotoped into a smooth submanifold. Thus the inequalities (1.1) and (1.2) hold when $f: S \to M$ is a locally flat piecewise linear embedding.

REMARK. Let M be as above and u any element of $H_2(M)$. Then there is a surface S and a smooth embedding $f: S \to M$ with $f_*[S] = u$. We give two proofs of this fact.

FIRST PROOF. Let v in $H^2(M)$ be dual to u and choose $f: M \to CP^2$ such that $f^*\gamma = v$, where γ is a generator for $H^2(CP^2)$. (Of course, CP^2 is the complex projective plane.) Deform f so as to be smooth and transverse regular to $CP(1) \subset CP(2)$ and let $S_0 = f^{-1}CP^1$. Now S_0 is a possibly disconnected 2-manifold in M with $f_*[S_0] = u$. We obtain S by connecting the components of S_0 with tubes.

SECOND PROOF. Using the Smale-Hirsch theory [8, Theorem D], [4, Theorem 8.3] and the Thom transverse regularity theory, we can find an immersion

$f: S^2 \to M$ of the 2-sphere S^2 with isolated self-intersection points satisfying $f_*[S^2] = u$. Let B be a small ball in M about the self-intersection point $p \in M$. Then $fS^2 \cap \partial B$ consists of two linked circles which we can span in B by an annulus. (See [**9**, Lemma 5].) Replacing $fS^2 \cap B$ by this annulus results in removing the self-intersection point and adding a handle to S^2. Continuing this process gives the result.

2. **Applications.** In this section, we apply Theorem 1.1 of the previous section to particular manifolds, comparing and contrasting our results with those already known.

Suppose M is a smooth simply connected four-manifold and $u \in H_2(M)$. We say *u can be represented by an embedded 2-sphere* if there is a locally flat piecewise linear embedding $f: S^2 \to M$ with $f_*[S^2] = u$. Of course, a smooth embedding is locally flat piecewise linear. Conversely, according to Theorem 2 of [**10**], any locally flat piecewise linear embedding can be replaced by a smooth embedding.

Let γ be a generator of $H_2(CP^2)$. It is known that $n\gamma$ for $|n| \leqq 2$ can be represented by embedded 2-spheres. (See [**5**] and [**9**, Lemma 5].) Our first result of this section states that no others can be so represented.

PROPOSITION 2.1 (TRISTRAM [**11**]). *The element $n\gamma$ in $H_2(CP^2)$ can be represented by an embedded 2-sphere if and only if $|n| < 3$.*

PROOF. Let $n = pq \geqq 3$, where p is prime. Then, according to (1.2), the genus g of a surface S realizing $n\gamma$ must satisfy

$$2g \geqq \frac{1}{p(p-1)}\left(n^2 \operatorname{cosec}^2 \frac{\pi}{p} - p^2\right) = \frac{p^2}{p(p-1)}\left(q^2 \operatorname{cosec}^2 \frac{\pi}{p} - 1\right).$$

Now, either $q = 1$ and $p \geqq 3$ or $q > 1$. In either case, the right side of the inequality is positive.

Now, let α and β be the standard generators for $H_2(S^2 \times S^2)$. It is known that $p\alpha + q\beta$ can be represented by an embedded sphere if either $|p| \leqq 1$ or $|q| \leqq 1$. (See [**5**] and [**9**, p. 132].)

PROPOSITION 2.2 (TRISTRAM [**11**]). *Let r and s be nonzero integers. Then, if r and s have a common factor greater than one, $r\alpha + s\beta \in H_2(S^2 \times S^2)$ cannot be represented by an embedded sphere.*

PROOF. If $u = r\alpha + s\beta$, $u^2 = 2rs$. Let p be a prime factor of both r and s. The inequality (1.2) becomes

$$2g \geqq \frac{1}{p(p-1)}\left(2rs \operatorname{cosec}^2 \frac{\pi}{p} - 2p^2\right).$$

Clearly, the right side of this inequality is positive unless $r = s = 2$. If $r = s = 2$, we use (1.1) which, in this case, becomes $2g \geqq 2$.

REMARK 1. In contrast to Propositions 2.1 and 2.2 above, Kervaire-Milnor [**5**, Theorem 2] have proved that, for $M = CP^2$ or $S^2 \times S^2$, any element of

$H_2(M)$ can be represented by a combinatorically embedded sphere. Of course, the embedding is not generally locally flat. To the best of our knowledge, this result is not known for arbitrary M.

REMARK 2. We should also mention the nonembeddability theorem of Kervaire-Milnor [**5**, Theorem 1]. Their result states that if $u \in H_2(M)$ is dual to $w_2(M)$ and represented by an embedded sphere, then $u^2 \equiv \sigma(M) \bmod 16$. When applied to $S^2 \times S^2$ and CP^2, this result gives far less information than our Theorem 1.1. However, this result is *not* a consequence of Theorem 1.1. For consider $M = CP^2 \# CP^2$ with standard generators γ_1 and γ_2 for $H_2(M)$. Then it follows from Theorem 1 of [**5**] that $u = 3\gamma_1 + 7\gamma_2$ cannot be represented by an embedded sphere (since $u^2 = 9 + 49 = 58$ which is not congruent to 2 mod 16). This can not be proved using Theorem 1.1.

REMARK 3. Using the embedding theorem of Wall [**9**, Theorem 3], Theorem 1 of [**5**], and Theorem 1.1 above, one can determine exactly which elements of the connected sum $S^2 \times S^2 \# S^2 \times S^2$ can be represented by an embedded sphere.

REMARK 4. Let M_k denote the connected sum of k copies of CP^2 and $\gamma \in H_2(M_k)$ the generator corresponding to the first summand. Boardman [**2**] has shown that $n\gamma$ can be represented by an embedded sphere in M_k for $k = \frac{1}{2}(n^2 - 3n + 4)$. (See [**5**] for the case $n = 3$.) Thus 3γ can be represented by an embedded sphere in M_2, 4γ in M_4, and 5γ in M_7. One can show, using Theorem 1.2, that neither 4γ nor 5γ can be represented by an embedded sphere in M_3.

3. **The proof of Theorem 1.1.** We now give the proof of Theorem 1.1.

Suppose $f: S \to M$ is a smooth embedding with $f_*[S] = u \in H_2(M)$ and let $\pi: N \to S$ be the normal disc bundle of f, $N \subset M$. Set $W = \text{closure}\,(M - N)$ and $d = \text{div}(u) > 1$.

LEMMA 3.1. *The group $H_1(W)$ is cyclic of order d.*

PROOF. By excision and Alexander duality, we have

$$H_1(W) \cong H^3(W, \partial W) \cong H^3(M, N) \cong H^3(M; S).$$

The exact sequence of the pair (M, S) gives

$$H^2(M) \xrightarrow{f^*} H^2(S) \xrightarrow{\delta^*} H^3(M; S) \longrightarrow 0$$

since $H^3(M) \cong H_1(M) = 0$ by Poincaré duality.

Let $u = dv$ and choose a base $v_1, \ldots, v_n$ for $H_2(M)$ (which is free abelian since $\pi_1 M = 0$) with $v_1 = v$. Let $g_1, \ldots, g_n$ be a dual base for $H^2(M) \cong \text{Hom}(H_2(M); Z)$ and $\mu \in H^2(S)$ dual to $[S] \in H_2(S)$. Then, since $f^* = \text{Hom}(f_*, 1)$,

$$\begin{aligned} f^*(g_i) &= d\mu \quad \text{if } i = 1, \\ &= 0 \quad \text{if } i > 1, \end{aligned}$$

so $H^3(M; S) \cong \text{coker}\, f^* \cong Z_d$.

Let p be a prime dividing div(u) and $\tilde{W} \to W$ the covering corresponding to the homomorphism

$$\pi_1(W) \to H_1(W) \cong Z_d \to Z_p.$$

Note that $\partial \tilde{W} \to \partial W$ is a connected covering since we have the Mayer-Vietoris sequence

$$\to H_1(\partial W) \to H_1(W) \oplus H_1(N) \to H_1(M) = 0$$

so that $H_1(\partial W) \to H_1(W)$ is onto.

Now, ∂W is a lens space (or $S \times S^1$ if $u^2 = 0$) and it is easy to see that we can complete $\tilde{W}$ to a p-fold branched covering $\pi: \tilde{M} \to M$, branched exactly along $S \subset M$. Thus we have an orientation preserving diffeomorphism $h: \tilde{M} \to \tilde{M}$ of period p with fixed point set $S \subset \tilde{M}$.

REMARK. For $M = CP^2$ and $u = 2\gamma$, $\tilde{M} = S^2 \times S^2$ with $h(x, y) = (y, x)$.

THEOREM 3.2. *The dimension of the real vector space* $H_2(\tilde{M}; R)$ *is* $2(p - 1)g + pb$, *where* g *is the genus of* S *and* $b = \dim H_2(M; R)$.

Before proving Theorem 3.2, we recall the G-Signature Theorem (in our situation) of Atiyah-Singer [**1**].

Let Z_p act on $\tilde{M}$ as above. Then we can decompose $H^2(\tilde{M}; R) = H_+ \oplus H_-$, where the cup product pairing of $\tilde{M}$ is positive definite on H_+, negative definite on H_- and H_+ and H_- are Z_p invariant. Proposition 6.18 of [**1**] now states that

$$\mathrm{Tr}(h^* \mid H_+) - \mathrm{Tr}(h^* \mid H_-) = \tilde{e}[S] \operatorname{cosec}^2(\pi/p),$$

where $\tilde{e} \in H^2(S)$ is the euler class of the normal bundle $S \subset \tilde{M}$ and Tr denotes trace. Since the euler class e of $S \subset M$ is given by $e[S] = u^2$, it follows easily that $\tilde{e}[S] = u^2/p$.

Thus we have

$$\mathrm{Tr}(h^* \mid H_+) - \mathrm{Tr}(h^* \mid H_-) = (u^2/p) \operatorname{cosec}^2(\pi/p). \tag{3.1}$$

To estimate the left side of (3.1), we first observe that $\pi^* H^2(M; R) \subset H^2(\tilde{M}; R)$ is exactly the $+1$ eigen space of h^* (see, for example [**3**, p. 38, Corollary 2.3]). Thus $\pi^* H^2(M; R)$ contributes exactly $\sigma(M)$ to the left side of (3.1). Now, on the remainder of $H^2(\tilde{M}; R)$ is a subspace of dimension $2(p - 1)g + pb - b = (p - 1)(2g + b)$ by Theorem 3.1, so the worst that could happen is that A_p could occur $(p - 1)(2g + b)$ times with a positive sign. Thus

$$\mathrm{Tr}(h^* \mid H_+) - \mathrm{Tr}(h^* \mid H_-) \leqq \sigma(M) + (p - 1)(2g + b)A_p. \tag{3.2}$$

Combining (3.1) and (3.2) gives the inequality of Theorem 1.1.

4. The proof of Theorem 3.2.

We begin with an algebraic lemma. We are indebted to D. M. Goldschmidt for supplying us with the proof.

LEMMA 4.1. *Let* G *be a finitely generated group with commutator subgroup* G'. *Suppose* G/G' *is cyclic of finite order and let* H *be a subgroup of* G *of prime power index with* $G' \subset H$. *Then* H/H' *is finite.*

The following is an immediate consequence of this lemma.

COROLLARY. *Let X be a complex with $\pi_1(X, x_0)$ finitely generated and $H_1(X)$ cyclic of finite order. Let p be a prime with p^r dividing the order of $H_1(x)$ and $\tilde{X} \to X$ the covering corresponding to the homomorphism*

$$\pi_1(X, x_0) \to H_1(X) \to Z_{p^r}.$$

Then $H_1(\tilde{X})$ is finite.

REMARK. Lemma 4.1 generalizes the result of Schenkman stated by Massey in [7, p. 148].

PROOF OF LEMMA 4.1. Suppose H/H' is infinite. Dividing through by H' and the torsion subgroup of H respectively, we can assume that H is finitely generated free abelian. Now, it is easily seen that, if P is a finite p-group with P/P' cyclic, then P is cyclic. (See, for example [6, p. 217].) Applying this to the group G/p^aH, where $p^r = [G:H]$ and a is any positive integer, we see that G/p^aH is cyclic. Thus $G' \subset p^aH$ for all positive integers a, so $G' = 0$. This contradicts the assumption that G/G' is finite.

For the remainder of this section all homology and cohomology groups will have real coefficients. We continue using the notation of the previous section.

LEMMA 4.2. (a) *If $u^2 \neq 0$, then* $\dim H_1(\partial W) = \dim H_2(\partial W) = 2g$.
(b) *If $u^2 = 0$, then* $\dim H_1(\partial W) = \dim H_2(\partial W) = 2g + 1$.

PROOF. Part (b) is trivial since $\partial W = \partial N = S \times S^1$. To prove part (a), consider the Gysin sequence for the S^1-bundle $\partial N \to S$,

$$0 \longrightarrow H^1(S) \longrightarrow H^1(\partial N) \longrightarrow H^0(S) \xrightarrow{\psi} H^2(S)$$
$$\longrightarrow H^2(\partial N) \longrightarrow H^1(S) \longrightarrow 0.$$

Since ψ is multiplication by the euler class $u^2 \cdot \mu$ of the bundle $\partial N \to S$ ($\mu \in H^2(S)$ is dual to $[S] \in H_2(S)$), it is an isomorphism and the result follows.

LEMMA 4.3. *The euler characteristic $\chi(W)$ of W is $b + 2g$.*

PROOF. It is easily seen that $H_1(W) = H_3(W) = H_4(W) = 0$. We compute $\dim H_2(W)$. Consider the Mayer-Vietoris sequence, for $M = W \cup N$,

$$0 \to H_2(\partial W) \to H_2(W) \oplus H_2(N) \to H_2(M)$$
$$\to H_1(\partial W) \to H_1(W) \oplus H_1(N) \to 0.$$

Suppose $u^2 \neq 0$. Then

$$\dim H_1(W) + \dim H_1(N) = \dim H_1(N) = 2g = \dim H_1(\partial W),$$

so the last map in the sequence is an isomorphism. Thus

$$\dim H_2(W) = \dim H_2(\partial W) + \dim H_2(M) - \dim H_2(N) = 2g + b - 1$$

and

$$\chi(W) = \dim H_0(W) + \dim H_2(W) = 2g + b.$$

If $u^2 = 0$, an entirely similar argument gives the same result.

As a consequence we have

COROLLARY. *The euler characteristic* $\chi(\tilde{W})$ *of* $\tilde{W}$ *is* $p(b + 2g)$.

LEMMA 4.4. *The dimension of* $H_2(\tilde{W})$ *is* $p(b + 2g) - 1$.

PROOF. Using Lemma 4.1 we see that $H_1(\tilde{W}) = 0$ and it follows easily that $H_3(\tilde{W}) = H_4(\tilde{W}) = 0$. Thus

$$\dim H_2(\tilde{W}) = \chi(\tilde{W}) - \dim H_0(\tilde{W}) = p(b + 2g) - 1.$$

To complete the proof of Theorem 3.2, let $\tilde{M} = \tilde{W} \cup \tilde{N}$ and consider the Mayer-Vietoris sequence

$$\begin{aligned} 0 &\to H_3(\tilde{M}) \to H_2(\partial\tilde{W}) \to H_2(\tilde{W}) \oplus H_2(\tilde{N}) \\ &\to H_2(\tilde{M}) \to H_1(\partial\tilde{W}) \to H_1(\tilde{W}) \oplus H_1(\tilde{N}) \to H_1(\tilde{M}) \to 0. \end{aligned}$$

Now

$$H_1(\partial\tilde{W}) = H_1(\partial\tilde{N}) \to H_1(\tilde{N}) \to H_1(\tilde{N}, \partial\tilde{N})$$

is exact and $H_1(\tilde{N}, \partial\tilde{N}) \cong H^3(\tilde{N}) \cong H^3(S) = 0$. Thus the last map in the sequence (4.1) is zero so that $H_1(\tilde{M}) = 0$ and $H_3(M) = 0$. The result now follows just as it did in the Proof of Lemma 4.3.

ADDED IN PROOF. It seems to us that a similar result was also obtained by V. A. Rohlin. (See Soviet Math. Dokl. **11** (1970), 318.)

REFERENCES

1. M. F. Atiyah and I. M. Singer, *The index of elliptic operators*. III, Ann. of Math. (2) **87** (1968), 546–604. MR **38** #5245.

2. J. M. Boardman, *Some embeddings of 2-spheres in 4-manifolds*, Proc. Cambridge Philos. Soc. **60** (1964), 354–356. MR **28** #3455.

3. E. E. Floyd, *Seminar on transformation groups*, Ann. of Math. Studies, no. 46, Princeton Univ. Press, Princeton, N.J., 1960.

4. M. W. Hirsch, *Immersions of manifolds*, Trans. Amer. Math. Soc. **93** (1959), 242–276. MR **22** # 9980.

5. M. A. Kervaire and J. W. Milnor, *On 2-spheres in 4-manifolds*, Proc. Nat. Acad. Sci. U.S.A. **47** (1961), 1651–1657. MR **24** #A2968.

6. A. G. Kuroš, *Theory of groups*, GITTL, Moscow, 1953; English transl., Vol. 2, Chelsea, New York, 1956. MR **15**, 501; MR **18**, 188.

7. W. S. Massey, *Proof of a conjecture of Whitney*, Pacific J. Math. **31** (1969), 143–156. MR **40** #3570.

8. S. Smale, *A classification of immersion of the two-sphere*, Trans. Amer. Math. Soc. **90** (1958), 281–290. MR **21** #2984.

9. C. T. C. Wall, *Diffeomorphisms of 4-manifolds*, J. London Math. Soc. **39** (1964), 131–140.

10. ———, *Locally flat P.L. submanifolds with codimension two*, Proc. Cambridge Philos. Soc. **63** (1967), 5–8. MR **37** #3577.

11. A. G. Tristram, *Some cobordism invariants for links*, Proc. Cambridge Philos. Soc. **66** (1969), 251–264.

YALE UNIVERSITY

ON CHARACTERISTIC CLASSES AND THE TOPOLOGICAL SCHUR LEMMA FROM THE TOPOLOGICAL TRANSFORMATION GROUPS VIEWPOINT

WU-YI HSIANG[1]

In this lecture, we shall discuss how to apply the method of algebraic topology to study topological transformation groups. Historically, this approach was originated in the work of L. E. J. Brouwer on periodic transformations and a little later the beautiful fixed point theorem of P. A. Smith for prime periodic maps on homology spheres. We refer to [1] for a systematic account of cohomology theory of topological transformation groups up to 1959. The most significant new feature in [1] is the setting of A. Borel which successfully brings together the modern theory of fibre bundles, spectral sequences and sheaves in a nice convenient way. However, it is rather regrettable that this excellent new setting was mainly used to look for the same type of fixed point theorem analogous to that of P. A. Smith, and consequently failed to reveal its actual potential. In fact, the effort to generalize P. A. Smith's theorem for general compact Lie groups eventually wound up with various puzzling counterexamples [2], [3], [7].

The purpose of this paper is to reformulate the Borel setting into a type of characteristic class theory for fibre bundles with a given G-space X as fibre. With such a new formulation, it is then not difficult to see that the most basic result in topological transformation groups will be a certain type of *splitting principle*. In the special case of linear transformation groups, Schur lemma shows that every representation of a torus group splits into a direct sum of rotations and this strong splittingness at *geometric level* for *linear* torus actions together with the

AMS 1970 *subject classifications*. Primary 57E10, 57E15.

[1] The author is an Alfred Sloan Fellow and is also partially supported by NSF Grant GP-8623.

maximal torus theorem of É. Cartan constitute the foundation of representation theory of compact Lie groups. In the general case of topological transformation groups, any kind of splittingness at *geometric level* is obviously out of the question. In view of the immense complexity of topological torus actions, it is a rather surprising, strong result that a kind of splittingness still holds at *characteristic class level*.

In §1, we introduce a concept of F-varieties which is an analog of Zariski closure for algebraic varieties. It seems to me that this is exactly the geometric setting that one needs in order to fully unfold the actual potential of Borel's setting. In §2, we shall indicate the proofs of topological Schur lemma and explain their geometric significance for topological torus actions on those testing spaces such as spheres, euclidean spaces, complex projective spaces, etc. In §3, we shall show that the topological Schur lemma, once established for a given space X, gives us such a strong grip of the geometric situation of topological torus actions on X that we can actually extend the central ideas of weight system of Schur, Cartan, and Weyl to study the geometric behavior of topological actions of compact Lie groups on X. A systematic development of this new approach will be appearing in forthcoming lecture notes by the author [**10**].

1. ***F*-varieties and the Borel setting.** (A) *The concept of F-varieties.* Let X be a topological space (resp. smooth manifold) and G be a compact Lie group. For the topological space X with a given topological (resp. differentiable) G-action Ψ, we introduce the following basic concepts:

DEFINITION. For a given point $x \in X$, let $G_x = \{g \in G; gx = x\}$ be the isotropy subgroup of x and G_x^0 be the identity (connected) component of G_x. We shall call G_x^0 the connected isotropy subgroup of x. Let the set of orbit types $\theta(\Psi)$ and connected orbit types $\theta^0(\Psi)$ of the given G-action Ψ on X be

$$\theta(\Psi) = \text{the conjugacy classes of subgroups of } \{G_x; x \in X\}$$

and

$$\theta^0(\Psi) = \text{the conjugacy classes of subgroups of } \{G_x^0; x \in X\}.$$

DEFINITION. For a point x on X with a *given* G-action Ψ, the fixed point set of G_x, $F(G_x, X)$, is called the F-variety spanned by x. For a subtle technical reason, it is usually more convenient to define the *connected F-variety spanned by x* to be the following set:

$$F^0(x) = \text{the connected component of } x \text{ of the set } F(G_x^0, X).$$

REMARKS. (i) In the special case that G is a torus, then G_x^0 is a subtorus and $F^0(x)$ is automatically an *invariant* subspace (since G_x^0 is a normal subgroup). This simple fact is one of the reasons that makes actions of torus groups easier to analyze than actions of general compact Lie groups.

(ii) The situation of G-spaces is quite analogous to that of algebraic varieties. The concept of F-varieties spanned by x is an analog of Zariski closure in the case of algebraic varieties. As one may expect, a great deal of important information of a given G-space is contained in the topological invariants of the network of F-varieties.

(B) *The Borel setting* [1]. Following A. Borel, we shall denote the twisted product of a G-space X and the total space of universal G-bundle E_G by X_G. Namely, X_G is the total space of the universal bundle

$$X \longrightarrow X_G \xrightarrow{\pi_1} B_G$$

with the given G-space X as fibre. Notice that there is another natural mapping $\pi_2 : X_G \to X/G$ with $\pi_2^{-1}(G(x)) = B_{G_x}$. One observes that this construction of X_G together with the two projections $\pi_1 : X_G \to B_G$ and $\pi_2 : X_G \to X/G$ from the given G-space X is clearly *functorial*. Hence, in the case that $G = T$ is a torus group, the collection of F-varieties of X consists of a canonical family of invariant subspaces, and consequently, the above construction of A. Borel will give us a network of spaces together with natural maps. One may then analyze this network of information from the traditional algebraic topology viewpoint. For example, one may analyze the network of commutative diagrams by passing to the ordinary cohomology theory. Then, the algebraic *invariants* one obtains can be viewed as the *characteristic classes* of the fibration

$$X \to X_G \to B_G.$$

Here, the Serre spectral sequence of $X_G \to B_G$ and the Larrey spectral sequence of the mapping $\pi_2 : X_G \to X/G$ offer a powerful tool in analyzing the algebraic relationships among the corresponding network of cohomology algebras.

(C) *Equivariant theories*. From the viewpoint of algebraic topology, it is rather natural to attach various equivariant theories to the category of G-spaces and equivariant mappings. Since the smallest entities in building a G-space are those homogeneous spaces of G, $\{G/H; H \text{ closed subgroups of } G\}$, it is clear that the *coefficient data* of a given equivariant theory $\mathscr{H}_G$ consists of a commutative network of algebras and homomorphisms, namely, to any conjugacy class of subgroups (H) of G, one has $\mathscr{H}_G(G/H)$ and to any pairs of subgroups $H_1 \subseteqq H_2 \subseteqq G$, one has the induced morphism $\mathscr{H}_G(G/H_2) \to \mathscr{H}_G(G/H_1)$. In fact, one may view $H^*(X_G; \cdot)$ as a kind of equivariant cohomology theory with $\{\mathscr{H}_G(G/H) = H^*(B_H; \cdot)$ and the usual induced mappings$\}$ as the coefficient data. Of course, the possible equivariant theories are many and, as usual, the choice mainly depends on the specific geometric problem that one intends to settle. However, there is an obvious natural choice for equivariant K-theory, namely the K_G-theory defined by Atiyah [9]. Especially in the study of differentiable actions, it seems to me K_G-theory is probably the most convenient natural setting to deal with. Hence, it is rather natural to choose our equivariant cohomology theory to be such that there is a nice character mapping:

$$\mathrm{Ch} : K_G^*(X) \to H_G^*(X).$$

It is not difficult to show that $H_G^*(X) = H^*(X_G)$ is the obvious candidate for our purpose. Hence, one may consider the Borel setting as a special kind of equivariant theory.

2. **Characteristic class theory and topological Schur lemma.** We shall only illustrate the central idea by explaining some results of [8], [10] in the simple testing cases:

(A) We recall the following well-known fact in the characteristic class theory of vector bundles. Let T be a torus group and Ψ be a complex linear action of T on $\mathbf{C}^n$, and Ψ be the induced T-action on $\mathbf{C}P^{n-1}$. Let $\alpha(\psi)$ be the associated $\mathbf{C}^n$-bundle over B_T and $\mathbf{C}P^{n-1} \to \mathbf{C}P_T^{n-1} \to B_T$ be the associated $\mathbf{C}P^{n-1}$-bundle of $\alpha(\psi)$. It is well known that

$$H^*(B_T; \mathbf{Z}) \cong \mathbf{Z}[t_1, \ldots, t_r] = \mathscr{A},$$

where $\{t_1, \ldots, t_r\}$ is a given basis of $H^2(B_T; \mathbf{Z}) \cong H^1(T; \mathbf{Z})$ and r is the rank of T. Furthermore,

$$H^*(\mathbf{C}P_T^{n-1}; \mathbf{Z}) \cong \mathscr{A}[\xi]/\{f(\xi) = 0\},$$

where $\xi \in H^2(\mathbf{C}P_T^{n-1}; \mathbf{Z})$ is the transgression of a generator of $H^1(S^1; \mathbf{Z}) \cong \mathbf{Z}$ in the fibration $S^1 \to S_T^{2n-1} \to \mathbf{C}P_T^{n-1}$, and

$$f(\xi) = \xi^n - C_1\xi^{n-1} + C_2\xi^{n-2} - \cdots + (-1)^n C_n,$$

with $C_i \in H^{2i}(B_T; \mathbf{Z})$ the ith Chern class of $\alpha(\psi)$. On the other hand, it follows from Schur lemma that the complex linear representation ψ splits into the direct sum of one-dimensional representations, say

$$\psi = \psi_1 \oplus \psi_2 \oplus \cdots \oplus \psi_n; \qquad \mathbf{C}^n = \mathbf{C} \oplus \mathbf{C} \oplus \cdots \oplus \mathbf{C},$$

where $\psi_j(t) = \exp(2\pi_i \cdot \omega_j(t))$ and $\omega_j \in H^1(T; \mathbf{Z})$ are integral linear functionals on T. Hence, by the so-called *splitting principle* of Chern classes, we have

$$\begin{aligned} f(\xi) &= \xi^n - C_1\xi^{n-1} + C_2\xi^{n-2} - \cdots + (-1)^n C_n \\ &= (\xi - \omega_1) \cdot (\xi - \omega_2) \cdot \cdots \cdot (\xi - \omega_n), \end{aligned}$$

where $\omega_j = C_1(\alpha(\psi_j))$ is the first Chern class of $\alpha(\psi_j)$ if we identify the integral linear functionals of T with elements of $H^2(B_T; \mathbf{Z})$ via the transgression $H^1(T; \mathbf{Z}) \cong H^2(B_T; \mathbf{Z})$. To summarize the above discussion, we see that the weight system $\Omega(\psi) = \{\omega_j\}$ of the (complex) linear T-action ψ is exactly the set of roots of the polynomial

$$f(\xi) = \xi^n - C_1\xi^{n-1} + C_2\xi^{n-2} - \cdots + (-1)^n C_n = 0$$

which, in turn, is the structural data of $H^*(P_T^{n-1}; \mathbf{Z})$ as an algebra over $H^*(B_T; \mathbf{Z})$.

(B) *Characteristic class theory for torus actions on cohomology* $\mathbf{C}P^n$. Let T be a torus group and X be a topological T-space with $H^*(X; \mathbf{Z}) \cong H^*(\mathbf{C}P^n; \mathbf{Z})$. Then the E_2-term of the Serre spectral sequence of the fibration $X \to X_T \to B_T$ is

$$E_2 = H^*(X; \mathbf{Z}) \otimes_{\mathbf{Z}} H^*(B_T; \mathbf{Z}).$$

Since there is no odd-dimensional element in E_2, it is obvious that $E_2 = E_\infty$ and hence $H^*(X_T; \mathbf{Z}) \to H^*(X; \mathbf{Z})$ is surjective. Let $\xi_0 \in H^2(X; \mathbf{Z})$ be a ring generator of $H^*(X; \mathbf{Z})$ and ξ be an arbitrarily fixed lifting of ξ_0 into $H^2(X_T; \mathbf{Z})$. It is easy to show that two different liftings differ by an element of $H^2(B_T; \mathbf{Z})$ and $H^*(X_T; \mathbf{Z})$ is generated by ξ as an algebra over $H^*(B_T; \mathbf{Z})$ with a single relation of the form

$$g(\xi) = \xi^{n+1} + C_1\xi^n + C_2\xi^{n-1} + \cdots + C_{n+1} = 0,$$

where $C_j \in H^{2j}(B_T; \mathbf{Z})$.

In the study of topological actions on *cohomology* complex projective spaces, the following splitting theorem of [7] is of fundamental importance.

THEOREM 1. *Let T be a torus group and X be a T-space with $H^*(X; \mathbf{Z}) \cong \mathbf{Z}[\xi_0]/\xi_0^{n+1} = 0$; $\deg \xi_0 = 2$. Let ξ be an arbitrary but fixed lifting of ξ_0 into $H^2(X_T; \mathbf{Z})$ and $g(\xi)$ be the defining relation of $H^*(X_T; \mathbf{Z})$ as an algebra over $H^*(B_T; \mathbf{Z})$. Then*

(i) *$g(\xi)$ is a splitting polynomial of $H^*(B_T; \mathbf{Z})[\xi]$, i.e. there exist distinct elements $\omega_1, \omega_2, \ldots, \omega_s \in H^2(B_T; \mathbf{Z})$ such that*

$$g(\xi) = (\xi - \omega_1)^{m_1} \cdot (\xi - \omega_2)^{m_2} \cdot \cdots \cdot (\xi - \omega_s)^{m_s},$$

and $\sum m_j = (n + 1)$.

(ii) *The fixed point set F consists of exactly s connected components, $F = F_1 + F_2 + \cdots + F_s$, and their cohomology algebras are isomorphic to the cohomology algebras of $\mathbf{C}P^{m_1-1}, \mathbf{C}P^{m_2-1}, \ldots, \mathbf{C}P^{m_s-1}$ respectively.*

(iii) *Let $\eta_j \in H^2(F_j; \mathbf{Z})$ be the image of $\xi_0 \in H^2(X; \mathbf{Z})$ under the induced mapping of inclusion $F_j \subseteq X$. Then, the induced mapping $j^*: H^*(X_T; \mathbf{Z}) \to H^*(F_T; \mathbf{Z}) \cong \Sigma \otimes H^*(B_T; \mathbf{Z}) \otimes H^*(F_k; \mathbf{Z})$ is given by*

$$j^*(\xi) = (\eta_1 + \omega_1, \eta_2 + \omega_2, \ldots, \eta_s + \omega_s).$$

(iv) *The set with multiplicities $\{\omega_j, m_j\}$ depends on the choice of the lifting ξ. Since two different liftings $\xi, \hat{\xi}$ are related by $\hat{\xi} = \xi + \beta$ for suitable $\beta \in H^2(B_T; \mathbf{Z})$, it is clear that the corresponding sets $\{\omega_j, m_j\}$ and $\{\hat{\omega}_j, m_j\}$ are related by $\hat{\omega}_j = \omega_j + \beta$, namely, only differ by a "translation."*

The above theorem completely determines the *structure of $H^*(X_T; \mathbf{Z})$ as an algebra over $H^*(B_T; \mathbf{Z})$*. This structural splittingness of the equivariant cohomology $H_G^*(X) = H^*(X_G; \mathbf{Z})$ has profound consequences on the geometric behavior of the given topological T-action on X. For example, the following corollary shows how the roots of the structural equation $g(\xi)$ determine the orbit types $\theta^0(\Psi)$ and the F-varieties structure of the given T-action Ψ on X.

COROLLARY 1. *Let Ψ be a topological T-action on a cohomology complex projective n-space X. Let $\Omega(\Psi) = \{\omega_j, m_j\}$ be the system of roots with multiplicities of the structural equation $g(\xi)$ as in Theorem 1. Then*

(i) *the set of connected orbit types $\theta^0(\Psi)$ consists of exactly those subtori given by $\omega_{j_1} = \omega_{j_2} = \cdots = \omega_{j_b}$, where $\{j_1, j_2, \ldots, j_b\}$ runs through all possible subsets of $\{1, 2, \ldots, s\}$;*

(ii) *to any point* $x \in X$, *let* $F_{j_1}, F_{j_2}, \ldots, F_{j_b}$ *be those connected components of* $F = F(T, X)$ *which are contained in* $F^0(x)$, *then* $F^0(x)$ *is the homological joint of* $F_{j_1}, F_{j_2}, \ldots, F_{j_b}$, *i.e.*

$$H^*(F^0(x); \mathbf{Z}) \cong H^*(\mathbf{C}P^m; \mathbf{Z}), \qquad m = m_{j_1} + \cdots + m_{j_b} - 1.$$

Roughly speaking, the above corollary shows that the orbit structure of a topological T-action on a CCP^nX with $\Omega(\Psi) = \{\omega_j, m_j\}$ has the same cohomological characteristics as the orbit structure of the linear T-action on $\mathbf{C}P^n$ with $\{\omega_j, m_j\}$ as its weight system. Hence, we shall call $\Omega(\Psi)$ the weight system of the topological action Ψ.

(C) *Torus actions on cohomology spheres*. In this subsection, we shall always assume X to be a cohomology n-sphere and Ψ to be a given T-action on X. Let F be the fixed point set of T which is well known to be also a cohomology sphere, i.e. $F \sim S^r$ for some r ($r = -1$ if F is empty). Then, one has

$$\begin{aligned} H^j(X - F; \mathbf{Z}) &= \mathbf{Z} \quad \text{if } j = n, r + 1, \\ &= 0 \quad \text{otherwise.} \end{aligned}$$

Hence, the Serre spectral sequence of the fibration $(X - F) \to (X - F)_T \to B_T$ consists of only two lines:

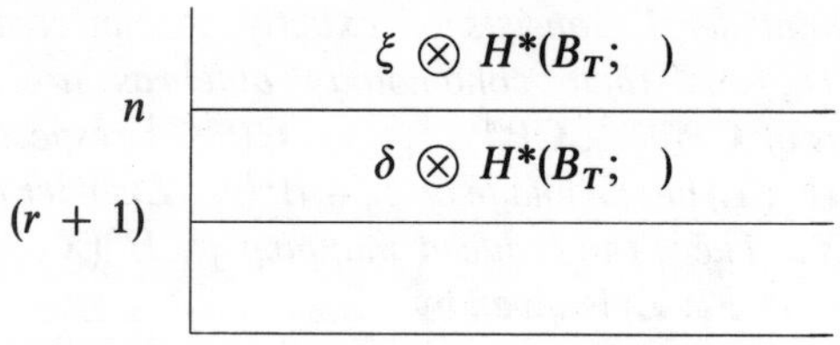

	$\xi \otimes H^*(B_T;\ )$
n	$\delta \otimes H^*(B_T;\ )$
$(r + 1)$	

On the other hand, since $(X - F)$ is by definition fixed point free, it is easy to show that the only nontrivial differential $d\xi = \delta \cdot a$ must be nonzero, i.e. $a \neq 0 \in H^{n-r}(B_T; \mathbf{Z})$. The topological Schur lemma in this case is the following refinement of a theorem of A. Borel [1, Chapter 13].

THEOREM 2. *The above homogeneous polynomial* $a \in H^{n-r}(B_T; \mathbf{Z})$ *splits into the product of linear forms. Namely, there exist suitable* $\omega_j \in H^2(B_T; \mathbf{Z})$ *such that*

$$a = l \cdot \omega_1^{m_1} \cdot \omega_2^{m_2} \cdot \cdots \cdot \omega_s^{m_s}, \qquad l \in \mathbf{Z},\ m_1 + m_2 + \cdots + m_s = (n - r)/2.$$

Again, it follows from the above splitting theorem that the orbit structure of Ψ strongly resembles (at cohomology level) the orbit structure of a linear T-action on S^n with

$$\{\omega_j, 2m_j\} \cup \{0, r + 1\}$$

as its weight system. To be precise, we state the following corollary.

COROLLARY 2. *Let* $x \in X$ *be an arbitrary point of* X. *Then, the connected isotropy subgroup* T_x^0 *is given by*

$$\omega_{j_1} = 0, \omega_{j_2} = 0, \ldots, \omega_{j_b} = 0$$

for a suitable subset of $\{j_1, j_2, \ldots, j_b\}$ *of* $\{1, 2, \ldots, s\}$. *Furthermore, the F-variety,* $F(x)$, *of* x *is a cohomology sphere of dimension* $2(m_{j_1} + m_{j_2} + \cdots + m_{j_b}) + r$.

REMARK. The above result for topological actions on cohomology spheres can easily be localized. This offers a workable substitute for the normal representations which is rather useful in the study of topological transformation groups. We refer to [7] for a systematic development of this idea.

ADDED IN PROOF. We quote here the following two theorems of fundamental theorems from **[10]**, which will be a useful tool to establish the topological Schur lemma for various spaces.

A FUNDAMENTAL FIXED THEOREM. *Let* $\{\xi_1, \ldots, \xi_k; v_1, \ldots, v_k\}$ *be a generator system of the* $\hat{R}$-*algebra* $\hat{H}^*(X_G; Q)$ *and* I *be the ideal of defining relations, namely*

$$I \to \hat{R}[x_1, \ldots, x_k] \otimes_{\hat{R}} \Lambda_{\hat{R}}[v_1, \ldots, v_h] \to \hat{H}^*(X_G; Q) \to 0$$

is an exact sequence of $\hat{R}$-*modules. Then*:

(i) *The radical of* I, $\sqrt{I}$, *decomposes into the intersection of* s *maximal ideals* $M_j = M(\alpha^{(j)})$ *whose variety is the point*

$$\alpha^{(j)} = (\alpha_1^{(j)}, \ldots, \alpha_k^{(j)}) \in \hat{R}^k, \quad i.e., \sqrt{I} = M_1 \cap \cdots \cap M_s.$$

(ii) *There is a one-to-one correspondence between the connected component of the fixed point set* $F = F^1 + \cdots + F^s$ *and the above* s *maximal ideals such that* $H^*(F^j; Q) \otimes \hat{R} \simeq A/I_j$ *where* $I_j = I_{M_j} \cap A$ (I_{M_j} *is the localization of* I *at* M_j).

(iii) $I = I_1 \cap \cdots \cap I_s = I_1 \cdot I_2 \cdots I_s$.

COROLLARY 1. *If* $\hat{H}^*(X_G; Q)$ *is generated by odd dimensional elements, then the fixed point set must be connected and* $H^*(F; Q)$ *is also generated by odd dimensional elements.*

THEOREM 3. *Let* J *be the kernel of* $\pi^* : H^*(B_G; Q) \to H^*(X_G; Q)$ *which is clearly a homogeneous ideal in* $R = H^*(B_G; Q) \simeq Q[t_1, \ldots, t_l]$. *Let* $\sqrt{J}$ *be the radical of* J *and* $\sqrt{J} = P_1 \cap \cdots \cap P_a$ *be the irreducible decomposition of* $\sqrt{J}$ *into its prime components. Then*:

(i) *Every* P_j *is integral linear in the sense that* P_j *is generated by integral linear elements of* R. *Namely, the variety of* P_j, $V(P_j)$, *is an integral linear subspace of* Q^r *and* P_j *is the kernel of* $R = H^*(B_G; Q) \to H^*(B_{H_j}; Q)$ *where* $H_j \subseteq G$ *is the subtorus with* $V(P_j)$ *as its Lie subalgebra.*

(ii) $\{H_j; j = 1, \ldots, a\}$ *are exactly those maximal elements of* $\theta^0(X)$.

(iii) *Let* $Y^j = F(H_j, X)$, $j = 1, \ldots, a$. *Then*

$$H^*(Y^j_G; Q) \simeq H^*(Y^j/G; Q) \otimes_Q H^*(B_{H_j}; Q),$$
$$H^*(X_G; Q)_{P_j} \simeq H^*(Y^j_G; Q)_{P_j} \simeq H^*(Y^j/G; Q) \otimes_Q \hat{R}_{H_j}$$

where $\hat{R}_{H_j} = H^*(B_{H_j}; Q)_{P_j}$ *is the quotient field of* $H^*(B_{H_j}; Q)$.

(iv) *Let* $k = \mathrm{Max}\{\mathrm{rk}\,(H_j);\ j = 1, \ldots, a\}$ *and* $H_1, \ldots, H_d$ *be those* H_j *with* $\mathrm{rk}\,(H_j) = k$. *Then the leading coefficient in the Laurent expansion of* $P(X_G, u)$ *at*

$u = 1$ *is given by*

$$a_{-k} = 2^{-k} \sum_{j=1}^{d} \dim_Q H^*(Y^j/G; Q).$$

3. **Geometric weight system and actions of general compact Lie groups.** Roughly speaking, the idea of geometric weight system of [**6**], [**7**] is to study the geometric behavior of a topological (resp. differentiable) G-action Ψ via its restriction Ψ/T to a maximal torus T of G. Heuristically, the maximal tori theorem of É. Cartan and the striking success of such an approach in the linear case strongly suggest that the restriction, Ψ/T, should be an "invariant" of dominant importance for the study of the G-action Ψ itself. Of course, such a simple-minded observation will be of little value if one still cannot understand their restrictions. Hence, the real reason that supports the idea of geometric weight system is the existence of deep theorems for torus actions, namely, the topological Schur lemmas for torus actions on spaces of various types. (Cf. §2.) Since the above-mentioned splitting theorems as well as the idea of geometric weight system are still at their very beginning stage, much work is yet to be carried out along this direction. In fact, even in the usually regarded as well-known case of linear actions, the understanding of the relationship between the weight system and the geometric behavior of linear representations are far from being reasonably explored.

So far, this approach is quite successful in settling several testing problems such as the classification of principal orbit types, regularity theorems for actions with simpler orbit structures, etc. We refer the reader to [**6**], [**7**], [**10**] for a discussion of those results.

References

1. A. Borel, *Seminar on transformation groups*, Ann. of Math. Studies, no. 46, Princeton Univ. Press, Princeton, N.J., 1960. MR **22** #7129.

2. G. Bredon, *Exotic actions on spheres*, Proc. Conf. on Transf. Groups (New Orleans, La., 1967), Springer-Verlag, Berlin and New York, 1968.

3. P. E. Conner and E. E. Floyd, *On the construction of periodic maps without fixed points*, Proc. Amer. Math. Soc. **10** (1959), 354–360. MR **21** #3860.

4. W. C. Hsiang and W. Y. Hsiang, *Differentiable actions of compact connected classical groups*. I, Amer. J. Math. **89** (1967), 705–786. MR **36** #304.

5. ———, *Differentiable actions of compact connected classical groups*. II, Chicago University, Chicago Ill., 1968 (mimeograph); Ann. of Math. (to appear).

6. W. Y. Hsiang, *On the geometric weight system of differentiable compact transformation groups on acyclic manifolds*, Invent. Math. (to appear).

7. ———, *On the geometric weight system of topological actions*. I, University of California, Berkeley, Calif., 1969. (mimeograph).

8. ———, *On generalizations of a theorem of A. Borel and their applications in the study of topological actions*, Proc. Topology Conf., Athens, Ga., 1969.

9. M. F. Atiyah, *Lecture notes on K-theory*, Benjamin, New York, 1969.

10. W. Y. Hsiang, *On some fundamental theorems in cohomology theory of topological transformation groups*, Taita J. Math. (1971), 61–87.

University of California, Berkeley

ON SPLITTINGS OF THE TANGENT BUNDLE OF A MANIFOLD

LEIF KRISTENSEN

1. **Introduction.** Let the tangent bundle τ of a manifold M split into a Whitney sum of two positive dimensional subbundles, $\tau = \alpha \oplus \beta$. If α is the trivial bundle then it is well known that certain of the characteristic classes of M will vanish, see e.g. Thomas [**9**]. It is the purpose of the present paper to consider the case when α is not necessarily trivial.

In §2 we shall indicate the proofs of the theorems stated below. §3 proves a certain property of the Thom class of an even dimensional nonsymplectic manifold which is needed in §2. In the proof of the theorems certain multiplicative properties of cohomology operations are needed. In §4 we give a general exposition of this subject. We focus on the general principles involved. The concrete properties needed are stated in a theorem without proof.

In the theorems stated below the Kervaire semicharacteristic and a new characteristic, "the third characteristic" are involved. Let us give the definitions. The semicharacteristic $\chi_{(2)}$ of an odd dimensional manifold M^n, $n = 2t + 1$, is defined by

(1)
$$\chi_{(2)} = \sum_{i=0}^{t} b_i \in Z_2, \qquad b_i = \dim(H_i(M; Z_2)).$$

The semicharacteristic is defined for an even dimensional manifold M^n, $n = 2t$, if $w_{2t}(M) = 0$. In this case we have

(2)
$$\chi_{(2)}(M) = \sum_{i=0}^{t-1} b_i + \tfrac{1}{2}b_t \in Z_2,$$

since b_t is even.

An even dimensional manifold M^n, $n = 2t$, is symplectic if the squaring map

(3)
$$Sq^t : H^t(M; Z_2) \to M^{2t}(M; Z_2) = Z_2$$

vanishes. For a symplectic manifold $w_{2t} = 0$ and $\chi_{(2)}$ is therefore defined.

AMS 1970 *subject classifications.* Primary 55F25, 57D20, 55G10, 55G20; Secondary 57D15.

Let us consider $H^*(M; Z_2)$ as a differential, graded algebra with Sq^1 as boundary operator.

Let

$$c_j = \dim(H_j(H^*(M); Sq^1)),$$

where $H_*(H^*(M); Sq^1)$ denotes the homology of $H^*(M)$ with respect to the boundary operator Sq^1. The third characteristic $\chi_{(3)}$ is defined if the integer $\sum_i c_{2i} \in Z$ is even. If this is the case then we define

$$\chi_{(3)}(M) = \tfrac{1}{2}\sum_i c_{2i} \in Z_2. \tag{4}$$

THEOREM 1. *Let M be an orientable manifold of dimension $2s$. Assume that $\tau = \alpha \oplus \beta$, with $\dim(\alpha) = p$ and $\dim(\beta) = q$. If $w_p(\alpha) = 0$, then*

$$\hat{\chi}_{(2)}(M) = w_{1,p}(\alpha) \cdot w_q(\beta) + w_{p-1}(\alpha) \cdot w_1(\beta) \cdot w_q(\beta),$$

with zero indeterminacy.

The notation here is as follows:

$$\hat{\chi}_{(2)}(M) = \chi_{(2)}(M) \cdot \mu(M) \in H^n(M),$$

where $\mu(M) \in H^n(M)$ is the generator. In order to define the characteristic classes $w_{i,n}$ we consider some secondary cohomology operations.

Let $\phi^{1,n}$ and $\phi^{2,n-1}$ be secondary operations associated with the relations (see (17) in §4)

$$Sq^1 Sq^n + \binom{n-1}{1} Sq^{n+1} = 0, \tag{5}$$

$$Sq^n Sq^1 + Sq^2 Sq^{n-1} + \binom{-n}{1} Sq^1 Sq^n + \binom{-n-1}{2} Sq^{n+1} = 0. \tag{6}$$

If the relation contains an unfactored term Sq^{n+1} the associated operation is of the unstable type which is only defined in dimensions $\leqq n$. For further information about these and other secondary operations see §4 below.

Let α be a p-vector bundle and let $T(\alpha)$ denote the associated Thom complex and $U = U(\alpha)$ the Thom class $U \in H^p(T(\alpha); Z_2)$. If $\phi^{i,s}$ is defined on U, let $w_{i,s}(\alpha) = w_{i,s} \subset H^{i+s-1}(B(\alpha))$, $i = 1, 2$, be such that $\phi^{i,s}(U) = w_{i,s} \cdot U$. It is clear that $w_{1,s}$ is defined if

$$w_s(\alpha) = 0 \quad \text{and,} \quad \text{for } s \text{ even,} \quad p = \dim(\alpha) \leqq s. \tag{7}$$

The indeterminacy is given by

$$\widetilde{Sq}^1(H^{s-1}(B(\alpha))), \tag{8}$$

where $\widetilde{Sq}^s$ is defined by

$$\widetilde{Sq}^s(x) = \sum_i Sq^i(x) \cdot w_{s-i}. \tag{9}$$

For $i = 2$ the situation is as follows: $w_{2,s}$ is defined provided

(10) $\quad w_1(\alpha) = 0, \quad w_s(\alpha) = 0, \quad p \leqq s + 1$ (if $s \equiv 0, 3 \pmod 4$).

The indeterminacy is given by

$$[\widetilde{Sq}^2 H^*(B(\alpha)) + \widetilde{Sq}^{s+1} H^*(B(\alpha))]^{s+1}.$$

THEOREM 2. *Let M be an orientable manifold of dimension $n = 2s$. Assume that $\tau = \alpha \oplus \beta$, with $\dim(\alpha) = p$ and $\dim(\beta) = q$. Assume that $w_i(\alpha) = 0$, $w_j(\beta) = 0$ for $i = p - 1, p$, $j = q - 1, q$. Then*

$$\chi_{(3)}(M) = w_{1 \cdot p}(\alpha) w_{1 \cdot q}(\beta).$$

The proof of this theorem (and a stronger version) we postpone to a later paper.

THEOREM 3. *Let M be a spin-manifold of dimension $4s + 1$ which is 1-connected (mod 2). Let $\tau = \alpha \oplus \beta$ with $w_p(\alpha) = 0$, $w_q(\beta) = 0$. Then with zero indeterminacy*

$$\hat{\chi}_2(M) = w_{2,p}(\alpha) \cdot w_{q-1}(\beta) + w_{p-1}(\alpha) \cdot w_{2,q}(\beta).$$

REMARK. i-connected (mod 2) means that $H_t(M; Z_2) = 0$ for $t \leqq i$.

An easy consequence of Theorem 3 is that $\chi_{(2)}(M) = 0$ if $w_{p-1}(\alpha) = 0$ and $w_{q-1}(\beta) = 0$. If α is trivial and $p \geqq 2$ we get the same conclusion.

The Theorems 1 and 3 above can be generalized to the case $n = 2^m(2s + 1) - 1$. Here, however, we make some stronger assumptions, and get a weaker result.

THEOREM 4. *Let M be an n-manifold, $n + 1 = 2^m(2s + 1)$, $m, s \geqq 1$, which is 2^{m-1}-connected (mod 2). Also assume that $w_i(M) = 0$ for $i = 2^m$ and $\chi_{(2)}(M) \neq 0$. Then the tangent bundle τ of M does not split as a Whitney sum $\tau = \alpha \oplus \beta$ of a p-dimensional bundle α and a q-dimensional bundle β satisfying $2^m \leqq p, q < 2^{m+1}s$ and*

$$w_i(\alpha) = 0 \quad \text{for } p - 2^m < i,$$
$$w_i(\beta) = 0 \quad \text{for } q - 2^m < i.$$

2. **Proof of theorems.** The proof of Theorems 1, 3, and 4 are quite similar. Here we shall prove Theorem 1. The proof of Theorem 2 will be given in a later paper.

The Whitney sum $\alpha \oplus \beta$ over M is the pull-back of $\alpha \times \beta$ over $M \times M$

$$\begin{array}{ccc} \alpha \oplus \beta & \longrightarrow & \alpha \times \beta \\ \downarrow & & \downarrow \\ M & \xrightarrow{\Delta} & M \times M \end{array}$$

where Δ is the diagonal map $\Delta(x) = (x, x)$. We use α, β, etc. to denote also the total space of the bundle. We have a map

$$\begin{array}{ccc} T(\alpha \oplus \beta) & \longrightarrow & T(\alpha \times \beta) \\ \| & & \| \\ T(\tau) & \xrightarrow{f} & T(\alpha) \wedge T(\beta) \end{array}$$

with the property

$$f^*(U(\alpha) \times U(\beta)) = U(\tau) \in H^n(T(\tau)),$$
$$U(\alpha) \times U(\beta) \in \tilde{H}^n(T(\alpha) \wedge T(\beta)) \xleftarrow{\cong} \tilde{H}^p(T(\alpha)) \otimes \tilde{H}^q(T(\beta)).$$

The classes $U(\alpha)$ and $U(\beta)$ satisfy the conditions of Proposition 7(v). Hence

$$\begin{aligned} &\phi(U(\alpha) \times U(\beta)) \\ &\quad = \phi^{1\cdot p}(U(\alpha)) \times U(\beta)^2 + U(\alpha) \cdot Sq^1(U(\alpha)) \times Sq^{q-1}(U(\beta)) \\ &\quad\quad + Sq^{p-1}(U(\alpha)) \times U(\beta) \cdot Sq^1(U(\beta)) \\ &\quad = w_{1\cdot p}(\alpha) \cdot U(\alpha) \times w_q(\beta) \cdot U(\beta) + w_{p-1}(\alpha) \cdot U(\alpha) \times w_1(\beta) w_q(\beta) U(\beta). \end{aligned}$$

After an application of f^* we get

$$\phi(U(\tau)) = (w_{1\cdot p}(\alpha) \cdot w_q(\beta) + w_{p-1}(\alpha) \cdot w_1(\beta) \cdot w_q(\beta)) \cdot U(\tau). \tag{1}$$

If $w_q(\beta) = 0$ we get $\phi(U(\tau)) = 0$ with zero indeterminacy. In order to complete the proof of Theorems 1 and 2 we need to show $\phi(U(\tau)) = \hat{\chi}_{(2)}$. The argument used for this is due to Thomas [**8**] in the case M is symplectic. In the non-symplectic case an additional argument is needed.

Pinching the complement of a tubular neighborhood of the diagonal to a point gives a map

$$g: M \times M \to T(\tau). \tag{2}$$

From §3 below we have that

$$g^*(U(\tau)) = A + t^*A + \lambda v_s \times v_s, \qquad \lambda \in Z_2,$$

where λ is zero if and only if M is symplectic. Also $t^*: H^*(M \times M) \to H^*(M \times M)$ denotes the transposition map. Further

$$\begin{aligned} A \cup t^*A &= \chi_{(2)}(M) \cdot \mu(M \times M), \\ A \cup (\lambda v_s \times v_s) = 0, &\qquad t^*A \cup (\lambda v_s \times v_s) = 0. \end{aligned} \tag{3}$$

If $\phi(A)$ and $\phi(t^*A)$ are defined with zero indeterminacy, then, by Proposition 7(ii) (see §4),

$$\begin{aligned} g^*(\phi(U(\tau))) &= \phi(A + t^*A + \lambda v_s \times v_s) \\ &= \phi(A) + t^*\phi(A) + \lambda\phi(v_s \times v_s) + A \cup t^*A \\ &\quad + A \cup (\lambda v_s \times v_s) + t^*A \cup (\lambda v_s \times v_s) \\ &= \chi_{(2)}(M) \cdot \mu(M \times M), \end{aligned}$$

since t^* is the identity in dimension $2n$, and, by Proposition 7 (§4),

(4) $\phi(v_s \times v_s) = \phi^{1 \cdot s}(v_s) \times v_2^2 + v_s \cdot Sq^1 v_s \times Sq^{s-1} v_s + Sq^{s-1} v_s \times v_s \cdot Sq^1 v_s = 0.$

This follows because $v_s^2 = Sq^s v_s = w_{2s} = 0$ by assumption and $v_s \cdot Sq^1 v_s = 0$ for dimensional reasons. Since g^* is an isomorphism in dimension $2n$ we get

$$\phi(U(\tau)) = \hat{\chi}_{(2)}(M) \cdot U(\tau). \tag{5}$$

A comparison between (1) and (4) gives the theorem.

In order to complete the proof we have to see that there is no indeterminacy in (1) and that $\phi(A)$ is defined. These, however, are both quite trivial computations which we prefer not to carry out here.

3. **The Thom class.** Thomas [8] has shown that if $\dim(M) = n$ is odd or if M is an even dimensional symplectic manifold then we can choose a basis $\{\alpha_1, \ldots, \alpha_d, \beta_d, \ldots, \beta_1\}$ for $H^*(M)$ such that if $\deg \alpha_i + \deg \beta_j = n$ then

$$\alpha_i \cup \beta_j = \delta_{ij} \cdot \mu(M), \qquad d \equiv \chi_{(2)}(M) \pmod 2.$$

Also if $A = \sum_1^d \alpha_i \otimes \beta_i \in H^*(M \times M)$ then (see (2) in §2)

$$g^*(U(\tau)) = A + t^*A, \tag{1}$$

where $t^*: H^*(M \times M) \to H^*(M \times M)$ denotes the transposition map, with

$$A \cup t^*A = \chi_{(2)}(M) \cdot \mu(M \times M).$$

Let us consider the case where $n = 2s$, $w_{2s} = 0$ and $Sq^s: H^s(M) \to H^{2s}(M) \cong Z_2$ is an epimorphism. The Wu-class v_s belongs to the kernel of Sq^s. Let $u \in H^s(M)$ be such that $u \cdot v_s \neq 0$. Then

$$u^2 = Sq^s u = v_s \cdot u = \mu.$$

An easy induction argument shows that there exists a basis $\{v_s, u, \alpha_1, \ldots, \alpha_d, \beta_d, \ldots, \beta_1\}$ for $H^*(M)$ with $\dim \alpha_i \leqq s$ satisfying

$$v_s^2 = 0, \qquad u \cdot v_s = u^2 = \mu, \qquad x \cdot \alpha_i = 0, \quad \text{if } x = v_s \text{ or } u$$

and $\dim(\alpha_i) = s$,

$$x \cdot \beta_i = 0, \quad \text{if } x = v_s \text{ or } u \tag{2}$$

and $\dim(\beta_i) = s$,

$$\alpha_i\alpha_j = 0, \qquad \beta_i\beta_j = 0, \quad \text{if } \dim(\alpha_i\alpha_j) = \dim(\beta_i\beta_j) = n,$$
$$\alpha_i\beta_j = \delta_{ij}, \quad \text{if } \dim(\alpha_i\beta_j) = n.$$

In his notes on characteristic classes Milnor shows that

$$g^*(U(\tau)) = Y^{-1}B,$$

where B is the column vector of basis elements $v_s, u, \alpha_1, \ldots, \beta_1$ and Y is the matrix with entries

$$\langle xy, \bar{\mu} \rangle, \qquad x, y \in \{v_s, u, \alpha_1, \ldots, \beta_1\}, \qquad \bar{\mu} \in H_n(M).$$

From (2) we get

$$g^*(U(\tau)) = A + t^*A + v_s \times v_s, \tag{3}$$

where

$$A = \sum_1^d \alpha_i \otimes \beta_i + v_s \times u. \tag{4}$$

We see that

$$\begin{gathered} A \cup t^*A = \chi_{(2)}(M) \cdot \mu(M \times M), \\ A \cup (v_s \times v_s) = 0, \qquad t^*A \cup (v_s \times v_s) = 0. \end{gathered} \tag{5}$$

4. **On product formulas.** In this section we shall give an exposition of a method to obtain exact (with smallest possible indeterminacy) Cartan formulas for secondary or for that matter higher order operations. This method is used to obtain the Cartan formulas needed for this paper. These are stated in Proposition 7 at the end of this section.

First let us recall that a secondary operation ϕ associated with a relation $\sum \bar{a}_\nu \bar{b}_\nu = 0$ in the Steenrod algebra can be considered as a Massey product $\langle A, B, \bar{x}\rangle$, where A is the row vector $(\bar{a}_1, \bar{a}_2, \ldots)$ and B is the column vector with entries $\bar{b}_\nu$. The Massey product $\phi(\bar{x}) = \langle A, B, \bar{x}\rangle$ is represented by the cocycle

$$\theta(x) + \sum a_\nu(w_\nu),$$

where

$$\nabla\theta = \sum a_\nu b_\nu,$$

a_ν and b_ν cochain operations representing $\bar{a}_\nu$ and $\bar{b}_\nu$, and w_ν cochains with $\delta w_\nu = b_\nu(x)$. In order to expand $\phi(\bar{x}\bar{y})$ we must be able to expand $\theta(x \cdot y)$. On the other hand, as soon as we know an expansion of $\theta(x \cdot y)$ it is a straightforward calculation to expand $\phi(\bar{x} \cdot \bar{y})$.

It is the purpose of this section to discuss expansions of this type.

Let us assume that we have given a system of elements

$$\alpha(t) \in \mathscr{A}, \qquad t \in T, \tag{1}$$

and a system of relations

$$r(s) \in \mathscr{A} \otimes \mathscr{A}, \qquad s \in S, \tag{2}$$

$$r(s) = \sum \rho(s; p, q)\alpha(p) \otimes \alpha(q), \qquad \rho(s; p, q) \in Z_2. \tag{3}$$

Here S and T are two indexing sets. Let the diagonal have the form

$$\psi(\alpha(s)) = \sum \delta(s; p, q)\alpha(p) \otimes \alpha(q), \qquad \delta \in Z_2, \tag{4}$$

$$\begin{aligned} \psi_{(2)}(r(s)) = & \sum \lambda(s, t; p, q) r(t) \otimes (\alpha(p) \otimes \alpha(q)) \\ & + \sum \mu(s, t; p, q)(\alpha(p) \otimes \alpha(q)) \otimes r(t), \end{aligned} \tag{5}$$

where

$$\psi_{(2)} : \mathscr{A} \otimes \mathscr{A} \to (\mathscr{A} \otimes \mathscr{A}) \otimes (\mathscr{A} \otimes \mathscr{A})$$

is given by $\psi_{(2)} = (1 \otimes T \otimes 1) \circ (\psi \otimes \psi)$. Let us assume that we have chosen cochain operations $a(p) \in Z\mathcal{O}$ representing $\alpha(p) \in \mathcal{A}$. Here Z denotes the kernel of the boundary operator $\nabla, \nabla\theta = \delta\theta + \theta\delta$, in $\mathcal{O}$—the set of all cochain operations. Then since $r(s)$ is a relation there exists $R(s) \in \mathcal{O}$ such that

$$\nabla R(s) = \sum \rho(s; p, q)a(p)a(q) = r(s). \tag{6}$$

Let $K(Z_2, 1)$ denote an Eilenberg-Mac Lane complex with the property that the coboundary operator in $C^*(K(Z_2, 1); Z_2)$ is the zero map. Let $u \in Z^1(K(Z_2, 1))$ denote the basic cocycle. Then

$$R(s)(u) = v_1(s)u^n, \qquad n = \deg(r(s)), \qquad v_1(s) \in Z_2, \tag{7}$$

$$R(s)(u^2) = v_2(s)u^{n+1}, \qquad v_2(s) \in Z_2. \tag{8}$$

Let us assume that fixed cochain operations T_p have been chosen with the property

$$\delta T_p(x, y) = a(p)(xy) + \sum \delta(s; p, q)a(p)(x) \cdot a(q)(y).$$

Properties of these cochain operations are given in the appendix. Let us assume that

$$T_p(u^{2^i}, u^{2^i}) = v_3(p, i) \cdot u^n, \qquad p \in T, \qquad v_3 \in Z_2, \tag{9}$$

$n = \deg(a(p)) + 2^{i+1} - 1$.

In the examples below not all values of v_3 are needed.

Let us consider the expression

$$\begin{aligned} A(s)(x, y) = R(s)(x \cdot y) + T_{r(s)}(x, y) &+ \sum \lambda(s, t; p, q)R(t)(x) \cdot a(p)a(q)(y) \\ &+ \sum \mu(s, t; p, q)a(p)a(q)(x) \cdot R(t)(y), \end{aligned} \tag{10}$$

where x and y are cocycles, and where $T_{r(s)}$ is a fixed cochain operation described in the appendix, with the property

$$\delta T_{r(s)}(x, y) = \sum \lambda(s, t; p, q)r(t)(x) \cdot a(p)a(q)(y) + \sum \mu(s, t; p, q)a(p)a(q)(x) \cdot r(t)(y).$$

The expression (10) is quite central in this section.

Let

$$F(\mathcal{A}): \bar{\mathcal{A}} \to \bar{\mathcal{A}} \otimes \bar{\mathcal{A}} \to \bar{\mathcal{A}} \otimes \bar{\mathcal{A}} \otimes \bar{\mathcal{A}} \to \cdots \tag{11}$$

denote the cobar resolution of the Steenrod algebra. The homology of the chain complex (11) is

$$H(F(\mathcal{A})) = \operatorname{Ext}_{\mathcal{A}^*}(Z_2, Z_2) = \Lambda\{Q_i; i \geqq 0\}, \tag{12}$$

where $Q_0 = Sq^1, Q_1 = Sq^{0,1}, \ldots, Q_i = Sq^{0,\ldots,0,1}$.

Using properties of the cochain operations T_p we get

$$A(s)(x, y) = 0 \tag{13}$$

whenever $x = 1$ or $y = 1$. Hence $A(s)$ (see (10)) determines an element $\bar{A}(s)$ in $\bar{\mathcal{A}} \otimes \bar{\mathcal{A}}$. The boundary of this element in the cobar resolution is given by

$$\delta(\bar{A}(s)) = \sum_{t \neq s} \lambda(s, t; p, q)\bar{A}(t) \otimes a(p)a(q) + \sum_{t \neq s} \mu(s, t; p, q)a(p)a(q) \otimes \bar{A}(t). \tag{14}$$

Also from (10) we get

$$A(s)(u, u) = v_4(s) \cdot u^n, \qquad v_4 \in Z_2, \qquad n = \deg(R(s)) + 2, \tag{15}$$

where $v_4(s)$ is expressible in terms of v_i, $i = 1, 2, 3$, and u is the basic 1-cocycle.

Now we look at the problem from a different angle. Let $\bar{A}(s) \in \bar{\mathscr{A}} \otimes \bar{\mathscr{A}}$ be a system of elements satisfying (13), (14), and (15). We then wish to show

THEOREM 5. *There exists a system of cochain operations* $\{R(s)\}$, $s \in S$, *satisfying* (6), (7), *and*

$$\begin{aligned} R(s)(xy) + A(s)(x, y) + T_{r(s)}(x, y) + \textstyle\sum \lambda(s, t; p, q)R(t)(x)a(p)a(q)(y) \\ + \textstyle\sum \mu(s, t; p, q)a(p)a(q)(x) \cdot R(t)(y) \sim 0 \end{aligned} \tag{16}$$

for all pairs of cocycles x and y.

PROOF. Let $\{R(s)\}$ be a system which satisfies (6) and (7). We show that if $\{R(s)\}$ also satisfies (16) whenever $\deg(R(s)) < k$ then there is another system $\{R'(s)\}$ which besides (6) and (7) satisfies (16) whenever $\deg(R'(s)) \leqq k$.

Let $R(s)$ be an operation of degree k. Let us denote the expression (16) by $B(x, y)$. Then B determines an element $\bar{B} \in \bar{\mathscr{A}} \otimes \bar{\mathscr{A}}$. An easy calculation using properties of $T_{r(s)}$ described in the appendix shows that $\bar{B}$ is a cycle in $F(\mathscr{A})$. If $\bar{B}$ determines a nonzero homology class in $H^*(F(\mathscr{A}))$ then there exists a cochain operation c such that, for all cocycles x and y,

$$B(x, y) \sim q_i(x) \cdot q_j(y) + c(xy) + xc(y) + c(x) \cdot y,$$

where q_i is a cochain operation representing Q_i. Putting $x = y = u$ we get a contradiction since $B(u, u) \sim 0$. Hence

$$B(x, y) \sim c(xy) + x \cdot c(y) + c(x) \cdot y.$$

If we replace $R(s)$ by $R'(s) = R(s) + c$ then (16) is fulfilled by $R'(s)$ also. This we do for all $R(s)$ of degree k. If degree $R(s)$ is different from k we put $R'(s) = R(s)$. Then $\{R'(s)\}$ satisfies (6). Since c can be chosen such that $c(u) = 0$, (7) is satisfied. Also (16) is satisfied in dimension $\leqq k$. This proves the theorem.

PROPOSITION 6. *If* $\{R(s)\}$ *and* $\{R'(s)\}$ *are systems satisfying* (6), (7), *and* (16) *then, for each cocycle* x, $R(s)(x) \sim R'(s)(x)$.

PROOF. We put $R(s) - R'(s) = P(s)$. Then (6), (7), and (16) implies

(i) $\nabla P(s) = 0$,

(ii) $P(s)(u) = 0$,

(iii) $P(s)(xy) \sim \sum \lambda(s, t; p, q)P(t)(x) \cdot a(p)a(q)(y) + \sum \mu(s, t; p, q)a(p)a(q)(x)P(t)(y)$.

From (i) we get an element $\bar{P}(s) \in \mathscr{A}$. Let us assume that $\bar{P}(s)$ is zero whenever $\deg(P(s)) < k$. Let $\bar{P}(s)$ have degree k. Then, by (iii), $\bar{P}(s)$ is primitive. By (ii), this primitive element must be zero. This completes the proof.

The method described is applicable to various convenient systems of relations. Here we mention the systems described in Kristensen [4].

If a, b, and $k \in Z$, then

$$r(a, b; k) = \sum c(j) Sq^{k-j} Sq^{j}, \qquad c(j) = \binom{b-1-j}{k+b-a-2j} + \binom{b-1-j}{j+b-a}$$

is a relation with

$$\psi(r(a, b; k)) = \sum r(a-2i, b-i; k-i-j) \otimes Sq^{j} Sq^{i} + \sum Sq^{j} Sq^{i} \otimes r(a-j, b-i; k-i-j).$$

We put

$$\bar{A}(a, b; k) = (Sq^{1} \otimes Sq^{0,1}) \psi\left(\sum c(j)(Sq^{k-j-3} Sq^{k-2} + Sq^{k-j-2} Sq^{j-3})\right).$$

Then with this system of $\bar{A}$'s, Theorem 5 and Proposition 6 hold true.

We can apply this to obtain information about the cohomology operations needed in §2.

Let us consider the relations

$$(17) \quad \begin{aligned} r(2^{m-1}(4s+1), 2^{m-1}; n+1) &: Sq(2^{m+1}s + 2^{m-1}) Sq(2^{m-1}) + Sq(2^{m}) Sq(2^{m+1}s) \\ &+ Sq(2^{m-1}) Sq(2^{m+1}s + 2^{m-1}) + Sq(0) Sq(2^{m+1}s + 2^{m}) = 0. \end{aligned}$$

The secondary operation ϕ associated with (7) is unstable. Putting $a = 2^{m-1}(4s+1)$, $b = 2^{m-1}$ and $n = a + b$, ϕ is defined on classes of dimension $\leqq n$ which are annihilated by Sq^{a}, Sq^{b}, and $Sq(2^{m+1}s)$. The indeterminacy is given by

$$(18) \qquad \text{Indet}^* = Sq^{a} H^*(-) + Sq^{2b} H^*(-) + Sq^{b} H^*(-).$$

PROPOSITION 7. *There is an unstable secondary operation ϕ associated with the relation* (17) *having the properties*

(i) *ϕ commutes with suspension whenever this makes sense (i.e. in dimensions $\leqq n$).*

(ii) *ϕ is additive in dimensions $<n$, and $\phi(x+y) = \phi(x) + \phi(y) + \{xy\}$ if x and y are n-dimensional classes.*

$$\text{(iii)} \qquad \begin{aligned} \phi &= 0 && \text{if } \dim(x) < 2^{m+1}s - 1, \\ \phi(x) &= \sum_{t<b} Sq^{t}(x) Sq^{2b-t}(x) && \text{if } \dim(x) = 2^{m+1}s - 1. \end{aligned}$$

(iv) *Let ϕ be defined on x and on y, $\dim(xy) \leqq n$. If*

$$\begin{aligned} Sq^{i}(x) &= 0 \quad \text{for } 1 \leqq i \leqq 2^{m-1} \text{ and } p - (2^{m} - 1) \leqq i, \\ Sq^{j}(y) &= 0 \quad \text{for } 1 \leqq j \leqq 2^{m-1} \text{ and } q - (2^{m} - 1) \leqq j, \end{aligned}$$

then ϕ is defined on xy and

$$\phi(xy) = x \cdot Sq(2^{m})(x) \cdot Sq(n - 2^{m} - p)(y) + Sq(n - 2^{m} - q)(x) \cdot y \cdot Sq(2^{m})(y),$$

and hence zero if $\dim(xy) < n$.

(v) *The case $m = 0$. Let $\dim(x) = p$, $\dim(y) = q$, $p + q = n$, and $x^2 = 0$.*

Then

$$\phi(xy) = \phi^{1,p}(x) \cdot y^2 + x \cdot Sq^1(x) \cdot Sq^{q-1}(y) + Sq^{p-1}(x) \cdot y \cdot Sq^1(y),$$

where $\phi^{1,p}$ *is as described in* §1.

(vi) *The case* $m = 1$. *Let* $\dim(x) = p$ *and* $\dim(y) = q$, $p + q = n$. *Further let* $x^2 = 0$, $y^2 = 0$, $Sq^1(x) = 0$, *and* $Sq^1(y) = 0$, *then*

$$\phi(xy) = \phi^{2\cdot p}(x) \cdot Sq^{q-1}(y) + Sq^{p-1}(x) \cdot \phi^{2\cdot q}(y) \\ + x \cdot Sq^2(x) \cdot Sq^{q-2}(y) + Sq^{p-2}(x) \cdot y \cdot Sq^2(y).$$

Operations of the unstable type were considered in Kristensen [3]. The properties (i), (ii), and (iii) in Proposition 7 were proved in that paper. Theorem 4.5 contains (i), Theorem 4.3 together with Corollary 3.6 contains (ii) and Theorem 4.6 contains (iii).

Appendix to §4. Here we shortly describe the properties of the cochain operations T_p needed for the proof in §4.

For each operation $a(s)$ in our system there is a cochain operation T_s in two variables satisfying

$$\text{(i)} \quad \nabla T_s(x, y) = a(s)(xy) + \sum \delta(s, p; q)a(p)(x) \cdot a(q)(y) \\ + d(a(s); \delta x \cdot y, x\delta y) + \deg(x) \cdot d(a(s); x\delta y, x\delta y),$$

where $d(a(s))$ is a cochain operation measuring the deviation from additivity of $a(s)$.

$$\text{(ii)} \quad T_s(x, 1) = 0 \quad \text{and} \quad T_s(1, y) = 0.$$

Further we wish the cochain operation chosen in such a way that, for cocycles x, y, and z,

$$\text{(iii)} \quad T_s(xy, z) + T_s(x, yz) + \sum \delta(s; p, q)a(p)(x) \cdot T_q(y, z) \\ + \sum \delta(s; t, q)T_p(x, y) \cdot a(q)(z) \sim 0.$$

There is a cochain operation $T_{p,q}$ satisfying (i) for the composition $a(p)a(q)$ such that, for cocycles x and y,

$$\text{(iv)} \quad T_{p,q}(x, y) = a(p)T_q(x, y) + \sum \delta(q; n, m)T_p(a(n)(x), a(m)(y)) \\ + d(a(p); a(q)(xy), \{\delta(q; n, m)a(n)(x) \cdot a(m)(y)\}) \\ + \sum \deg(a(n))\deg(y)\mathscr{H}(a(p))(\delta(q; n, m)a(n)(x) \cdot a(m)(y)) \\ + \mathscr{H}(a(p))a(q)(xy).$$

The operation $T_{r(s)}$ described in §4 is defined by

$$\text{(v)} \quad T_{r(s)} = \sum \rho(s; p, q)T_{p,q}.$$

It has the following property needed in the proof

$$
\text{(vi)}\quad
\begin{aligned}
T_{r(s)}(xy, z) + T_{r(s)}(x, yz) \sim {} & \sum \lambda(s, t; p, q) T_{r(t)}(x, y) \cdot a(p)a(q)(z) \\
& + \sum \mu(s, t; p, q) \cdot a(p)a(q)(x) T_{r(t)}(y, z) \\
& + \sum \lambda(s, t; p, q) \cdot r(t)(x) \cdot T_{p,q}(y, z) \\
& + \sum \mu(s, t; p, q) T_{p,q}(x, y) \cdot r(t)(z).
\end{aligned}
$$

Property (vi) makes it possible to determine the boundary in the cobar resolution of the element under consideration.

REFERENCES

1. J. F. Adams, *Vector fields on spheres*, Ann. of Math. (2) **75** (1962), 603–632. MR **25** #2614.

2. D. Frank and E. Thomas, *A generalization of the Steenrod-Whitehead vector field theorem*, Topology **7** (1968), 311–316. MR **37** #4827.

3. L. Kristensen, *On secondary cohomology operations*, Math. Scand. **12** (1963), 57–82. MR **28** #2550.

4. ———, *On secondary cohomology operations*. II, Conference on Algebraic Topology (University of Illinois, Chicago Circle, Chicago, Ill., 1968), Univ. of Illinois, Chicago Circle, Ill., 1969, pp. 117–133. MR **40** #3539.

5. ———, *Fields of k-planes on manifolds*, Aarhus Univ. Preprint Series #10, 1969/70.

6. N. Steenrod and J. H. C. Whitehead, *Vector fields on the n-sphere*, Proc. Nat. Acad. Sci. U.S.A. **37** (1951), 58–63. MR **12**, 847.

7. N. Steenrod, *The topology of fibre bundles*, Princeton Math. Series, vol. 14, Princeton Univ. Press, Princeton, N.J., 1951. MR **12**, 522.

8. E. Thomas, *The index of a tangent 2-field*, Comment. Math. Helv. **42** (1967), 86–110. MR **35** #6158.

9. ———, *Vector fields on manifolds*, Bull. Amer. Math. Soc. **75** (1969), 643–683. MR **39** #3522.

AARHUS UNIVERSITY

COBORDISM AND CLASSIFYING SPACES

PETER S. LANDWEBER[1]

Let G be a *finite* group and let BG be a classifying space for G-bundles.

If $h^*(\ )$ is a cohomology theory, then elements of $h^*(BG)$ assign characteristic classes to principal G-bundles $P \to X$, taking values in $h^*(X)$. And if $h_*(\ )$ is a homology theory represented by a Thom spectrum then one knows how to interpret $h_*(BG)$ as bordism classes of free G-actions on suitable closed manifolds.

Today I will talk about the complex bordism $\Omega_*^U(BG)$ and complex cobordism $\Omega_U^*(BG)$ modules of classifying spaces. These theories are represented by the unitary Thom spectrum MU, and the appropriate manifolds are those whose stable normal bundle has a complex structure. The coefficient ring is a polynomial ring

$$\Omega_*^U = \Omega_U^* = Z[x_1, x_2, \ldots],$$

where $x_i = [M^{2i}] \in \Omega_{2i}^U = \Omega_U^{-2i}$ for a suitable manifold M^{2i}.

Recall that one often studies $\Omega_*^U(X)$ with the help of the Thom homomorphism

$$\mu : \Omega_*^U(X) \to H_*(X; Z),$$

which sends a bordism class $[M^n, f]$ into the image of the fundamental class under the induced homomorphism $f_* : H_n(M^n; Z) \to H_n(X; Z)$.

PROBLEM 1. *For which G is the Thom homomorphism $\mu : \Omega_*^U(BG) \to H_*(BG; Z)$ onto?*

It is easy to see that G has periodic cohomology $\Rightarrow H^{od}(BG; Z) = 0 \Rightarrow \mu$ maps onto $H_*(BG; Z)$ and so it is reasonable to ask if in fact these are equivalences.

REMARKS. The groups with periodic cohomology are the groups which do not contain a subgroup of the form $Z_p \times Z_p$. Equivalently they are the groups whose Sylow subgroups are either cyclic or (if $p = 2$) generalized quaternion

AMS 1970 *subject classifications*. Primary 57D90, 55F40, 18H10; Secondary 55B20, 20J05.

[1] Research supported in part by NSF grant GP-9452 at Yale University.

[**2**, Chapter XII, §11]. For these groups a computation shows that $H^{od}(BG; Z) = 0$, i.e. $\tilde{H}_{ev}(BG; Z) = 0$; and from this it follows that the spectral sequence

$$H_*(BG; \Omega_*^U) \Rightarrow \Omega_*^U(BG)$$

collapses, hence μ (its edge homomorphism) is onto.

THEOREM 1 [**6**]. *The Thom homomorphism $\mu : \Omega_*^U(BG) \to H_*(BG; Z)$ is onto if and only if G has periodic cohomology.*

The proof that $H^{od}(BG; Z) = 0 \Rightarrow Z_p \times Z_p \nleq G$ was shown to me by R. Swan [**9**], and was obtained earlier by M. Atiyah and J. Tate. One notes first that the spectral sequence of Atiyah

$$H^*(BG; Z) \Rightarrow K^*(BG) = \widehat{R(G)}$$

collapses, hence $H^{2n}(BG; Z) \cong R_{2n}(G)/R_{2n+2}(G)$, where $\{R_{2n}(G)\}$ is the topological filtration of $R(G)$. If G has order g and if $R(G)$ is free abelian of rank r, then g kills $H^{2n}(BG; Z)$ and $R_{2n}(G)$ is free abelian of rank $r - 1$ for $n > 0$, hence $|H^k(BG; Z)| \leqq g^{r-1}$ for $k > 0$ and so G has "bounded cohomology". The following lemma shows that each subgroup of G also has bounded cohomology; the Künneth formula shows that $Z_p \times Z_p$ does not have this property, so $Z_p \times Z_p \nleq G$.

LEMMA. *Let K be a closed subgroup of a compact Lie group H. Then $H^*(BH; Z)$ is a finitely generated ring and $H^*(BK; Z)$ is a finitely generated module over $H^*(BH; Z)$.*

Here is a sketch of the proof given by E.-A. Weiss in [**11**]. Embed H in a unitary group $U(n)$; it is enough to show that $H^*(BH; Z)$ is finitely generated over $H^*(BU(n); Z) = Z[c_1, \ldots, c_n]$. The fibration

$$U(n)/H \to BH \to BU(n)$$

leads to a spectral sequence

$$H^*(BU(n)) \otimes H^*(U(n)/H) \Rightarrow H^*(BH).$$

Each E_r is finitely generated over the Noetherian ring $H^*(BU(n))$, hence for some $r < \infty$, $E_r = E_\infty$. This implies the result.

REMARK. Quillen proves the following more general result. If G is a finite group and if p is a prime, then the largest n such that $(Z_p)^n \leqq G$ (the p-rank of G) is also the least integer m such that

$$\lim_{k\to\infty} \frac{\dim H^k(G; Z_p)}{k^m} = 0.$$

Next we argue that $H^{od}(BG; Z) = 0$ *if* $\mu : \Omega_*^U(BG) \to H_*(BG; Z)$ *is onto.* Let $K_*(\ )$ be the homology theory dual to K-theory. J. Vick [**10**] has recently shown that $K_1(BG)$ is isomorphic to the character group of the compact group $\tilde{K}^0(BG)$, and that also $\tilde{K}_0(BG) = 0$ (recall $K^1(BG) = 0$). Our assumption is simply that the spectral sequence $H_*(BG; \Omega_*^U) \Rightarrow \Omega_*^U(BG)$ collapses. Now there is a natural transformation $\mu_c : \Omega_*^U(\ ) \rightarrow K_*(\ )$ which on coefficient rings becomes simply the Todd genus $\mu_c : \Omega_*^U \rightarrow Z$. It follows that the spectral sequence $H_*(BG; Z) \Rightarrow K_*(BG)$ also collapses. Since $\tilde{K}_0(BG) = 0$ this implies that $\tilde{H}_{ev}(BG; Z) = 0$, hence $H^{od}(BG; Z) = 0$. Q.E.D.

PROBLEM 2. This is the *conjecture* that *for each finite* G, *the* Ω_*^U*-module* $\Omega_*^U(BG)$ *has finite projective dimension, equal to the rank of* G, by which we mean the largest n such that $(Z_p)^n \leqq G$ for some prime p.

REMARK. We can put Theorem 1 into the following form: if $G \neq \{1\}$ then

$$\text{proj dim}_{\Omega_*^U} \Omega_*^U(BG) = 1 \Leftrightarrow \text{rank } G = 1.$$

This requires that some results of P. Conner and L. Smith [4] on finite complexes be extended to infinite complexes, which can be done if one builds resolutions in Boardman's stable category as in Adams' Seattle lectures [1].

REMARK. Results of Conner and Smith [4, §5] imply that the projective dimension of $\Omega_*^U(BG)$ is at least n for $G = (Z_p)^n$, p a prime. However, not much more is presently known about this conjecture. The following several problems arose while searching for techniques to handle the conjecture.

PROBLEM 3. *If X is an infinite complex and if for some $r < \infty$ we have $E_r = E_\infty$ in the spectral sequence*

$$H^*(X; Z) \Rightarrow K^*(X),$$

must this also be the case for the spectral sequence

$$H^*(X; \Omega_U^*) \Rightarrow \Omega_U^*(X)?$$

If $E_2 = E_\infty$ in the K-theory spectral sequence, then also the complex cobordism spectral sequence collapses (use the Hattori-Stong theorem). For $X = BH$ and H a compact Lie group, the Noetherian argument of [11] shows that the K-theory spectral sequence does terminate.

PROBLEM 4. If X is a finite complex then $\Omega_*^U(X)$ is a *coherent* Ω_*^U-module [8]. This means that $\Omega_*^U(X)$ is finitely generated and that each finitely generated submodule is finitely presented. Moreover [4, §1] each coherent Ω_*^U-module is obtained by extending scalars from a finitely generated module over a subring $Z[x_1, \ldots, x_n] \subset \Omega_*^U$, hence has finite projective dimension. We ask the question: *For which infinite complexes X with finite skeletons is $\Omega_*^U(X)$ a pseudo-coherent module*? Following Bourbaki, this means that each finitely generated submodule is finitely presented. The next two simple results motivate this question.

PROPOSITION 1. *If $\Omega_*^U(X)$ is pseudo-coherent then the annihilator ideal of each element of $\Omega_*^U(X)$ is finitely generated.*

PROPOSITION 2. *If X has finite skeletons, then $\Omega_*^U(X)$ is pseudo-coherent $\Leftrightarrow$ for each n $\exists$ m, $n \leqq m < \infty$, such that*

$$\text{Ker}\{\Omega_*^U(X^n) \rightarrow \Omega_*^U(X)\} = \text{Ker}\{\Omega_*^U(X^n) \rightarrow \Omega_*^U(X^m)\}$$

$\Leftrightarrow$ *in the spectral sequence* $H_*(X; \Omega_*^U) \Rightarrow \Omega_*^U(X)$, *for each* p, $\exists\, r(p) < \infty$ *such that* $E_{p,*}^{\infty} = E_{p,*}^{r(p)}$.

In particular, are $K(Z_p, n)$ and BG for $G = (Z_p)^n$ of this type? However for the *spectra* $K(Z_p)$, $\Omega_*^U(K(Z_p))$ is not pseudo-coherent since the annihilator ideal of the "unit" element is not finitely generated.

PROBLEM 5. *Compute the complex cobordism rings* $\Omega_U^*(BG)$ and apply this information to the harder problem of studying the Ω_*^U-modules $\Omega_*^U(BG)$.

In fact I am able to compute $\Omega_U^*(BG)$ if G is *finite abelian*. The result is not surprising, and is a sort of Künneth formula, but there is some new algebra in the proof.

We first review the structure of $\Omega_U^*(BZ_k)$. Let $\eta \to CP^\infty$ be the canonical line bundle and $C = E(\eta) \in \Omega_U^2(CP^\infty)$ its Euler class, so that $\Omega_U^*(CP^\infty) = \Omega_U^*[[C]]$. Then Stong has observed that the sphere bundle $S(\eta^k)$ is a classifying space for Z_k, and so we can find $\Omega_U^*(BZ_k)$ from the Gysin sequence. Let $[k](C)$ be the power series

$$[k](C) = E(\eta^k) = kC + \text{higher terms}.$$

Then

$$\Omega_U^*(BZ_k) = \Omega_U^*[[C]]/([k](C)).$$

THEOREM 2. *If* $G = Z_{k_1} \times \cdots \times Z_{k_r}$ *then*

$$\Omega_U^*(BG) \cong \frac{\Omega_U^*[[C_1, \ldots, C_r]]}{([k_1](C_1), \ldots, [k_r](C_r))}.$$

Here is some of the algebra involved in the proof; a full report will appear in [7]. Since this is a completed Künneth formula, it is reasonable to try to establish a flatness property for $\Omega_U^*(BZ_k)$.

LEMMA. $\Omega_U^*[[C]]$ *is a flat* Ω_U^**-module.*

For over a coherent ring, products of flat modules are flat [3].

Call $f \in \Omega_U^*[[C]]$ *regular* if no proper finitely generated ideal in Ω_U^* contains all coefficients of f.

LEMMA. f *is regular* $\Leftrightarrow$ *for all primes* p *and* $n \geq 0$, $f \to \bar{f} \neq 0$ *under the coefficient homomorphism*

$$\Omega_U^* \to Z_p[x_{n+1}, x_{n+2}, \ldots].$$

LEMMA. *For all* k, *the power series* $[k](C)$ *is regular.*

We can now state the main algebraic result.

THEOREM 3. *If* f *is a regular power series and* M *is a coherent* Ω_U^**-module then*

$$\operatorname{Tor}_1^{\Omega_U^*}(\Omega_U^*[[C]]/(f), M) = 0.$$

In fact this implies that $\Omega_U^*[[C]]/(f)$ *is a flat* Ω_U^**-module.*

COROLLARY 1. *If X is a finite complex then*

$$\Omega_U^*(BZ_k \times X) \cong \Omega_U^*(BZ_k) \otimes_{\Omega_U^*} \Omega_U^*(X) \cong \Omega_U^*(X)[[C]]/([k](C)).$$

COROLLARY 2. *If X is a complex with finite skeletons such that $\Omega_U^*(X)$ has no elements of infinite filtration, then*

$$\Omega_U^*(BZ_k \times X) \cong \Omega_U^*(X)[[C]]/([k](C)).$$

For a clarification of the hypothesis of Corollary 2 see [**5**, §1]. In particular, we may take X to be BG with G finite and then Theorem 2 follows by induction.

REFERENCES

1. J. F. Adams, *Lectures on generalized homology theories*, Lecture Notes in Math., no. 99, Springer-Verlag, Berlin and New York, 1969, pp. 1–138.
2. H. Cartan and S. Eilenberg, *Homological algebra*, Princeton Univ. Press, Princeton, N.J., 1956. MR **17**, 1040.
3. S. U. Chase, *Direct products of modules*, Trans. Amer. Math. Soc. **97** (1960), 457–473. MR **22** #11017.
4. P. E. Conner and L. Smith, *On the complex bordism of finite complexes*, Inst. Hautes Études Sci. Publ. Math. No. 37 (1970), 117–212.
5. P. S. Landweber, *On the complex bordism and cobordism of infinite complexes*, Bull. Amer. Math. Soc. **76** (1970), 650–654.
6. ———, *Complex bordism of classifying spaces*, Proc. Amer. Math. Soc. **27** (1971), 175–179.
7. ———, *Coherence, flatness, and cobordism of classifying spaces*, Proc. Aarhus Summer Institute on Algebraic Topology, 1970, 256–269.
8. L. Smith, *On the finite generation of $\Omega_*^U(X)$*, J. Math. Mech. **18** (1968/69), 1017–1023. MR **39** #6300.
9. R. G. Swan, *Groups with no odd dimensional cohomology*, J. Algebra (to appear).
10. J. W. Vick, *Pontrjagin duality in K-theory*, Proc. Amer. Math. Soc. **24** (1970), 611–616.
11. E.-A. Weiss, *Kohomologiering und Darstellungsring endlicher Gruppen*, Bonn. Math. Schr. No. 36 (1969). MR **40** #4375.

RUTGERS UNIVERSITY

THE IMMERSION APPROACH TO TRIANGULATION AND SMOOTHING

R. LASHOF

PART I. LEES' IMMERSION THEOREM

Chapter 1. The tangent bundle.

DEFINITION.[1] An n-dimensional *microbundle* ε over B is a diagram

$$\varepsilon : B \xrightarrow{s} E \xrightarrow{p} B$$

of spaces and maps such that $ps = 1_B$ and, for each $b \in B$, there exists neighborhoods U of b in B and V of $s(b)$ in E and a homeomorphism $h : V \to U \times R^n$, so that

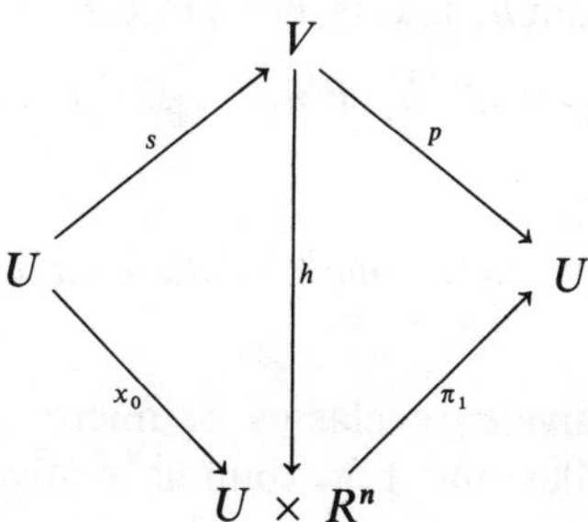

commutes.

Two microbundles over B,

$$\varepsilon_i : B \xrightarrow{s_i} E_i \xrightarrow{p_i} B, \qquad i = 1, 2,$$

AMS 1970 *subject classifications*. Primary 57D05, 57D10, 57D40, 57D55, 57A35, 57A55, 57C15, 57C25, 57C30, 57C35.

[1] Due to J. Milnor [7, Part II].

are *equivalent* if there exist neighborhoods E_i^0 of $s(B)$ in E_i, and a homeomorphism $h: E_1^0 \to E_2^0$ such that

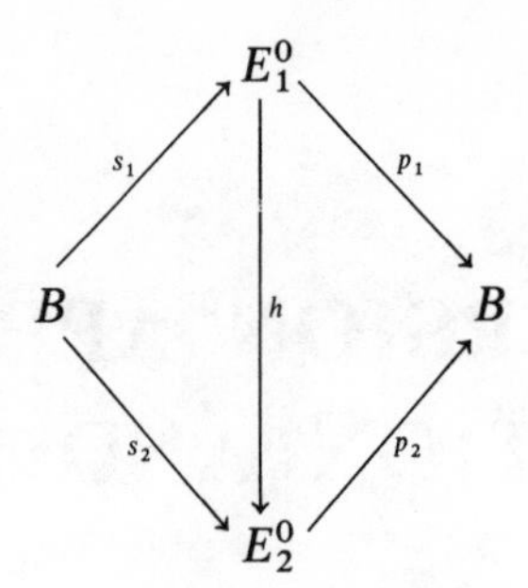

commutes.

EXAMPLE 1. Let M be a topological n-manifold, then

$$\tau: M \xrightarrow{\Delta} M \times M \xrightarrow{\pi_1} M$$

is a microbundle called the *tangent* microbundle of M.

EXAMPLE 2. A fibre bundle with fibre R^n is an n-dimensional microbundle. Equivalent fibre bundles are equivalent as microbundles.

DEFINITION. We will say that an n-dimensional microbundle $\varepsilon: B \xrightarrow{s} E \xrightarrow{p} B$ *contains* an R^n-fibre bundle, if there exists an R^n-bundle η and a homeomorphism h of the total space $E(\eta)$ of η onto a neighborhood E^0 of $s(B)$, such that $h: E(\eta) \to E^0$ is a microbundle equivalence.

Two fibre bundles (η_i, h_i), $i = 1, 2$, contained in ε are called *equivalent* if $h_1: E(\eta_1) \to E_1^0 \subset E$ is isotopic (i.e., homotopic through microbundle equivalences) to a homeomorphism $\bar{h}_1$ taking $E(\eta_1)$ into $E_2^0 \subset E$ with

$$h_2^{-1}\bar{h}_1: E(\eta_1) \to E(\eta_2)$$

a fibre bundle equivalence.

THEOREM (KISTER [3]). *Every microbundle contains a fibre bundle, unique up to equivalence.*

Thus in studying equivalence classes of microbundles, one might as well restrict to the fibre bundles and fibre bundle equivalences. It is nevertheless convenient to keep the notion of microbundle around, since the tangent microbundle of a manifold is a natural object, whereas the tangent fibre bundle is defined only up to equivalence.

DEFINITION. An n-dimensional microbundle ε *reduces* to a *vector bundle* if it contains an n-dimensional vector bundle (η, h). Two such reductions (η_i, h_i), $i = 1, 2$, are called *equivalent* if $h_1: E(\eta_1) \to E_1^0 \subset E$ is isotopic to a homeomorphism

$$\bar{h}_1: E(\eta_1) \xrightarrow{\text{onto}} E_2^0 \subset E$$

and $h_2^{-1}\bar{h}_1: E(\eta_1) \to E(\eta_2)$ is a vector bundle equivalence.

DEFINITION. An R^n-fibre bundle ε reduces to a *vector bundle* if there exists a vector bundle η, and a fibre bundle equivalence $h: E(\eta) \to E(\varepsilon)$. Two such reductions (η_i, h_i), $i = 1, 2$, are called *equivalent* if

$$h_1 : E(\eta_1) \to E(\varepsilon)$$

is isotopic (i.e., homotopic through fibre bundle equivalences) to

$$\bar{h}_1 : E(\eta_1) \to E(\varepsilon)$$

such that $h_2^{-1}\bar{h}_1 : E(\eta_1) \to E(\eta_2)$ is a vector bundle equivalence.

LEMMA. *An equivalence class of reductions of a microbundle ε to a vector bundle defines an equivalence class of reductions of any fibre bundle contained in ε, and conversely.*

PROOF. Let (μ, g) be a fibre bundle contained in ε. By Kister's theorem a reduction

$$h: E(\eta) \to E^0 \subset E$$

of ε to a vector bundle is isotopic to

$$\bar{h}: E(\eta) \xrightarrow{\text{onto}} g(E(\mu))$$

with $g^{-1}\bar{h}: E(\eta) \to E(\mu)$ a fibre bundle equivalence. To show equivalent reductions of ε give equivalent reductions of μ, one applies a relative version of Kister's theorem to $\varepsilon \times I$, I the unit interval. The converse is trivial.

The group for R^n-fibre bundles is the group of homeomorphisms with the C-O topology, denoted Top_n. For vector bundles we can always take the group to be the orthogonal group, O_n. The inclusion, $O_n \subset \mathrm{Top}_n$ induces a map

$$BO_n \to B\,\mathrm{Top}_n$$

of the classifying spaces, which we can consider to be a fibre bundle projection with fibre Top_n/O_n.

PROPOSITION. *The equivalence classes of reductions of an R^n-fibre bundle ε over B to a vector bundle are in one to one correspondence with the isotopy classes of lifts of a classifying map $f: B \to B\,\mathrm{Top}_n$ to BO_n.*

PROOF. Let γ be the universal n-dimensional vector bundle over BO_n, τ the universal R^n-bundle over $B\,\mathrm{Top}_n$. Then

$$\pi: BO_n \to B\,\mathrm{Top}_n$$

is covered by an R^n-bundle map $\bar{\pi}: \gamma \to \tau$. Choose a fixed bundle map $\bar{f}: \varepsilon \to \tau$ covering f. $\bar{f}$ induces a bundle equivalence $h: \varepsilon \to f^*(\tau)$.

Let $\tilde{f}: B \to BO_n$ be a lift of f, i.e. $\pi\tilde{f} = f$.

Let $\mu = \tilde{f}^*(\gamma)$. The bundle map $\bar{\pi}$ induces a bundle map

$$q: \mu = \tilde{f}^*(\gamma) \to f^*(\tau).$$

Define a reduction (μ, g) of ε by

$$g: \mu \xrightarrow{q} f^*(\tau) \xrightarrow{h^{-1}} \varepsilon.$$

If $\tilde{f}_1$ is a lift isotopic to $\tilde{f}$, let

$$\tilde{F}: B \times I \to BO_n$$

be the isotopy; then $\pi\tilde{F} = fp_1$. Let $F = fp_1$, i.e.

$$F = \pi\tilde{F}: B \times I \xrightarrow{p_1} B \xrightarrow{f} B\,\mathrm{Top}_n.$$

F is covered by

$$\bar{F}: \varepsilon \times I \xrightarrow{p_1} \varepsilon \xrightarrow{\bar{f}} \tau.$$

$\bar{F}$ induces the bundle equivalence

$$h^{-1} \times 1: f^*\tau \times I \to \varepsilon \times I.$$

We define

$$G: \tilde{F}^*(\gamma) \longrightarrow F^*(\tau) = f^*\tau \times I \xrightarrow{h^{-1} \times 1} \varepsilon \times I.$$

Then G over F_0 is g and G over F_1 is the reduction

$$g_1: \mu_1 = \tilde{F}^*(\gamma) \mid B \times 1 \to \varepsilon$$

corresponding to f_1. Let $k: \mu_1 \times I \to \tilde{F}^*(\gamma)$ be a vector bundle equivalence such that $k_1 =$ identity. Then $k_0: \mu_1 \to \mu$ is a vector bundle equivalence. Hence $G \circ k$ gives an equivalence between (μ, g) and (μ_1, g_1).

Conversely, given a reduction (μ, g) of ε, let $\bar{a}: \mu \to \gamma$ be a vector bundle map over $a: B \to BO_n$. Then $\pi\bar{a}: B \to B\,\mathrm{Top}_n$.

Since $B\,\mathrm{Top}_n$ is a classifying space, πa is homotopic to f, where the homotopy is covered by a bundle homotopy of $\bar{f}$ to $\pi\bar{a}g^{-1}$, and hence $\bar{a}$ is homotopic to a lift $\tilde{f}$ of f, by a covering homotopy.

If f_1 is a lift corresponding to an equivalent reduction (μ_1, g_1) then we claim f_1 is isotopic to f. In fact, if $\bar{a}_1: \mu_1 \to \gamma$ is a classifying map for μ_1, then $\bar{a}$ is a homotopic through vector bundle maps to $\bar{a}_1 g_1^{-1}\bar{g}$, where $\bar{g}: \mu \to \varepsilon$ is bundle homotopic over the identity to g and $g_1^{-1}\bar{g}: \mu \to \mu_1$ is a vector bundle map.

Hence $\bar{\pi}\bar{a}g^{-1}$ is homotopic through bundle maps to $\bar{\pi}\bar{a}_1 g_1^{-1}$, covering a homotopy of πa to πa_1. The homotopy of f to πa, is covered by a bundle homotopy of $\bar{f}$ to $\bar{\pi}\bar{a}g^{-1}$, by assumption.

Similarly, the homotopy πa_1 to f is covered by a bundle homotopy of $\bar{\pi}\bar{a}_1 g_1^{-1}$ to $\bar{f}$. Let

$$F: B \times I \to B\,\mathrm{Top}_n$$

be the homotopy f to πa to πa_1 to f; and

$$\bar{F}: \varepsilon \times I \to \tau$$

be the bundle homotopy $\bar{f}$ to $\bar{\pi}\bar{a}g^{-1}$ to $\bar{\pi}\bar{a}_1 g_1^{-1}$ to $\bar{f}$ covering it. Then since τ is universal (see lemma below), F is homotopic to $f \circ p_1$, $p_1: B \times I \to B$ the projection, rel end points. But then the homotopy

$$\tilde{F}: B \times I \to BO_n$$

$\tilde{f}$ to a to a_1 to $\tilde{f}_1$, is homotopic rel end points to a lift of $f \circ p_1$, i.e. an isotopy of $\tilde{f}$ to $\tilde{f}_1$.

LEMMA. *Let G be a topological group, BG the universal base space, and τ the universal bundle over BG. Let ε be a G-bundle over a paracompact B, and let $f_0, f_1 : B \to BG$. Then any two homotopies*

$$F, \bar{F} : B \times I \to BG$$

between f_0 and f_1, which are covered by bundle maps $\varepsilon \times I \to \tau$ which agree over f_0 and f_1, are homotopic rel end points.

REMARK. All the above definitions and propositions hold for the categories PL and Diff and the reductions of Top to PL, and PL to Diff, mod technical modifications.

Chapter 2. Lees' Immersion Theorem. In [2], Haefliger and Poenaru give an immersion theory which holds in both the PL and Diff categories. Their proof holds in the Top category once an isotopy extension theorem is proved in this category. This was established by Lees [3] for the case needed in triangulation and smoothing theory, using a lemma of Kirby. The general isotopy extension theorem was established by Edwards and Kirby [1]. Haefliger and Poenaru assume the manifold being immersed has a handlebody decomposition (always the case in PL and Diff categories); the appropriate modifications in the proof necessary for general topological manifolds was given essentially, in [4]. See also [1].

DEFINITION. Let M^m and Q^m be m-dimensional topological manifolds, M^m an open manifold. A map $f : M \to Q$ is called an *immersion* if f is a local homeomorphism, i.e., for each $x \in M$ there is a neighborhood U_x such that $f \mid U_x$ is a homeomorphism into a neighborhood of $f(x)$.

An *embedding* is an immersion $f : M^m \to Q^m$ which is one to one.

A *regular homotopy* between two immersions $f_0, f_1 : M \to Q$ is an immersion $F : I \times M \to I \times Q$ commuting with projection onto the unit interval I with $f \mid 0 \times M = f_0$, $f \mid 1 \times M = f_1$. An *isotopy* between two embeddings f_0, $f_1 : M \to Q$ is a regular homotopy f, which is an embedding of $I \times M$ in $I \times Q$.

More generally, let Δ be a k-simplex. An immersion $f : \Delta \times M \to \Delta \times Q$ which commutes with projection onto Δ, is called a *regular k-homotopy*. If f is also one to one, it is called a *k-isotopy*.

An *ambient* k-isotopy, is a k-isotopy $f : \Delta \times Q \to \Delta \times Q$, such that f is onto and f_0 is the identity, where f_0 is f restricted to the 0-vertex of Δ.

Finally, let X be an arbitrary subset of M. A regular k-homotopy (k-isotopy) of X in M, is a regular k-homotopy (k-isotopy) $f : \Delta \times U \to \Delta \times Q$, where U is an open neighborhood of X in M. Two such regular k-homotopies (k-isotopies)

$$f_i : \Delta \times U_i \to \Delta \times Q, \qquad i = 1, 2,$$

are identified, if there is a neighborhood U of X, $U \subset U_1 \cap U_2$, such that

$$f_1 \mid \Delta \times U = f_2 \mid \Delta \times U.$$

DEFINITION. Given a map $f: M \to Q$, we call

$$(f, f): M \times M \to Q \times Q$$

the differential of f and denote it df. Note that df commutes with π_1.

If f is an immersion, then $df: \tau M \to \tau Q$ is a microbundle map of the tangent microbundle of M into the tangent microbundle of Q. It follows from Kister's theorem (§1), applied to the induced microbundle $f^*\tau Q$ that if TM is any R^m-bundle in τM and TQ is any R^m-bundle in Q, then df induces a unique isotopy class of R^m-bundle maps (again denoted df) of TM into TQ.

More generally, if $f: \Delta \times M \to \Delta \times Q$ is a regular k-homotopy, then $df: \Delta \times \tau M \to \Delta \times \tau Q$ is the microbundle map $df_t = f_t \times f_t, t \in \Delta$.

DEFINITION. Let M^m and Q^m be manifolds and X a subset of M. Let $g: U \to Q$ be an immersion of X in Q. We denote by $\vartheta_g(M, Q)$ the semisimplicial complex whose k-simplices are regular k-homotopies

$$f: \Delta^k \times M \to \Delta^k \times Q,$$

which agree with $1 \times g$ on some neighborhood of $\Delta^k \times X$. The face and degeneracy operators are defined in the usual way.

We denote by $R_{dg}(\tau M, \tau Q)$ the semisimplicial complex whose k-simplices are microbundle maps

$$\varphi: \Delta^k \times \tau M \to \Delta^k \times \tau Q,$$

commuting with projection on Δ^k and such that φ agrees with $1 \times dg$ on some neighborhood of X.

LEMMA. *$\vartheta_g(M, Q)$ and $R_{dg}(\tau M, \tau Q)$ are Kan complexes.*

PROOF. Let Λ denote the union of all faces of the k-simplex Δ except one. Denote these faces by $\delta_1, \ldots, \delta_k$. Suppose given regular k-isotopies

$$f_j: \delta_j \times M \to \delta_j \times Q$$

satisfying the compatibility conditions, i.e., so as to define

$$h: \Lambda \times M \to \Lambda \times Q.$$

Let $r: \Delta \to \Lambda$ be the standard retraction. Define $f: \Delta \times M \to \Delta \times Q$ by

$$\pi_1 \circ f = \pi_1, \qquad \pi_2 \circ f = \pi_2 \circ h \circ (r \times 1).$$

Then it is easy to check that f is a regular k-isotopy satisfying the Kan condition, i.e.

$$f \mid \delta_j \times M = f_j.$$

This shows that $\vartheta_g(M, Q)$ is a Kan complex. A similar argument works for $R_{dg}(\tau M, \tau Q)$.

LEMMA. *The differential* $d:\vartheta_g(M, Q) \to R_{dg}(\tau M, \tau Q)$ *is a semisimplicial map.*

Let M^m and Q^m be manifolds without boundary, M open. Let M' be a closed, locally flat submanifold of M of the same dimension as M.

Let $R(\tau M \mid M', \tau Q)$ be the s.s. complex of microbundle maps of $\tau M \mid M'$ in τQ.

Let $R(\tau M', \tau Q)$ be the s.s. complex of microbundle maps of τU in τQ, where U is a neighborhood of M' in M, and two microbundle maps

$$f_i:\Delta \times \tau U \to \Delta \times \tau Q, \qquad i = 1, 2,$$

are identified if they agree on a smaller neighborhood $U \subset U_1 \cap U_2$.

LEMMA. *The restriction map* $R(\tau M', \tau Q) \to R(\tau M \mid M', \tau Q)$ *is a homotopy equivalence.*

PROOF. This follows from the fact that M' is a deformation retract of a collar neighborhood of M' in M, and the covering homotopy theorem for microbundle maps.

THEOREM A. *The restriction map* $R(\tau M, \tau Q) \to R(\tau M', \tau Q)$ *is a Kan fibration.*

PROOF. Let $\varphi:\Delta \times \tau U \to \Delta \times \tau Q$ be in $R(\tau M', \tau Q)$ and suppose

$$\psi:\Lambda \times \tau M \to \Lambda \times \tau Q$$

agrees with φ on $\Lambda \times \tau U$. We need to show that φ extends to $\Delta \times \tau M$ so that its restriction to some neighborhood of M' agrees with φ.

But there is a deformation of $\Delta \times M$ into $\Lambda \times M \cup \Delta \times U$ which is the identity on $\Lambda \times M$ and $\Delta \times U_1$, where U_1 is a closed collar neighborhood of M' in U. The result then follows from the covering homotopy theorem for microbundle maps.

Let X be a closed subset of M, where either X is compact or a closed locally flat submanifold. Further assume each component of $M - X$ has noncompact closure.

Let $\vartheta(X, Q)$ be the s.s. complex of regular k-isotopies of U in Q, where U is a neighborhood of X; two such isotopies being identified if they agree in a smaller neighborhood.

THEOREM B. *The restriction map* $\vartheta(M, Q) \to \vartheta(X, Q)$ *is a Kan fibration.*

We will prove Theorem B in the next section. Here we will state the main immersion theorem and show how it follows from Theorems A and B.

LEES' IMMERSION THEOREM. *Let* M^m *and* Q^m *be m-dimensional manifolds without boundary, and let* X *be a closed subset of* M *as above (possibly empty). Let* $g:U \to Q$ *be an immersion of a neighborhood of* X *in* M*, into* Q*. Then*

$$d:\vartheta_g(M, Q) \to R_{dg}(M, Q)$$

is a homotopy equivalence.

Manifolds with boundary. If $\partial M \neq \varnothing$, we may apply Lees' Immersion

Theorem to $M \cup \partial M \times [0, 1)$. If M' is a closed locally flat submanifold of an open M of the same dimension, it is easy to check that the restriction map

$$\vartheta(M' \cup \partial M' \times [0, 1), Q) \to \vartheta(M', Q)$$

is a homotopy equivalence. Consequently, we will use the notation $\vartheta(M, Q)$ for manifolds with boundary to mean $\vartheta(M \cup \partial M \times [0, 1), Q)$ and if $M' \subset M$ as above $\vartheta(M', Q)$ will stand for the earlier definition or for $\vartheta(M' \cup \partial M' \times [0, 1), Q)$, if this leads to no difficulties.

PROOF OF LEES' IMMERSION THEOREM FOR HANDLE BODIES. We assume M^m is a handle body and X^m is a subhandle body, and M is obtained from X by adding handles of dimension less than m.

The scheme of the proof is as follows:

1. Prove theorem for $M^m = D^m$, the m-disc, and $X = \varnothing$, i.e.,

$$d : \vartheta(D^m, Q) \to R(\tau D^m, \tau Q)$$

is a homotopy equivalence.

2. Prove inductively for $M^m = D^k \times D^{m-k}$, $X = \partial D^k \times D^{m-k}$, i.e.

$$d : \vartheta_g(D^k \times D^{m-k}, Q) \to R_{dg}(\tau D^k \times D^{m-k}, \tau Q)$$

is a homotopy equivalence for $k < m$.

3. Prove inductively for M by adding a handle at a time.

PROOF OF 1. Let $0 \in D^m$ be the center of the disc. Let $\vartheta(0, Q)$ be the s.s. complex at immersions of a neighborhood U of 0 in D^m into Q, two such being identified if they agree on a smaller neighborhood. Let $R(\tau D^m \mid 0, \tau Q)$ be the s.s. complex of bundle maps $\tau D^m \mid 0$ into τQ.

(a) The restriction $r : \vartheta(D^m, Q) \to \vartheta(0, Q)$ is a homotopy equivalence.

Let $f : S^k \times U \to S^k \times Q$ represent a homotopy class α of $\pi_k \vartheta(0, Q)$. Now U contains a disc D^m of radius ε about 0. Let $h : D \to D_\varepsilon$ be a homeomorphism which is the identity on $D_{\varepsilon/2}$. Extend h to a homeomorphism of an open collar neighborhood $\tilde{D}$ of D onto an open collar neighborhood $\tilde{D}_\varepsilon$ of D_ε. Then

$$f \circ (1 \times h) : S^k \times \tilde{D} \to S^k \times Q$$

represents an element of $\pi_k \vartheta(D^m, Q)$ whose image is α. This shows r_* is onto.

Now let $f : S^k \times \tilde{D} \to S^k \times Q$ represent a class $\beta \in \pi_k(\vartheta(D^m, Q))$ such that $r_* \beta$ is trivial. But this means that $f \mid S^k \times \tilde{D}_\varepsilon$, for some ε, extends to a regular $k + 1$-homotopy

$$g : D^{k+1} \times \tilde{D}_\varepsilon \to D^{k+1} \times Q.$$

Let h_t, $0 \leqq t \leqq \frac{1}{2}$, be an isotopy of the identity map of $\tilde{D}$ to a shrinking of $\tilde{D}$ into $\tilde{D}_\varepsilon$. Then $f \circ (1 \times h_t)$ and $g \circ 1 \times h_1$ combine to extend f to a regular $k + 1$-homotopy, showing that β is trivial and hence r_* is monic.

(b) The restriction $r : R(\tau D^m, \tau Q) \to R(\tau D^m \mid 0, \tau Q)$ is a homotopy equivalence.

This follows from the covering homotopy theorem and the fact that 0 is a deformation retract of D^m.

(c) $d : \vartheta(0, Q) \to R(\tau D^m \mid 0, \tau Q)$ is an isomorphism.

Let $f:\Delta \times U \to \Delta \times Q$ be a k-simplex of $\vartheta(0, Q)$. Since f is an immersion and we can replace U by a smaller neighborhood, we may assume f is an embedding. Then $df \mid 0$ is given by

$$df(t, 0, x) = (t, f_t(0), f_t(x)).$$

On the other hand, a k-simplex $\varphi \in R(\tau D^m \mid 0, \tau Q)$ is an embedding

$$\varphi:\Delta \times 0 \times U \to \Delta \times Q \times Q$$

which preserves the diagonal and commutes with projection. In both complexes, two such maps are identified if they agree in a smaller neighborhood of 0 in D^m.

Now φ defines an element $s(\varphi)$ of $\vartheta(0, Q)$ by setting

$$s(\varphi)(t, x) = \pi_{1,3}\varphi(t, 0, x).$$

Then sd and ds are both the identity.

(d) Since

$$\begin{array}{ccc} \vartheta(D^m, Q) & \xrightarrow{r} & \vartheta(0, Q) \\ \big\downarrow d & & \big\downarrow d \\ R(\tau D^m, \tau Q) & \xrightarrow{r} & R(\tau D^m \mid 0, \tau Q) \end{array}$$

commutes, (a), (b), (c) imply $d:\vartheta(D^m, Q) \to R(\tau D^m, \tau Q)$ is a homotopy equivalence.

PROOF OF 2. Consider the fibrations

$$\begin{array}{ccc} \vartheta(S^k \times D^{m-k}, Q) & \xrightarrow{d} & R(\tau S^k \times D^{m-k}, \tau Q) \\ \big\downarrow r_1 & & \big\downarrow r_1 \\ \vartheta(D^k_- \times D^{m-k}, Q) & \xrightarrow{d} & R(\tau D^k_- \times D^{m-k}, \tau Q) \end{array}$$

and

$$\begin{array}{ccc} \vartheta(D^k \times D^{m-k}, Q) & \xrightarrow{d} & R(\tau D^k \times D^{m-k}, \tau Q) \\ \big\downarrow r_2 & & \big\downarrow r_2 \\ \vartheta(S^{k-1} \times D^{m-k+1}, Q) & \xrightarrow{d} & R(\tau S^{k-1} \times D^{m-k+1}, \tau Q), \end{array}$$

where

(a) D^k_- is the lower hemisphere of S^k and r_1 is the obvious restriction.

(b) $S^{k-1} \times D^{m-k+1} = S^{k-1} \times I \times D^{m-k}$ and $S^{k-1} \times I$ is a collar neighborhood of $S^{k-1} = \partial D^k$ in D^k; r_2 the obvious restriction.

(c) By definition $\vartheta(S^k \times D^{m-k}, Q)$ means immersions of $S^k \times \tilde{D}^{m-k}$, where $\tilde{D}^{m-k}$ is an open collar neighborhood of D^{m-k}. Similarly, $\vartheta(D^k_- \times D^{m-k}, Q)$ means immersions of $\tilde{D}^k_- \times \tilde{D}^{m-k}$, where $\tilde{D}^k_-$ is an open collar of D^k_- in S^k. This satisfies the hypothesis of Theorem B with $M = S^k \times \tilde{D}^{m-k}$ and $M' = D^k_- \times \tilde{D}^{m-k}$ provided $k < m$.

(d) Similarly, the second fibration satisfies the hypothesis of Theorem B with

$$M = \tilde{D}^k \times \tilde{D}^{m-k}, \qquad \tilde{D}^k = D^k \times S^{k-1} \cup [0, 1)$$

and

$$M' = S^{k-1} \times (I \cup [0, 1)) \times \tilde{D}^{m-k},$$

provided $k < m$.

(e) The fibres for both fibrations are $\vartheta_g(D^k \times D^{m-k}, Q)$ and $R_{dg}(\tau D^k \times D^{m-k}, \tau Q)$ where $g : U \to Q$, and:

For the first fibration $D^k = D^k_+$ and

$$U = (\text{neigh } \partial D^k_+ \times [0, 1) \text{ in } \tilde{D}^k_+) \times \tilde{D}^{m-k}.$$

For the second fibration,

$$U = (\text{neigh } S^{k-1} \times (I \cup [0, 1)) \text{ in } \tilde{D}^k) \times \tilde{D}^{m-k}.$$

By step 1, the bottom d in the first fibration and the top d in the second fibration are homotopy equivalences.

Inductive argument: $d : \vartheta_g(D^k \times D^{m-k}, Q) \to R_{dg}(\tau D^k \times D^{m-k}, \tau Q)$ is a h.e. $k = 0$. For $k = 0$, $U = \varnothing$; and $\vartheta_g(D^k \times D^{m-k}, Q)$ is simply $\vartheta(D^m, Q)$ and result holds by step 1. Assume for $k - 1$: Then in fibration one for $k - 1$, d is a h.e. both on the fibres and bottom line, hence

$$d : \vartheta(S^{k-1} \times D^{m-k+1}, Q) \to R(\tau S^{k-1} \times D^{m-k+1}, \tau Q)$$

and hence d on the fibres is a h.e., giving the desired result for k.

PROOF OF 3. M is obtained by adding a finite number of handles to X:

$$X = M_0 \subset M_1 \subset \cdots \subset M_r = M, \qquad M_{i+1} = M_i \cup H, \qquad H = D^k \times D^{m-k}, \quad k < m.$$

Consider the commutative diagram

$$\begin{array}{ccc} \vartheta_h(M_{i+1}, Q) & \xrightarrow{d} & R_{dh}(\tau M_{i+1}, \tau Q) \\ \downarrow r & & \downarrow r \\ \vartheta_h(M_i, Q) & \xrightarrow{d} & R_{dh}(\tau M_i, \tau Q), \end{array}$$

$h : U \to Q$, U a neigh of X.

(a) r is a fibration, since it is the restriction of the fibration

$$r : \vartheta(M_{i+1}, Q) \to \vartheta(M_i, Q)$$

to the part over the subcomplex $\vartheta_h(M_i, Q)$. Similarly, for the right side.

(b) The fibre of r is $\vartheta_g(D^k \times D^{m-k}, Q)$, since it is immersions of

$$\tilde{M}_{i+1} = M_{i+1} \cup \partial M_{i+1} \times [0, 1)$$

which are fixed on $\tilde{M}_i = M_i \cup \partial M_i \times [0, 1)$.

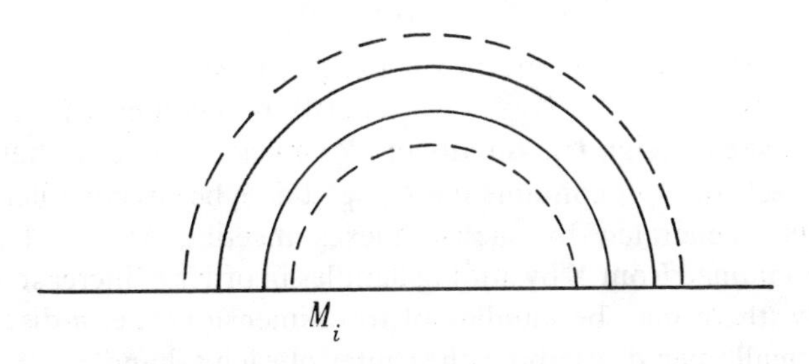

(c) For $i = 0$, $\vartheta_h(M_0, Q)$ is the point complex corresponding to $h:U \to Q$ and $d:\vartheta_h(M_0, Q) \to R_{dh}(\tau M_0, \tau Q)$ is an isomorphism.

(d) Assume $d:\vartheta_h(M_i, Q) \to R_{dh}(\tau M_i, \tau Q)$ is a h.e. Then since d is a h.e. on the fibres, we get

$$d:\vartheta_h(M_{i+1}, Q) \to R_{dh}(\tau M_{i+1}, \tau Q)$$

is a h.e.

LEES' IMMERSION THEOREM (GENERAL CASE). *$d:\vartheta_g(M, Q) \to R_{dg}(\tau M, \tau Q)$ is a h.e. X closed in M, $g:U \to Q$ an immersion of a neigh of X in M. Every component of $M - X$ has noncompact closure, X as above.*

PROOF. Cover M by a locally finite countable number of coord. neighs $U_i, i = 1, 2, \ldots$, and let $\{V_i\}$ be a shrinking of this cover such that $M \subset \bigcup \operatorname{Int} V_i$, V_i compact, $V_i \subset U_i$.

Given $f_i:S^p \to R_{dg}(\tau M, \tau Q)$ we will inductively construct maps

$$f_i:S^p \to \vartheta_g(M_i, Q), \qquad M_i = \bigcup_{j \leq i} (V_j - P_j) \cup X,$$

where $P_j \subset V_j$ is a finite set of disjoint open discs in the interior of $V_j - M_{j-1}$, with the properties:

(a) f_i restricts to f_{i-1}.

(b) There is a homotopy h_i of df_i to φ_i (φ_i restriction of φ to $R_{dg}(M_i, Q)$) and h_i restricts to h_{i-1}.

(c) There is a fixed neigh $U_1 \subset U$ of X on which all the f_i agree with g.

Assuming this for the moment, the f_i's determine a map

$$\hat{f}:S^p \to \vartheta_g\left(M - \bigcup_{j=1}^{\infty} P_j, Q\right)$$

such that there is a homotopy h of $d\hat{f}$ to $\hat{\varphi}$, $\hat{\varphi}$ the restriction of φ to $M - \bigcup_{j=1}^{\infty} P_j$.

We will then show that there is an isotopy ψ_t of M into $M - \bigcup \bar{P}_j$, fixed on U_1. If we let f be the restriction of $\hat{f}$ to $\psi_1(M)$, then df is homotopic to g.

1. The assertion is true for $M_0 = X$.

2. Assume true for $i-1$. I.e. we are given

$$f_{i-1}: S^p \times N_{i-1} \to S^p \times Q,$$

N_{i-1} a neigh of M_{i-1} and a homotopy h_i of df_{i-1} to φ_{i-1}.

3. Consider $M_{i-1} \cup V_i \subset N_{i-1} \cup U_i$. Let W_i be a compact PL neighborhood of V_i in the Euclidean neigh U_i. Subdivide W_i sufficiently fine so that any simplex of W_i, which meets M_{i-1} is contained in N_{i-1}. Let Y_i be a second derived neigh of the subcomplex generated by the simplexes meeting M_{i-1}. Then $Y_i \subset W_i$. Further W_i is obtained from Y_i by adding handles in order of increasing dimension. Unfortunately, there may be handles of top dimension; i.e. n-discs. Let P_j be the union of small open discs about the center of each n-handle.

4. Now f_{i-1} is already defined on a neigh of Y_i, and so it is only necessary to extend f_{i-1} inductively over each handle of dimension less than n so that the differential is homotopic to φ_i by a homotopy extending that already obtained on a neigh of the boundary. To do this we use Theorems A and B for a handle and its boundary and the following lemma (see Phillips [**6**, Appendix]).

LEMMA. *Let*

$$p: A_1 \to A_2 \quad \textit{and} \quad p: B_1 \to B_2$$

be fibrations and let

$$d_1: A_1 \to B_1 \quad \textit{and} \quad d_2: A_2 \to B_2$$

be h.e. such that

$$\begin{array}{ccc} A_1 & \xrightarrow{d_1} & B_1 \\ \Big\downarrow{\scriptstyle p} & & \Big\downarrow{\scriptstyle p} \\ A_2 & \xrightarrow{d_2} & B_2 \end{array}$$

commutes.

Then given

$$f_2: S^p \to A_2 \quad \textit{and} \quad \varphi_1: S^p \to B_1$$

and a homotopy h_2 of $d_2 f_2$ to $p\varphi_1$, there exists $f_1: S^p \to A$ and a homotopy h_1 of $d_1 g_1$ to φ_1 such that

$$pg_2 = g_1 \quad \textit{and} \quad ph_1 = h_2.$$

5. Note that f_i is defined on the same neighborhood of X as f_{i-1}, except near V_i. Since the $\{U_i\}$ are locally finite, on any compact set, the neigh of X is reduced only a finite number of times. Hence there is a neighborhood U_1 of X on which all the f_i agree with g.

6. The isotopy ψ_t of M into $M - \bigcup \bar{P}_j$: The center points of the discs in $\bigcup P_j$ form a countable sequence of points,

$$p_1, p_2, \ldots \in M - X.$$

Each p_i is connected to infinity by a proper path in $M - X$. Given any compact subset K of M, we claim only a finite number of the p_i are in K, and only a finite number of the corresponding paths need meet K. The first statement follows since only a finite number of U_i meet K. For the second statement consider

Case 1. X compact. Only finite number of U_i meet $X \cup K$, say $U_1, \ldots, U_r$. Let

$$\Lambda_{r+1} = U_{r+1} \cup U_{r+2} \cup \cdots.$$

Let G be a component of X_{r+1}. Since U_i is connected, G is the union of a subcollection of the U_i in Λ_{r+1}. G is open, and $bG \cap \Lambda_{r+1} = \emptyset$ since G is closed in Λ_{r+1} (being complement of union of other components). Either G is a component of $M - X$ or $bG \neq \emptyset$, since $bG = \emptyset$ implies G is open and closed. If $bG \neq \emptyset$. some U_j in G meets $U_1 \cup \cdots \cup U_r$; hence only a finite number of such components, and therefore only a finite number of bounded components.

Now there is an n, such that $p_i \in \Lambda_{r+1}$, $i > n$. Further, only a finite number can be in the bounded components. For the others, there is a proper path to infinity not meeting $X \cup K$.

Case 2. X closed, locally flat submanifold. If $\dim X \leqq m - 2$, apply above argument with K in place of $X \cup K$. Then either $G = M$ or $bG \neq 0$. Hence again only a finite number of bounded components in Λ_{r+1}. If p_i is in a nonbounded component, the path to infinity in this component can be pushed off of X by general position arguments.

If $\dim X = m - 1$, we can replace X by a product neigh $X \times [0, 1]$ of X in M, missing all the points p_i. The closure of the components of the complement will again be noncompact. So now assume $\dim X = n$.

We can construct a locally finite cover $\{U_i\}$, where U_i is either an open coord. neigh of $M - X$ or the interior of X, or U_i is a half-space with boundary on ∂X.

Then the closure of the components of $M - X$ are unions of the U_i. Since only a finite number of the U_i meet K, only a finite number of the components have closures that meet K.

Now apply the argument of Case 1 to the closure of each component that meets K, with K in place of $X \cup K$. Further note that a path in the closure of a component of $M - X$ which misses K can also be pushed off of X, since the closure of a component is a manifold with boundary, with boundary the intersection with X.

7. If $\dim M \geqq 3$, we can inductively construct locally flat paths from the p_i to infinity, which do not intersect. Further we can construct normal tubes which are disjoint. The normal tube is trivial and hence homeomorphic to the half-space H^m, and contains a small disc about the point in its interior. There is an obvious isotopy of H^m, fixed in a neigh of the boundary which deforms H^m into H^m-disc. Since only a finite number of tubes meet any compact set, the topology of the union of the tubes in M is that of a disjoint union. Hence we may simultaneously perform the isotopy on all the tubes at once. This gives the desired isotopy ψ_t of M into an open subset of M missing $\bigcup P_j$.

8. If $\dim M < 3$, then M is triangulable by classical results, and M is an infinite handle body, with handles of dim $< n$. Hence no points need be omitted.

This proves that d_* is an epimorphism on homotopy. The argument that d_* is a monomorphism is similar. See Phillips [6].

Chapter 3. Proof of Theorem B. In order to convert the regular homotopy extension problem to an isotopy extension problem, we prove (see [2]):

PROPOSITION. *Let $f:M^n \to Q^n$ be an immersion. Then with respect to any metric on Q, there is an $\varepsilon:M \to R^+$ such that if $g:M \to Q$ is an immersion within ε of f, it may be factored $g = fh$, where $h:M \to M$ is a local homeomorphism and h depends continuously on g.*

PROOF. Choose a metric on M. Let K be a compact subset of M. Since f is a local homeomorphism, there is a $\delta > 0$, such that if $x_1 \in K$ and x_2 is within δ of x_1, then

$$f(x_1) \neq f(x_2) \qquad \text{if } x_1 \neq x_2.$$

Consider $f(N_\delta(x))$, $x \in K$. There is an $\varepsilon > 0$, such that

$$N_\varepsilon(f(x)) \subset f(N_\delta(x)), \qquad \text{all } x \in K.$$

Then if $y \in Q$ is within ε of $f(x)$ for some $x \in K$, there is a unique x' within δ of x for which $f(x') = y$.

Since M is σ-compact, say $M = \bigcup K_n$, there is an $\varepsilon:M \to R^+$, such that if $x \in K_n$, $\varepsilon(x)$ is less than the corresponding ε_n for K_n. Now given g within ε of f, let $y = g(x)$ and $h(x) = x'$ as above. Then $g = f \circ h$, h is a local homeomorphism, since f and g are, and $h(x)$ depends continuously on $g(x)$. I.e. if we write h_g for the h corresponding to g, $h_g(x)$ depends continuously on g and x.

REMARK. There is a $\delta:M \to R^+$ and an $\varepsilon:M \to R^+$ such that if g is within δ of f, then h may be chosen within δ of the identity and as such is unique. In particular, the h corresponding to f is the identity.

COROLLARY. *Let $f:\Delta \times M \to \Delta \times Q$ be a regular k homotopy. Let $K \subset M$ be compact. Then for any $t_0 \in \Delta$, there is an $\varepsilon > 0$, and a neighborhood L of K, such that $f_t \mid L$ can be factored*

$$f_t = f_{t_0} \circ h_t \quad \textit{for } |t - t_0| < \varepsilon.$$

Further h_{t_0} is the inclusion and $h_t:L \to M$ is an isotopy.

PROOF. The existence of h_t follows from the proposition. Further

$$h:N_\varepsilon(t_0) \times L \to N_\varepsilon(t_0) \times M$$

is a local homeomorphism since f and $\mathrm{id} \times f_{t_0}$ are. By the remark above, we may assume h_{t_0} is the inclusion. If we take L compact, there is a neighborhood of $t_0 \times L$ in $N_\varepsilon(t_0) \times L$ on which h is a homeomorphism. This neighborhood contains one of the form $N_{\varepsilon'}(t_0) \times L$. Thus after restricting ε we get that h_t is an isotopy.

The proof of Theorem B will depend on Lees' Isotopy Extension Theorem (to be proved in Chapter 4).

Let C be a compact subset of a topological manifold M without boundary. Let U be a neighborhood of C in M.

THEOREM C. *Let $h:\Delta \times U \to \Delta \times M$ be a k-isotopy. Then there is an ambient k-isotopy*

$$g:\Delta \times M \to \Delta \times M$$

such that g is fixed outside a compact set and gh_0 agrees with h on a neighborhood of C.

COROLLARY. *Let V be an open set in M such that $h_t(C) \subset V$ all $t \in \Delta$. By replacing M by V, we get that g is fixed outside V.*

LEMMA. *The proof of Theorem B can be reduced to the case where*

(1) M is a handle $D^p \times D^{n-p}$ and X is its boundary $(\partial D^p) \times D^{n-p}$ and $p < n$. That is, if we let

$$\tilde{M} = 2\mathring{D}^p \times 2\mathring{D}^{n-p} \quad \text{and} \quad \tilde{X} = (2\mathring{D}^p - \mathring{D}^p) \times 2\mathring{D}^{n-p},$$

we need to show: Let

$$f:\Lambda \times \tilde{M} \to \Lambda \times Q \quad \text{and} \quad g:\Delta \times U \to \Delta \times Q$$

be regular homotopies with U a neigh of X in M, say

$$U = (2\mathring{D}^p - \tfrac{1}{2}D^p) \times 2\mathring{D}^{n-p},$$

with f and g agreeing on $\Lambda \times$ (neigh X). Then f extends to $\Delta \times$ (neigh M), with f agreeing with g in a neigh of X.

(2) We are given an isotopy $h:\Delta \times V \to \Delta \times U$, where V is a compact neigh of X in U; such that $g_t = g_{t_0} \circ h_t$ for some $t_0 \in \Delta$ and $h_{t_0} =$ inclusion. Say

$$V = (\tfrac{3}{2}D^p - \tfrac{3}{4}\mathring{D}^p) \times \tfrac{5}{4}D^p.$$

PROOF. (1) f may be extended to Δ, coordinate neigh by coordinate neigh as in §2. In each coordinate neigh the extension is handle by handle. We may omit top dimensional handles as before, under our assumptions on M and X.

(2) (Δ, Λ) is topologically equivalent to a cube and its front face; i.e., (I^k, I^{k-1}), if Δ is a k-simplex. Now given the regular homotopy

$$g:I^k \times U \to I^k \times Q;$$

for each $t_0 \in I^k$ there is an isotopy

$$h_t:V \to Q,$$

for t in a neigh of t_0, so that $g_t = g_{t_0} \circ h_t$. Subdivide I^k uniformly into sufficiently small cubes I_i^k, so that each small cube is in such a neighborhood.

By proceeding little cube by little cube over the bottom layer, etc., we reduce the situation to (2). That is, when we come to any little cube I_i^k, the extension f of having been made over the previous ones, we will have the extension defined over certain faces of I_i^k, whose union is always homeomorphic to I^{k-1} in I^k (see Phillips [6]).

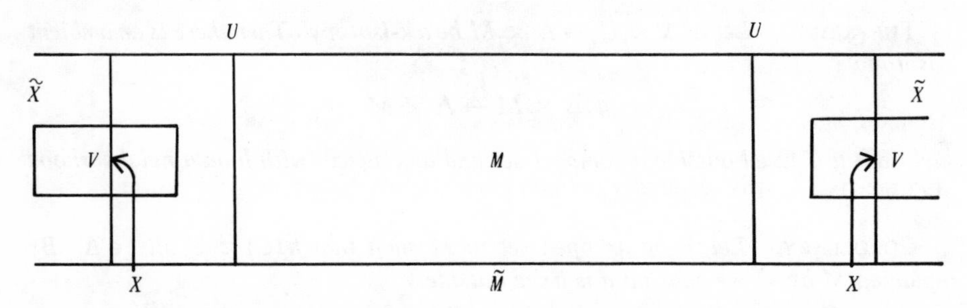

NOTATION. As in proof of (2) above, we will replace (Δ, Λ) by (I^k, I^{k-1}) in the remainder of the proof of Theorem B.

With U, V, M, X as above, let f be a regular $(k-1)$-homotopy of M in Q, and

$$g: I^k \times V \to I^k \times Q$$

a regular k-homotopy of X in Q, which agree on $I^{k-1} \times V_1$,

$$V_1 = (\tfrac{3}{2}D^p - \tfrac{7}{8}\mathring{D}^p) \times \tfrac{5}{7}D^{n-p}.$$

DEFINITION. We will say that f is in *good position* with respect to U, if there is an ambient isotopy

$$k: I^k \times U \to I^k \times U,$$

a $(k-1)$-isotopy

$$h: I^{k-1} \times V \to I^{k-1} \times U,$$

and an immersion

$$\varphi: U \to Q$$

such that

(a) $k \mid I^k \times V_1 = g \mid I^k \times V_1$,

(b) $\varphi h = f \mid I^{k-1} \times V$,

(c) $k \mid I^{k-1} \times V_1 = h \mid I^{k-1} \times V_1$,

(d) there exists a neigh $U_1 \subset U$ with $\overline{U}_1$ compact in U, such that k is the identity outside U_1, and $h_t(\dot{V}) \subset U - \overline{U}_1$, where $\dot{V} = (\partial\tfrac{3}{4}D^p) \times \tfrac{5}{4}D^{n-p}$.

CONCLUSION OF PROOF OF THEOREM B, ASSUMING GOOD POSITION. If $t = (t_1, \ldots, t_k)$, let $\bar{t} = (t_1, \ldots, t_{k-1}, 0) \in I^{k-1}$. Then we can extend f to I^k on the neighborhood M_1 of M, where $M_1 = \tfrac{3}{2}D^p \times \tfrac{5}{4}D^{n-p}$, by setting

(a) $f_t = \varphi \circ k_t \circ k_{\bar{t}}^{-1} \circ h_{\bar{t}}$ on V,

(b) $f_t = f_{\bar{t}}$ on $M_1 - V$.

Note that

(1) For $t \in I^{k-1}$,

$$k_t k_{\bar{t}}^{-1} = \text{identity}$$

and

$$\varphi h_t \circ k_{\bar{t}}^{-1} h_{\bar{t}} = \varphi \circ h_t \quad \text{on } I^{k-1} \times V.$$

Hence extended f agrees with original f on $I^{k-1} \times V$.

(2) Outside U_1, k_t and $k_{\bar{t}}^{-1}$ are the identity, so that $f_t = f_{\bar{t}}$ on a neigh of $\dot{V}$, for $t \in I^k$. Hence f is well defined.

(3) On V_1,

$$\varphi \circ k_t \circ k_{\bar{t}}^{-1} h_{\bar{t}} = \varphi k_t = g_t.$$

Hence f_t agrees with g_t on V_1.

Putting f in good position. Let f and g be as in the lemma above. We may assume that h_t was originally defined on a slightly larger neighborhood $V_2 \supset V$, say

$$V_2 = (\tfrac{3}{2}D^p - \tfrac{5}{8}\mathring{D}^p) \times \tfrac{5}{4}D.$$

Let U_1 open in U, such that $h_t(V) \subset U_1$, all $t \in I$, $\bar{U}_1$ compact in U.

(1) There exists $\varepsilon > 0$ and an ambient isotopy

$$J: I^{k-1} \times [0, \varepsilon] \times U \to I^{k-1} \times [0, \varepsilon] \times U$$

fixed on $U - U_1$, such that $J_t = h_t$ on V_1 and $J_t(\dot{V}_2) = J_{\bar{t}}(\dot{V}_2)$. (See §4.)

(2) There is an ambient isotopy

$$q: [0, \varepsilon] \times U \to [0, \varepsilon] \times U$$

which is fixed on V_1 and a neighborhood of $\dot{V}_2$ in V_2 and which squeezes $\dot{V}$ vertically (i.e. in a D^{n-p} direction) into $U - \bar{U}_1$. This is where we need $p < n$.

(3) Let $\varphi = g_{t_0}$ and define f_t, $0 \leqq t_k \leqq \varepsilon$ by

$$\begin{aligned} f_t &= \varphi \circ J_t \circ 1 \times q_{t_k} && \text{on } V_2, \\ &= f_{\bar{t}} && \text{on } M_1 - V_2. \end{aligned}$$

Then at $I^{k-1} \times \varepsilon$, f and g agree on V_1.

Let $k: I^k \times U \to I^k \times U$ be an ambient isotopy extending h and fixed outside U_1.

Let $J \circ 1 \times q \mid I^{k-1} \times \varepsilon \times V$ be the new h. Note that it agrees with the old h on V_1.

Chapter 4. The Isotopy Extension Theorem. Kirby and Edwards [1] have written up the isotopy extension theorem for 1-isotopies. Since the k-isotopy theorem follows easily from the main theorem of their paper, we will content ourselves with showing how to obtain the k-isotopy extension result from their theorem.

MAIN THEOREM OF [1]. *Let M be a topological manifold without boundary and let C and U be subsets of M such that C is compact and U is a neighborhood of C. Given any neighborhood R of the inclusion $\eta: U \subset M$ in $I(U; M)$, there is a neighborhood S of η in $I(U; M)$ and a deformation*

$$\phi: S \times I \to R$$

of S into $I(U; C; M)$. Furthermore, ϕ is modulo the complement of a compact *neighborhood of C in U and*

$$\phi(\eta, t) = \eta \quad \textit{for all } t \in I.$$

Here $I(U; M)$ is the space of embeddings of U in M in the $C - O$ topology, and $I(U; C; M)$ is the subspace of embeddings which are the inclusion on C.

k-ISOTOPY EXTENSION THEOREM. *Let*

$$h: I^k \times U \to I^k \times M, \qquad h_0 = \text{inclusion},$$

be a k-isotopy. There is an ambient isotopy

$$g: I^k \times M \to I^k \times M$$

such that g is fixed outside a compact set and $g \mid I^k \times C = h \mid I^k \times C$.

COROLLARY. *By replacing C by a compact neighborhood of C in U, we get that g agrees with h on a neighborhood of C.*

We first prove:

PROPOSITION. *Let C, U, M be as in the main theorem. There is a neighborhood S of the inclusion* $\eta: U \subset M$ *in* $I(U; M)$ *and a continuous map*

$$\psi: S \to H(M)$$

such that $\psi(\eta) = 1_M$ *and, for all* $h \in S$,

(a) $\psi(h) \mid C = h \mid C$,

(b) $\psi(h)$ *is the identity outside a compact neighborhood of C.*

PROOF. Let S and ϕ be as in the main theorem. Let

$$\begin{aligned} \psi(h) &= h\phi(h, 1)^{-1} && \text{on } h(U), \\ &= \text{identity} && \text{on } M - h(U). \end{aligned}$$

That

$$\psi(\eta) = 1_M \quad \text{and} \quad \psi(h) \mid C = h \mid C$$

is immediate from the properties of Φ. For property (b), let N be the compact neigh of C such that

$$\Phi(h, t) \mid U - N = h \mid U - N$$

according to the main theorem. Let V be a compact neigh of N, and restrict S so that $h(N) \subset V$, all $h \in S$. Then $\psi(h)$ is the identity outside V.

COROLLARY. *We may assume* $\psi(h)$ *and h agree on a neigh of C.*

PROOF OF THE k-ISOTOPY EXTENSION THEOREM. The proof is by induction on k. $k = 1$: Let $h: I \times U \to I \times M$ be an isotopy. Let $s \in I$, and identify for the moment U with $h_s(U)$. Then we may apply the proposition with $h_s = \eta$; to obtain

$$g_t^s: M \to M,$$

t in a neigh of s; an ambient isotopy with $g_s^s = 1_M$, extending h_t on a neigh of C.

Choose $0 = t_0 < t_1 < \cdots < t_r = 1$, so that the neighborhoods of t_i for which $g^i = g^{t_i}$ is defined, cover I. Choose

$$t_i < s_i < t_{i+1}, \qquad i = 0, 1, \ldots, r - 1,$$

so that $g^i_{s_i}$ and $g^{i+1}_{s_i}$ are both defined. Set $g_t = g^0_t$ for $0 \leqq t \leqq s_0$. Assume g has been defined for $0 \leqq t \leqq s_i$. Set

$$g_t = g_{s_i}(g^{i+1}_{s_i})^{-1} g^{i+1}_t, \qquad s_i \leqq t \leqq s_{i+1}.$$

Assume theorem for $k - 1$: Let

$$h: I^k \times U \to I^k \times M$$

be a k-isotopy. Write $I^k = I^{k-1} \times I$. For any $s \in I$, (identifying $h_{0,s}(U)$ with U) let

$$f^s: I^{k-1} \times M \to I^{k-1} \times M$$

be a $(k - 1)$-isotopy extending $h \mid I^{k-1} \times s \times U_1$, U_1 a neigh of C. Consider

$$h^s = (f^s \times 1)^{-1} \circ h: I^k \times U \to I^k \times M.$$

Then $h^s_t: U_1 \to M$ is the inclusion for $t \in I^{k-1} \times s$, and is in S for $t \in I^{k-1} \times s'$, s' in a neigh of s. By the proposition, there is an ambient isotopy

$$\bar{g}^s_t: M \to M, \qquad t \in I^{k-1} \times (\text{neigh of } s)$$

extending h^s_t on a neigh of C. Let $g^s = (f^s \times 1) \circ \bar{g}^s$. Then g^s_t extends h_t on a neigh of C, $t \in I^{k-1} \times$ (neigh of s). Proceed as above to obtain an ambient isotopy $g: I^k \times M \to I^k \times M$ extending h on a neighborhood of C.

Note that g is fixed outside of a compact set, by (b) of the proposition, and the inductive assumption.

References

1. Robert D. Edwards and Robion C. Kirby, *Deformations of spaces of imbeddings* (mimeograph).

2. Andre Haefliger and V. Poenaru, *La classification des immersions combinatoires*, Inst. Hautes Études Sci. Publ. Math. No. 23 (1964), 75–91. MR **30** #2515.

3. J. M. Kister, *Microbundles are fibre bundles*, Ann. of Math. (2) **80** (1964), 190–199. MR **31** #5216.

4. R. Lashof, *Lees' immersion theorem and the triangulation of manifolds*, Bull. Amer. Math. Soc. **75** (1969), 335–538. MR **39** #960.

5. J. Lees, *Immersions and surgeries on topological manifolds*, Bull. Amer. Math. Soc. **75** (1969), 529-534. MR **39** #959.

6. A. Phillips, *Submersions of open manifolds*, Topology **6** (1967), 171–206. MR **34** #8420.

7. D. Gauld, *Mersions of topological manifolds*, Trans. Amer. Math. Soc. **149** (1970), 539–560.

Part II. Triangulations and Smoothings

Chapter 5. Structures on microbundles. Let X be a topological space. If X is homotopy equivalent to a smooth manifold (PL space) M, we wish to define a smooth (PL) microbundle over X to be a smooth (PL) microbundle over M.

Given two such, say M_1 and M_2 homotopy equivalent to X, we want to define when ε_1 over M_1 is equivalent to ε_2 over M_2. For this we need to build the homotopy equivalence into the definition. (We speak only of the smooth case in all that follows; the PL case being identical.)

DEFINITION 1. Let X be a topological space. A *smooth microbundle over* X is a pair (ε, f) where ε is a smooth microbundle over a smooth manifold M, and $f:M \to X$ is a homotopy equivalence.

Given two smooth microbundles (ε_1, f_1) and (ε_2, f_2) over X, consider

$$M_1 \xrightarrow{f_1} X \xrightarrow{g_2} M_2,$$

where g_2 is a homotopy inverse of f_2. Then $g_2 \circ f_1$ is homotopic to a smooth homotopy equivalence $h:M_1 \to M_2$.

The equivalence class of $h^*\varepsilon_2$ is independent of the choice of g_2 and h, since homotopic smooth maps are connected by a smooth homotopy. Denote the class of $h^*\varepsilon_2$ by $(f_2^{-1}f_1)^*(\varepsilon_2)$. Define (ε_1, f_1) to be *equivalent* to (ε_2, f_2) if $\varepsilon_1 \in (f_1 f_2^{-1})^*(\varepsilon_2)$. It is easy to check that this is an equivalence relation.

Given a topological microbundle τ over X, we would like to define a smooth reduction of τ to be a smooth microbundle (ε, f) over τ and a topological microbundle map $\varphi:\varepsilon \to \tau$ over f. In order to define where two such reductions are equivalent, we need f to be a special kind of homotopy equivalence.

DEFINITION 2. A *strong homotopy equivalence* (f, g) between topological spaces M and X consists of:

(a) Maps $f:M \to X, g:X \to M$.

(b) Homotopies $\rho:I \times M \to I \times M$ between gf and 1_M, $\sigma:I \times X \to I \times X$ between fg and 1_X satisfying

(1) The resultant two homotopies of fgf to f are homotopic rel endpoints.

(2) The resultant two homotopies of gfg to g are homotopic rel endpoints.

DEFINITION 3. A *smooth reduction* of τ over X, is a quadruple $(\varphi, \varepsilon; f, g)$, where ε is a smooth microbundle over a smooth manifold M, (f, g) is a strong homotopy equivalence between M and X, and $\varphi:\varepsilon \to \tau$ is a topological microbundle map over f.

If $(\varphi', \varepsilon'; f, g)$ is another smooth reduction with φ' over M, and ε' over f, we say it is equivalent to $(\varphi, \varepsilon; f, g)$ if there is a smooth microbundle equivalence

$$\lambda:\varepsilon \to \varepsilon'$$

such that $\varphi'\lambda \sim_f \varphi$.

Given two smooth reductions $(\varphi_1, \varepsilon_1; f_1, g_1)$ and $(\varphi_2, \varepsilon_2; f_2, g_2)$ of τ over X, consider

$$M_1 \xrightarrow{f_1} X \xrightarrow{g_2} M_2.$$

Let $h:M_1 \to M_2$ be a smooth homotopy equivalence homotopic to $g_2 \circ f_1$. Then we have the topological microbundle map

$$\varphi_2 h_* :h^*\varepsilon_2 \to \tau,$$

where $h_* : h^*\varepsilon_2 \to \varepsilon_2$ is the natural projection. Now $\varphi_2 \circ h_*$ covers $f_2 \circ h$ which is homotopic to $f_2 \circ g_2 \circ f_1$ and hence homotopic to f_1. This homotopy is covered by a homotopy of $\varphi_2 \circ h_*$ through microbundle maps to a microbundle map ψ of $h^*\varepsilon_2$ into τ over f_1. The equivalence class of $(h^*\varepsilon_2, \psi; f_1, g_1)$ is independent of the choice of h and ψ provided we use the given homotopies to produce ψ. Consequently we will denote it by $(g_2 f_1)^*((\varphi_2, \varepsilon_2); f_1, g_1)$.

We will say that $(\varphi_1, \varepsilon_1; f_1, g_1)$ is equivalent to $(\varphi_2, \varepsilon_2; f_2, g_2)$ if

$$(\varphi_1, \varepsilon_1; f_1, g_1) \in ((g_2 f_1)^*(\varphi_2, \varepsilon_2); f_1, g_1).$$

This is an equivalence relation:

(a) $$(\varphi, \varepsilon; f, g) \in ((gf)^*(\varphi, \varepsilon); f, g).$$

PROOF. In fact, we may choose h to be 1_M by using the homotopy ρ of gf to 1_M. Then $\varphi \circ h_* : h^*\varepsilon \to \tau$ is the equivalent to $\varphi : \varepsilon \to \tau$.

On the other hand, to produce ψ we are supposed to use the homotopy of $f \circ h$ to f by the homotopies

$$f \circ 1_M \overset{\rho}{\sim} f \circ g \circ f \overset{\sigma}{\sim} 1_X \circ f.$$

Since (f, g) is a strong homotopy equivalence, this homotopy is itself homotopic to the constant homotopy rel endpoints. Hence

$$(gf)^*(\varphi, \varepsilon) = (\varphi, \varepsilon).$$

Abbreviate $(\varphi, \varepsilon; f, g)$ by (φ, ε) if (f, g) is understood.

(b) $$(\varphi_1, \varepsilon_1) = (g_2 f_1)^*(\varphi_2, \varepsilon_2) \Rightarrow (\varphi_2, \varepsilon_2) = (g_1 f_2)^*(\varphi_1, \varepsilon_1).$$

PROOF. Consider

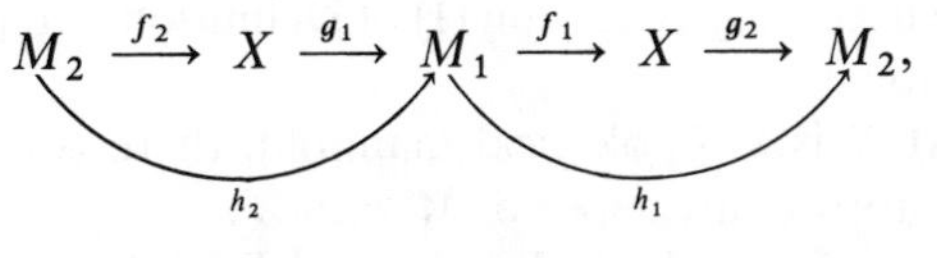
$M_2 \xrightarrow{f_2} X \xrightarrow{g_1} M_1 \xrightarrow{f_1} X \xrightarrow{g_2} M_2,$
h_2
h_1

where h_2 and h_1 are smooth maps homotopic to $g_1 f_2$ and $g_2 f_1$ resp.

The composition

$$(g_1 f_2)^*((g_2 f_1)^*(\varphi_2, \varepsilon_2)) = (g_1 f_2)^*(\varphi_1, \varepsilon_1)$$

corresponds to the homotopy

(i) $$f_2 \circ h_1 \circ h_2 \sim (f_2 \circ g_2) \circ f_1 \circ h_2 \sim f_1 \circ h_2 \sim (f_1 \circ g_1) \circ f_2 \sim f_2.$$

On the other hand, $(g_2 f_2)^*(\varphi_2, \varepsilon_2) = (\varphi_2, \varepsilon_2)$ corresponds to the homotopy

(ii) $$f_2 \circ h_1 \circ h_2 \sim f_2 \circ g_2 \circ (f_1 \circ g_1) \circ f_2 \sim (f_2 \circ g_2) \circ f_2 \sim f_2.$$

Now (i) is homotopic rel endpoints to

(i′) $$f_2 \circ h_1 \circ h_2 \sim (f_2 \circ g_2) \circ f_1 \circ h_2 \sim (f_2 \circ g_2) \circ (f_1 \circ g_1) \circ f_2 \sim (f_2 \circ g_2) \circ f_2 \sim f_2.$$

And (ii) is homotopic rel endpoints to

(ii′) $f_2 \circ h_1 \circ h_2 \sim f_2 \circ g_2 \circ f_1 \circ h_2 \sim f_2 \circ g_2 \circ (f_1 \circ g_1) \circ f_2 \sim f_2 \circ (g_2 \circ f_2) \sim f_2.$

By condition (1) of Definition 2, (i′) and (ii′) are homotopic rel endpoints. Hence

$$(\varphi_2, \varepsilon_2) = (g_2 f_2)^*(\varphi_2, \varepsilon_2) = (g_1 f_2)^*(g_2 f_1)^*(\varphi_2, \varepsilon_2) = (g_1 f_2)^*(\varphi_1, \varepsilon_1).$$

(c) $$(\varphi_1, \varepsilon_1) = (g_2 f_1)^*(\varphi_2, \varepsilon_2) \quad \text{and} \quad (\varphi_2, \varepsilon_2) = (g_3 f_2)^*(\varphi_3, \varepsilon_3) \Rightarrow (\varphi_1, \varepsilon_1) = (g_3 f_1)^*(\varphi_3, \varepsilon_3).$$

PROOF. Consider

$$M_1 \xrightarrow{f_1} X \xrightarrow{g_2} M_2 \xrightarrow{f_2} X \xrightarrow{g_3} M_3,$$

with $h_1: M_1 \to M_2$ and $h_2: M_2 \to M_3$,

where h_1 and h_2 are smooth maps homotopic to $g_2 f_1$ and $g_3 f_2$ respectively.

The composition

$$(g_2 f_1)^*(g_3 f_2)^*(\varphi_3, \varepsilon_3) = (\varphi_1, \varepsilon_1)$$

corresponds to the homotopy

(i) $$f_3 \circ h_2 \circ h_1 \sim (f_3 \circ g_3) \circ f_2 \circ h_1 \sim f_2 \circ h_1 \sim (f_2 \circ g_2) \circ f_1 \sim f_1.$$

On the other hand $(g_3 f_1)^*(\varphi_3, \varepsilon_3)$ corresponds to

(ii) $$f_3 \circ h_2 \circ h_1 \sim f_3 \circ g_3 \circ (f_2 \circ g_2) \circ f_1 \sim (f_3 \circ g_3) \circ f_1 \sim f_1.$$

Since (i) and (ii) are homotopic rel endpoints, (c) follows.

REMARK. We need only condition (1) of Definition 2 to prove the equivalence relation of Definition 3.

EXAMPLE 1. If X is a topological manifold, there is a smooth manifold M and a strong homotopy equivalence of M with X:

Embed X in a sufficient high dimensional Euclidean space so that it has a normal microbundle ν, see [7]. The total space of ν contains a R^n-bundle with total space E an open subset of Euclidean space and hence smooth.

It follows from Kister's theorem that the zero section $X \subset E$ is a deformation retract by a fibrewise deformation. It is easy to check that this implies (p, i) is a strong homotopy equivalence

$$p: E \to X, \quad \text{the projection,}$$
$$i: X \to E, \quad \text{the zero section.}$$

If Y is an open subset of X, then $E \mid Y$ has the same relation to Y as E does to X. It follows that a reduction $(\varphi, \varepsilon; p, i)$ of τ over X restricts to a reduction $(\varphi \mid p^{-1}Y, \varepsilon \mid p^{-1}Y; p \mid p^{-1}Y, i \mid Y)$ of $\tau \mid Y$.

EXAMPLE 2. If X has a smoothing (M, f), M smooth, $f:M \to X$ a homeomorphism, then (f, f^{-1}) is a strong homotopy equivalence.

Let M_α and M_β be two smooth manifolds with underlying space X. Let $(\varphi_\alpha, \varepsilon_\alpha)$, $(\varphi_\beta, \varepsilon_\beta)$ be reductions of τ, ε_α and ε_β differentiable microbundles over M_α and M_β, resp. Note that

$$\varphi_\alpha:\varepsilon_\alpha \to \tau, \qquad \varphi_\beta:\varepsilon_\beta \to \tau$$

are topological microbundle equivalences (i.e. microbundle maps over the identity). Then the condition that $(\varphi_\alpha, \varepsilon_\alpha)$ be equivalent to $(\varphi_\beta, \varepsilon_\beta)$ simplifies to $\varphi_\beta^{-1}\varphi_\alpha:\varepsilon_\alpha \to \varepsilon_\beta$ is homotopic through microbundle maps to a smooth microbundle map.

Let M be a smooth manifold with underlying space X. Let (φ, ε) over M be a reduction of τ; and let $(\psi, \eta; p, i)$ over E be a reduction of τ. Now $\psi:\eta \to \tau$ covering p, has a left inverse ψ^{-1} as a topological microbundle map, ψ^{-1} covering i. The condition that these two reductions be equivalent is:

$$\psi^{-1}\varphi:\varepsilon \to \eta$$

is homotopic to a smooth microbundle map. (Note that the smooth microbundle map and homotopy may be chosen to cover maps $X \to E$ arbitrarily close to the inclusion i.)

All the above holds for PL microbundles as well as smooth microbundles. The remarks below are for smooth microbundles only.

Just as we talked about R^n-fibre bundles contained in topological microbundles; we have that every smooth microbundle contains a vector bundle unique up to equivalence. Then a smooth reduction $(\varphi, \varepsilon; f, g)$ of τ over X defines a unique equivalence class of vector bundle reductions, and conversely if X is a topological manifold: In fact, given $(\varphi, \varepsilon; f, g)$, let ε_0 be a vector bundle in ε. Then since

$$X \xrightarrow{g} M \xrightarrow{f} X$$

is homotopic to the identity, there is a topological microbundle equivalence of $g^*\varepsilon_0$ with τ, over 1_X. Conversely, given a vector bundle ε contained in τ, let $p:E \to X$ be as in Example 1, then

$$p_*:p^*\varepsilon \to \varepsilon \subset \tau$$

is the smooth microbundle reduction; since every vector bundle over a smooth manifold is a smooth microbundle.

More generally, if there exists a smooth M strongly homotopy equivalent to X, then the equivalence classes of vector bundle reductions of τ are in one to one correspondence with the equivalence classes of smooth microbundle reductions of τ.

NOTE. R. Vogt has shown that if (f, g) is a homotopy equivalence, $\rho:gf \sim 1$, $\sigma:fg \sim 1$; then one may choose σ', so that $(f, g; \rho, \sigma')$ is a strong homotopy equivalence.

Chapter 6. The existence theorem.

DEFINITION. A topological manifold M *admits a smoothing* if there exists a smooth manifold V and a homeomorphism $h: V \to M$. Two such smoothings (V_1, h_1) and (V_2, h_2) are called *equivalent* if h_1 is isotopic through onto homeomorphisms to a homeomorphism $\bar{h}_1$ such that $h_2^{-1}\bar{h}_1$ is a diffeomorphism. (Two such smoothings are called *identical* if $h_2^{-1}h_1$ is a diffeomorphism.)

LEMMA. *Let* $f: M \xrightarrow{n} N^n$ *be a topological immersion,* M, N *without boundary,* M *open. Let* (W, k) *be a smoothing of* N*, then there is a unique smoothing* (V, h) *of* M *such that*

$$k^{-1}fh: V \to W$$

is a smooth immersion.

PROOF. First we show that two such smoothings (V_1, h_1) and (V_2, h_2) are identical.

Take an open covering $\{U_\alpha\}$ of M such that $f \mid U_\alpha$ is a homeomorphism. Then $\{h_1^{-1}U_\alpha\}$ and $\{h_2^{-1}U_\alpha\}$ are open coverings of V_1 and V_2 respectively. Now $h_2^{-1}h_1$ is a diffeomorphism, since

$$h_2^{-1}h_1 \mid h_1^{-1}U_\alpha = (k^{-1}fh_2)^{-1}k^{-1}fh_1 \mid h_1^{-1}U_\alpha$$

is a diffeomorphism.

Second we will show that f defines a smooth manifold V with underlying space M, so that (V, identity) will be the required smoothing.

Now $f^{-1}k \mid k^{-1}fU_\alpha$ defines a smoothing (V_α, h_α) of U_α,

$$V_\alpha = k^{-1}fU_\alpha, \qquad h_\alpha = f^{-1}k \mid k^{-1}fU_\alpha.$$

Consider the overlap $U_\alpha \cap U_\beta$ of two such neighborhoods. Then

$$h_\beta^{-1}h_\alpha \mid h_\alpha^{-1}(U_\alpha \cap U_\beta) = (f^{-1}k)^{-1}f^{-1}k \mid k^{-1}f(U_\alpha \cap U_\beta)$$

is the identity, and hence a diffeomorphism. Thus $\{V_\alpha, h_\alpha\}$ defines a smooth structure (V, identity) on M. Since

$$k^{-1}fh_\alpha: V_\alpha \to W$$

is

$$1_{V_\alpha}: k^{-1}fU_\alpha \to k^{-1}fU_\alpha \subset W,$$

$k^{-1} \circ f \circ$ identity is a smooth immersion.

NOTATEM. We will call (V, h) the *smoothing induced by* f from (W, k).

PROPOSITION 1. *An equivalence class of smoothings of a topological manifold* M*, defines an equivalence class of smooth reductions of* τM.

PROOF. Let (V, h) be a smoothing of M. Then $(dh, \tau V)$ is a smooth reduction of τM,

$$dh = (h, h): V \times V \to M \times M.$$

Now let (V_α, g_α), (V_β, g_β) be equivalent smoothings. Identify the underlying spaces of V_α and V_β with M via g_α and g_β and denote the smoothings by M_α and M_β resp. By assumption there is an ambient isotopy h_t of M,

$$h_0 = 1_M; \qquad h_1 : M_\alpha \to M_\beta,$$

a diffeomorphism. Then

$$dh_t : \tau M_\alpha \to \tau M_\beta$$

is a homotopy through microbundle maps from

$$dh_0 = 1_{\tau M} = (dg_\beta)^{-1} dg_\alpha$$

to a smooth microbundle map.

LEMMA. *Let K^n_α be a smooth compact manifold with boundary embedded topologically in Euclidean n-space E^n as a flat submanifold. Note that the smooth structure K_α on K extends to an open collar neighborhood $\tilde{K}$ of K. Suppose the smooth reduction of $\tau\tilde{K}$ given by $\tilde{K}_\alpha$ is equivalent to one that extends to a smooth reduction of τE^n. Then the smooth structure K_α extends to a smooth structure on $E^n - P$, where P is a finite subset.*

PROOF. Let $\varphi : \varepsilon \to \tau E^n$ be the smooth reduction such that $(\varphi, \varepsilon) \mid \tilde{K} \cong (\tau\tilde{K}_\alpha, 1)$, where ε is a smooth bundle over the standard smoothing E^n_0 of E^n, and φ covers the identity map. Then there is a homotopy $\lambda_t : \tau\tilde{K}_\alpha \to \varepsilon \mid \tilde{K}$, $\lambda_0 = \varphi^{-1} \mid \tilde{K}$, λ_1 a smooth microbundle map.

Since E^n is contractible, ε is trivial, and we get a smooth bundle equivalence

$$\rho : \varepsilon \to \tau E^n_0.$$

Then $\rho\lambda_1 : \tau\tilde{K}_\alpha \to \tau E^n_0 \mid \tilde{K} = \tau\tilde{K}_0$ induces a smooth immersion

$$f : \tilde{K}_\alpha \to \tilde{K}_0 \subset E^n_0,$$

with $df \sim \rho\lambda_1$.

In order to apply Lees' theorem we need each component of $E^n - K$ to have noncompact closure. If this is not the case, remove a point from each bounded component. (There are only a finite number, by Alexander duality.) Let P be this finite set.

As a topological microbundle map

$$df \sim \rho\lambda_1 \sim \rho\lambda_0 = \rho\varphi^{-1} \mid K.$$

By Theorem A of §2, df extends to a microbundle map

$$\sigma : \tau(E^n - P) \to \tau(E^n - P), \qquad \sigma \sim \rho\varphi^{-1}.$$

Then we get a topological immersion

$$\bar{f} : E^n - P \to (E^n - P)_0$$

extending f on a neighborhood of K, and with $d\bar{f} \sim \sigma \sim \rho\varphi^{-1}$, $\bar{f}$ induces a smoothing $(E^n - P)_\alpha$ on $E^n - P$, extending K_α.

ADDENDUM. *The homotopy* λ_t *extends to a homotopy*

$$\bar{\lambda}_t : \tau(E^n - P)_\alpha \to \varepsilon \mid p^{-1}(E^n - P)$$

with

$$\bar{\lambda}_0 = \varphi^{-1} \mid (E^n - P), \quad \text{and} \quad \bar{\lambda}_1 \text{ smooth.}$$

In particular, the reduction of $\tau(E^n - P)$ *given by* $\tau(E^n - P)_\alpha$ *is equivalent to* $(\varphi, \varepsilon) \mid (E^n - P)$.

PROOF. The homotopy

$$\bar{u}_t : \tau(E^n - P)_\alpha \to \tau(E^n - P)_0$$

composed of the homotopies

$$\rho\varphi^{-1} \sim \sigma \sim d\bar{f}$$

extends the homotopy u_t, $0 \leqq t \leqq 2$, composed of

$$\rho\varphi^{-1} \mid \tilde{K} \sim \rho\lambda_0 \sim \rho\lambda_1 \sim df.$$

Consider $\rho^{-1}\bar{u}_t$ and $\rho^{-1}u_t$. Now $\rho^{-1}u_t$ is λ_t, $0 \leqq t \leqq 1$, and is a smooth homotopy, $1 \leqq t \leqq 2$. This smooth homotopy may be extended to a smooth homotopy

$$v_t : \tau(E^n - P)_\alpha \to \varepsilon \mid p^{-1}(E^n - P), \qquad 1 \leqq t \leqq 2,$$

with $v_2 = d\bar{f}$. Since $\rho^{-1}u_t$ followed by $\rho^{-1}u_{4-t}$, $2 \leqq t \leqq 3$, is homotopic rel endpoints to λ_t, $\rho^{-1}\bar{u}_t$ followed by v_{4-t} is homotopic rel endpoints to an extension $\bar{\lambda}_t$ of λ_t.

THEOREM. *Let* M^n *be an open topological manifold (without boundary), and let* (ε, φ) *be a smooth reduction of* τM^n. *Then* M^n *admits a smoothing, such that the reduction of* τM *given by the smoothing is equivalent to* (ε, φ).

PROOF. Cover M by a countable locally finite family $\{U_\alpha\}$, $\alpha = 1, 2, \ldots$, of coordinate neighborhoods, and let $\{V_\alpha\}$, $\bar{V}_\alpha$ compact, be a shrinking of this cover. Let $M_k = \bigcup_{\alpha=1}^k \bar{V}_\alpha$. We will inductively define a smoothing on a neighborhood of $M_k - P_k$, P_k a finite subset, so that the reduction of the tangent bundle defined by the smoothing is equivalent to the given one.

First apply the above lemma to $U_1 = E^n$, with $K = \emptyset$. We get a smoothing of U_1 with the desired property ($P_1 = \emptyset$ since there are no bounded components).

Now suppose we have a smoothing of a neighborhood W_k of M_k, minus a finite number of points P_k, defining a reduction of the tangent bundle equivalent to $(\varepsilon, \varphi) \mid W_k$. Let $\tilde{P}_k$ be an open collar neighborhood of P_k; i.e. a union of small open balls, centered about each point of P_k. Let $M_k^0 = M_k - \tilde{P}_k$. Then M_k^0 is compact, and $W_k^0 = W_k - P_k$ is a neighborhood of M_k^0.

Consider $\bar{V}_{k+1} \cap M_k^0$. This is a compact subset of the smooth manifold $U_{k+1} \cap W_k^0$. Let K be a compact smooth submanifold of $U_{k+1} \cap W_k^0$ containing $\bar{V}_{k+1} \cap M_k^0$ in its interior. Now apply the above lemma to $U_{k+1} = E^n$ and K. We get a smoothing of U_{k+1} minus a finite number of points, which agrees with the given smoothing on a neighborhood of K. Choose sufficiently small neighborhoods U'_{k+1} of $\bar{V}_{k+1}$ in U_{k+1}, and W'_k of M^0_k in W_k, so that

$$U'_{k+1} \cap W'_k \subset \operatorname{Int} K.$$

Let

$$W_{k+1} = U'_{k+1} \cup W'_k \cup \tilde{P}_k.$$

Since $(W'_k \cup \tilde{P}_k) - P_k$ can be isotoped into W'_k, rel a neighborhood of M^0_k, we have a smoothing of

$$W^0_{k+1} = W_{k+1} - P_{k+1},$$

on a neighborhood of M_k minus P_k.

We need to show that the smoothing α of W^0_{k+1} defines a reduction of τW^0_{k+1} equivalent to $(\varepsilon, \varphi) \mid W^0_{k+1}$. Let N be the topological normal bundle of M^n in a large Euclidean space, so that ε is a smooth microbundle over N and $\varphi : \varepsilon \to \tau M$ covers $p : N \to M$.

By the induction assumption, there is a homotopy

$$u_t : \tau(W^0_k)_\alpha \to \varepsilon \mid p^{-1}(W^0_k)$$

from $\varphi^{-1} \mid W^0_k$ to a smooth microbundle map. On the other hand, the reduction $(\varphi_{k+1}, \varepsilon_{k+1})$ over $U_{k+1} = E^n_0$ is given by taking a homotopy

$$f_t : U_{k+1} \to p^{-1}(U_{k+1})$$

of $i \mid U_{k+1}$ to a smooth map, and then covering f_t by a microbundle homotopy

$$\sigma_t : f^*_1 \varepsilon \to \tau U_{k+1}$$

with $\sigma_1 = \varphi f_{1*}$. Then $\varepsilon_{k+1} = f^*_1$ and $\varphi_{k+1} = \sigma_0$.

Let $h_t : \tilde{K}_\alpha \to \tilde{K}_0$ be a homotopy of the identity to a smooth map. Then

$$f_t \circ h_t : \tilde{K}_\alpha \to p^{-1}(U_{k+1})$$

is a homotopy of $i \mid \tilde{K}$ to a smooth map. Since $u_t \mid \tilde{K}_\alpha$ also covers a homotopy from $i \mid \tilde{K}$ to a smooth map into $p^{-1}(W^0_{k+1})$, we can deform u_t so that $u_t \mid \tilde{K}$ covers $f_t h_t$, changing u_t only over an arbitrarily small neighborhood of K. Then $u_t \mid \tilde{K}$ can be factored,

$$u_t \mid \tilde{K} = f_{t*} \circ u'_t$$

and we get the commutative diagram

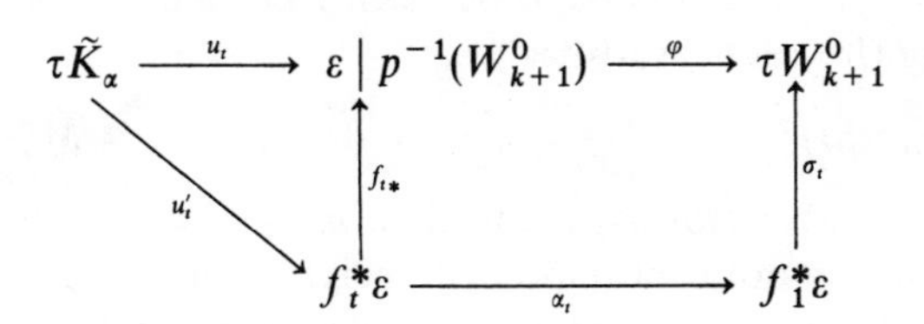

where α_t is a microbundle equivalence with $\alpha_1 = 1$.

Let $\lambda_t = \alpha_t u_t'$. Then

$$\varphi \circ u_t = \sigma_t \lambda_t \quad \text{on } \tau \tilde{K}_\alpha.$$

Thus

$$\sigma_0 \lambda_0 = \varphi \circ u_0 = 1 \quad \text{and} \quad \lambda_0 = \varphi_{k+1}^{-1} \mid \tilde{K}.$$

Also

$$\lambda_1 = \alpha_1 u_1' = u_1',$$

a smooth microbundle map. By the addendum to the lemma, λ_t extends to

$$\bar{\lambda}_t : \tau(U_{k+1}^0)_\alpha \to f_1^* \varepsilon, \qquad \bar{\lambda}_0 = \varphi_{k+1}^{-1}, \quad \bar{\lambda}_1 \text{ smooth}.$$

Extend u_t to W_{k+1}^0 by setting,

$$u_t = f_{t*} \alpha_t^{-1} \bar{\lambda}_t \quad \text{on } U_{k+1}^0.$$

This completes the inductive step.

Thus we get a smoothing of $M - P_\infty$, where P_∞ is a countable subset, and only a finite number of points of which are in any compact set. Further, the reduction of the tangent bundle of $M - P_\infty$ given by the smoothing is equivalent to $(\varepsilon, \varphi) \mid M - P_\infty$. As in §2 of Part I, there is an isotopy of the identity map of M, deforming M into $M - P_\infty$, $n \geqq 3$. This gives a smoothing of M which defines a reduction of the tangent bundle equivalent to (ε, φ). (Actually, the argument works for $n = 2$, using say the two-dimensional annulus theorem. For $n = 1$, the theorem is trivial.)

Chapter 7. The uniqueness theorem.

DEFINITION. Let M^n be a topological manifold. Then two smoothings (V_α, g_α) and (V_β, g_β) are called *sliced concordant* if there exists a smoothing (V, h) of $I \times M$ such that

(a) $p_1 \circ h$ is regular and hence $h^{-1}(t \times M)$ is a smooth submanifold of V for each $t \in I$.

(b) $(h_0^{-1}(M), h_0)$ and $(h_1^{-1}(M), h_1)$ are equivalent to (V_α, g_α) and (V_β, g_β) resp.

PROPOSITION 1. *Let (V_α, g_α) and (V_β, g_β) be two smoothings of an open topological manifold M which define equivalent reductions of τM; then (V_α, g_α) and (V_β, g_β) are sliced concordant.*

PROOF. Identify the underlying spaces of V_α and V_β with M by g_α and g_β resp; and denote the smoothings by M_α and M_β. Then our assumption implies $1_{\tau M} : \tau M_\alpha \to \tau M_\beta$ is homotopic to a smooth microbundle map φ. Then φ induces a smooth immersion $f : M_\alpha \to M_\beta$ such that $df \sim \varphi$. Since $\varphi \sim 1_{\tau M}$, there is a regular homotopy $h : I \times M \to I \times M$ between f and 1_M.

The smoothing $(I \times M)_{h*\beta}$ satisfies

$$h_0^{-1}(M) = f^{-1}(M) = M_\alpha \quad \text{and} \quad h_1^{-1}(M) = M_\beta.$$

Since $h_t : M \to M_\beta$ is a topological immersion, $h_t^{-1}(M)$ is a smooth submanifold. In fact, $h_t^{-1}(M)$ has smooth local product neighborhoods in $(I \times M)_{h*\beta}$, since this is true for $t \times M_\beta$ in $I \times M$ and h is a local homeomorphism commuting with projections on I.

REMARK. It is easy to show the converse is true: If (V_α, g_α) and (V_β, g_β) are sliced concordant smoothings of M, then they define equivalent reductions of τM.

DEFINITION. Let M_α and M_β be smoothings of M. Two homotopies $\theta_t, \psi_t : \tau M_\alpha \to \tau M_\beta$ from smooth microbundle maps to the identity $1_{\tau M}$, will be called *strongly homotopic* if the two homotopies are themselves homotopic through such homotopies.

REMARK. Let $K \subset M$ and let $\theta_t, \psi_t : \tau \tilde{K}_\alpha \to \tau \tilde{K}_\beta$ be two homotopies from smooth microbundle maps to the identity, which are strongly homotopic. If θ_t extends over M_α with the same property, then θ_t over M_α is strongly homotopic to an extension of ψ_t over M_α.

ADDENDUM TO PROPOSITION 1. *Let $\theta_t : \tau M_\alpha \to \tau M_\beta$ be the homotopy from a smooth microbundle map to 1_M given by the equivalence of the reductions induced by M_α and M_β. Then θ_t is strongly homotopic to dh_t.*

PROOF. Let ψ_t be the homotopy

$$df \sim \varphi = \theta_1 \sim \theta_0 = 1_{\tau M}.$$

The first homotopy is smooth and hence ψ_t is strongly homotopic to θ_t. On the other hand, dh_t is homotopic rel endpoints to ψ_t.

LEMMA. *Let M^n, $n \geqq 5$, be a smooth (PL) manifold. Let M be homeomorphic to $K \times R$, where K is a closed connected topological manifold. If there is a smooth (locally flat PL) submanifold $N^{n-1} \subset M$, such that N is a deformation retract of M, then M is diffeomorphic (PL equivalent) to $N \times R$, with N corresponding to $N \times 0$.*

PROOF. First consider the PL case:

By Siebenmann [**8**], there are arbitrarily small $(n - 3)$ neighborhoods V of each end $\pm\varepsilon$ of M. Then, in particular, $\pi_1(\varepsilon) \cong \pi_1(V) \cong \pi_1(M)$. Also, for some $t \in R$, we have $K \times t \subset V \subset M$. Hence $\pi_i(V) \to \pi_i(M)$ is onto, all i. From the exact homotopy sequence of (M, V) we see that $\pi_j(M, V) = 0$ for $j \leqq 2$.

Now N splits M into two PL submanifolds M_i, $i = 1, 2$, with boundary N and one end. It is easy to check that each M_i is homotopy equivalent to M; and hence for V in M_i, $\pi_j(M_i, V) = 0, j \leqq 2$. It now follows by the Stallings engulfing argument [**11**], that

$$M_1 = N \times [0, \infty), \qquad M_2 = N \times [0, -\infty)$$

and hence $M = N \times R$.

The smooth case follows from the PL case by applying the Cairns-Hirsch Theorem [**2**].

REMARK. The above argument shows that it is sufficient to assume M is the proper homotopy type of $K \times R$ to obtain the above result.

COROLLARY. *Let M_α and M_β be smoothings of a connected topological manifold M^n, $n \geqq 5$. Suppose that M is homeomorphic to the open collar neighborhood of a compact submanifold N^n. Suppose also that M_α has a smooth collar neighborhood* *of its end. Let h be a homeomorphism of M_α onto M_β which is smooth on a neighborhood N_1 of N. Then h is isotopic to a diffeomorphism, rel a neighborhood of N.*

PROOF. $M = N \cup \partial N \times [0, 1)$. Let the smooth collar neighborhood of the end of M_α be

$$W \cong \partial W \times [0, 1).$$

There is a smooth copy of ∂W in $(\partial N \times (0, 1))_\alpha$, and by the lemma, $(\partial N \times (0, 1))_\alpha$ is diffeomorphic to $\partial W \times (0, 1)$. Consequently, there is a smooth copy $\partial W \times \varepsilon$ of ∂W in $N_1 \cap (\partial N \times (0, 1))_\alpha$. Again by the lemma, $h(\partial N \times (0, 1))$ is diffeomorphic to $\partial W \times (0, 1)$ so that $h(\partial W \times \varepsilon)$ corresponds to $\partial W \times \varepsilon$. By the uniqueness of topological collars [1], h is isotopic to a diffeomorphism, rel a neighborhood of N.

PROPOSITION 2. *Let (V_α, g_α) and (V_β, g_β) be two smoothings of M^n, $n \geqq 5$, which define equivalent reductions of M. Suppose that V_α is diffeomorphic to the open collar neighborhood of a compact smooth submanifold N_α. Then the two smoothings are equivalent.*

Further, if $g_\beta^{-1} g_\alpha$ is already smooth in a neighborhood of a compact smooth submanifold N_α^0 in the interior of N, then the isotopy of g_α to $\bar{g}_\alpha$ so that $g_\beta^{-1}\bar{g}_\alpha$ is smooth, can be taken fixed on a neighborhood of N^0, if (1) the equivalence of the reductions of τM extends the equivalence over N^0 given by $g_\beta^{-1} g_\alpha$ being smooth on a neighborhood of N^0, (2) every component of $M - N^0$ has noncompact closure.

PROOF. Again identify V_α and V_β with M by g_α and g_β, and denote the smoothings by M_α and M_β. Then by Proposition 1, M_α is sliced concordant to M_β, and the sliced concordance comes from a regular homotopy

$$h: I \times M \to I \times M_\beta.$$

Further, we may assume h_t = identity on a neighborhood of N^0 since the smooth immersion $f: M_\alpha \to M_\beta$ of Proposition 1 may be taken with $f = g_\beta^{-1} g_\alpha$ (i.e. the identity) on a neighborhood of N^0.

By the corollary to the Proposition of §3, for any $s \in I$, there is an $\varepsilon > 0$, and a neighborhood N_1 of N, such that $h_t \mid N_1$ can be factored

$$h_t = h_s \circ k_t \quad \text{for } |t - s| < \varepsilon;$$

where $k_t: N_1 \to M$ is an isotopy and k_s is the inclusion. By the isotopy extension theorem, §4, we may assume k_t is an ambient isotopy of M such that $h_t = h_s \circ k_t$ on N_1 and k_s is the identity; i.e.

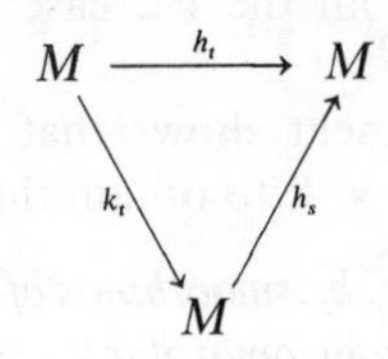

commutes on N_1.

Now

$$k_t : M_{h_t^* \beta} \to M_{h_s^* \beta}$$

is smooth on N_1. If either $M_{h_t^* \beta}$ or $M_{h_s^* \beta}$ is equivalent to M_α, then k_t is isotopic to a diffeomorphism rel a neighborhood of N, by the corollary of the lemma.

Cover I, by a finite number of such intervals. Then we can choose $0 = s_0 < t_0 < s_1 < t_1 \cdots < t_r < s_{r+1} = 1$ with ambient isotopies k^i, such that

$k^i_{s_i}$ = identity and

k^i_t is defined for $t_{i-1} \leqq t \leqq t_i$, with $h_t = h_{s_i} \circ k^i_t$ on N_1.

By the above,

$$M_\alpha = M_{h^*_{s_0}} \cong M_{h^*_{t_0}} \cong M_{h^*_{s_1}} \cong \cdots \cong M_{h^*_{s_{r+1}}} = M_\beta.$$

Further, each succeeding isotopy is fixed on a neighborhood of N^0, and so therefore the composite isotopy.

ADDENDUM 1 TO PROPOSITION 2. *Let*

$$\theta_t : \tau M_\alpha \to M_\beta$$

be the homotopy of a smooth microbundle map to $1_{\tau M}$ given by the equivalence of the reductions induced by M_α and M_β. Then if j is the ambient isotopy of M from a diffeomorphism of M_α onto M_β to the identity given by Proposition 1, d_{j_t} is strongly isotopic to θ_t.

PROOF. Proposition 2 shows that we can choose a finite sequence $0 = t_0 < t_1 < \cdots < t_n = 1$, and ambient isotopic k^i from a diffeomorphism of M_{t_i} onto $M_{t_{i+1}}$ to the identity, such that:

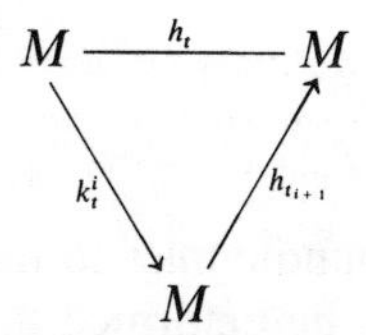

commutes on $N_i \subset M$, $t_i \leqq t \leqq t_{i+1}$, where M_{t_i} is the smoothing on M induced from M_β by h_{t_i}, and N_i is a deformation retract of M.

Inductively, define j^i, an ambient isotopy from a diffeomorphism of M_{t_i} onto $M_{t_n} = M_\beta$ to the identity, by

$$j^i_t = j^{i+1}_{t_{i+1}} \circ k^i_t, \qquad t_i \leqq t \leqq t_{i+1},$$

and

$$j^i_t = j^{i+1}_t \quad \text{for } t \geqq t_{i+1}.$$

Then $j = j^0$. Also

$$j_t = j^i_t \quad \text{for } t \geqq t_i.$$

Then

$$dh_t = dh_{t_{i+1}} \circ dk^i_t = dh_{t_{i+1}} \circ (dj^{i+1}_{t_{i+1}})^{-1} \circ df_t$$

on N_i, for $t_i \leqq t \leqq t_{i+1}$.

Also, for $t = t_i$,

$$dh_{t_i}:\tau M_{t_i} \to \tau M_\beta \quad \text{and} \quad dj_{t_i}:\tau M_{t_i} \to \tau M_\beta$$

are smooth, and, for $t = t_{i+1}$,

$$dh_{t_{i+1}}:\tau M_{t_{i+1}} \to \tau M_\beta \quad \text{and} \quad dj_{t_{i+1}}:\tau M_{t_{i+1}} \to \tau M_\beta$$

are smooth. Therefore

$$dh_{t_i}(dj_{t_i})^{-1}:\tau M_\beta \to \tau M_\beta$$

is a smooth bundle map which equals $dh_{t_{i+1}} \circ (dj^{i+1}_{t_{i+1}})^{-1}$ on N_i, and hence these two bundle maps are smoothly homotopic rel N_i. Similarly at t_{i+1}. Again since N_i is a deformation retract of M, we can deform d_{j_t}, $t_i \leqq t \leqq t_{i+1}$, rel endpoints, to γ_t such that $dh_t = \alpha_t \circ \gamma_t$, where $\alpha_t:\tau M_\beta \to \tau M_\beta$ is a smooth bundle map.

Doing this between each t_i and t_{i+1}, we can deform j_t, $0 \leqq t \leqq 1$, rel endpoints to γ_t, where $dh_t = \alpha_t \circ \gamma_t$ and $\alpha_t:\tau M_\beta \to \tau M_\beta$ is smooth. Since $dh_1 = dj_1 = 1_{\tau M}$ α_t may be itself homotoped to the identity homotopy by a smooth deformation fixed at $t = 1$. Therefore j_t is strongly homotopic to h_t.

ADDENDUM 2. *Let M_α and M_β be two smoothings of M^n which induce equivalent reductions of τM. Let $N^n_\alpha \subset M^n_\alpha$ be a compact smooth submanifold. Let $\tilde{N}$ be an open collar neighborhood of N. Then $\tilde{N}_\alpha$ and $\tilde{N}_\beta$ induce equivalent reductions of $\tau\tilde{N}$. Let j be the ambient isotopy of $\tilde{N}$ from a diffeomorphism of $\tilde{N}_\alpha$ onto $\tilde{N}_B$ to the identity, given by Proposition 2. Extend j to an ambient isotopy of M, and let $M_\gamma = M_{j_0^*\beta}$. Then M_γ satisfies:*

(a) $\tilde{N}_\alpha = \tilde{N}_\gamma$.

(b) *There is an isotopy $\psi_t:\tau M \to \tau M_\gamma$ from a smooth bundle map to the identity so that $\psi_t \mid \tilde{N}$ is the identity.*

PROOF. Let

$$\theta_t:\tau M_\alpha \to \tau M_\beta$$

be a homotopy from a smooth bundle map to the identity. Then if $h_t:M_\alpha \to M_\beta$ is the map covered by θ_t, h_t may be deformed to h'_t so that $h'_t(\tilde{N}) \subset (\tilde{N})$, and the deformation from h_0 to h'_0 is smooth and from h_1 to h'_1 is fixed. Then we can cover this deformation by a strong homotopy of θ_t to θ'_t. But then

$$\theta'_t \mid \tau\tilde{N}:\tau\tilde{N}_\alpha \to \tau\tilde{N}_\beta$$

shows that $\tilde{N}_\alpha$ and $\tilde{N}_\beta$ induce equivalent reductions of the tangent bundle.

M_γ is equivalent to M_β by the extended j. Then $dj_t \mid \tilde{N}$ is strongly isotopic to $\theta'_t \mid \tilde{N}$ by Addendum 1, and we can deform θ'_t to θ''_t so that it agrees with dj_t on $\tilde{N}$. Then

$$\psi_t = \theta''_t \circ dj_t^{-1}$$

is a homotopy from a smooth bundle map of τM_α into τM_γ to the identity, with

$$\psi_t \mid \tilde{N} = \text{identity}.$$

THEOREM 1. *Let (V_α, g_α) and (V_β, g_β) be two smoothings of an open topological manifold M^n, $n \geqq 5$, which define equivalent reductions of τM. Then the two smoothings are equivalent.*

PROOF. An open smooth manifold $V = \bigcup_{i=1}^{\infty} N_\alpha^i$, where N_α^i is a compact smooth n-dimensional submanifold, and $N_\alpha^i \subset \operatorname{Int} N_\alpha^{i+1}$. Further, we can assume that if V_α^i is an open smooth collar neighborhood of N_α^i in $\operatorname{Int} N_\alpha^{i+1}$, then every component of $V_\alpha^{i+1} - N_\alpha^i$ has noncompact closure. The result then follows by induction using Proposition 2 and addendums.

THEOREM 2. *Let M^n be a closed topological manifold $n \geqq 5$ and let (ε, φ) be a reduction of τM to a vector bundle. Then M^n admits a smoothing, such that the reduction of τM, given by the smoothing is equivalent to (ε, φ).*

PROOF. Apply the theorem of §6 to $M - p$, p a point in M. We then get a smoothing $(M - p)_\alpha$ defining a reduction of $\tau M \mid M - p$ equivalent to $(\varepsilon, \varphi) \mid M - p$. Let U be a coordinate neighborhood of p. The reduction of $\tau(U - p)$ given by α, extends to a reduction of τU. This gives a smoothing U_β of U by the theorem of §6, such that the reduction of $\tau(U - p)$ given by β is equivalent to that given by α. By Theorem 1, $(U - p)_\alpha$ is equivalent to $(U - p)_\beta$.

Thus there is a homeomorphism h of U into U_β which is a diffeomorphism of $(U - p)_\alpha$ into $(U - p)_\beta$ when restricted to $U - p$. Thus the smoothing of $M - p$ extends to a smoothing of M. It follows by Addendum 2 of Proposition 2 that this smoothing gives the desired equivalence class of reductions of τM.

THEOREM 3. *Let (V_α, g_α) and (V_β, g_β) be two smoothings of a closed topological manifold M^n, $n \geqq 5$, which define equivalent reductions of τM. Then the two smoothings are equivalent.*

PROOF. Apply Theorem 1 to $M - p$, then $(M - p)_\alpha$ is equivalent to $(M - p)_\beta$. Let U be a coordinate neighborhood of p. Consider $I \times U$. Put the smoothing α on $0 \times U$, and β on $1 \times U$. Using the equivalence of $(U - p)_\alpha$ with $(U - p)_\beta$ we may extend this smoothing to a smoothing on $I \times U - q$, $q = \frac{1}{2} \times p$. Also, the reduction of the tangent bundle of $I \times U - q$ given by the smoothing extends to $I \times U$. By the argument of Theorem 2, we can extend our smoothing to a smoothing of $I \times U$. By the h-cobordism theorem, the diffeomorphism of $(U - p)_\alpha$ into $(U - p)_\beta$ extends to a diffeomorphism of U_α into U_β. Since any two homeomorphisms of a disc which agree on the boundary are isotopic, the identity map of M is isotopic to a diffeomorphism of M_α onto M_β.

REMARK. For $n = 4$, it can be shown that if the tangent bundle τM of a closed M^4 reduces to a smooth microbundle, then $M \# k(S^2 \times S^2)$ is smoothable, (i.e. the connected sum with k copies of $S^2 \times S^2$) some k. Further, two smoothings M_α, M_β of a closed M^4 which induce equivalent reductions of τM, satisfy: For some k, $M_\alpha \# k(S^2 \times S^2)$ is equivalent to $M_\beta \# k(S^2 \times S^2)$.

(These results will appear in a forthcoming paper with J. Shaneson.)

It remains to show when τM reduces to a smooth or PL bundle. Since the equivalence classes of reductions correspond to homotopy classes of lifts of the classifying map

$$M \to B\,\mathrm{Top}_n$$

to BO_n or $B\,\mathrm{PL}_n$, they are determined by the homotopy groups of Top_n/O_n or $\mathrm{Top}_n/\mathrm{PL}_n$.

THEOREM.

$$\pi_i(\mathrm{Top}_n/\mathrm{PL}_n) = 0 \quad \text{for } i \leqq n, i \neq 3, n \geqq 5.$$

$$\pi_3(\mathrm{Top}_n/\mathrm{PL}_n) = Z_2 \quad \text{for } n \geqq 5.$$

This result depends on work of Hsiang and Shaneson [3], Wall [9] and Kirby and Siebenmann [4]. The argument is outlined in [4] and [6].

In general (loc. cit.),

THEOREM.

$$\pi_i(\mathrm{Top}_n/\mathrm{PL}_n) \cong \pi_i(\mathrm{Top}/\mathrm{PL}), \qquad i \leqq n, n \geqq 5.$$

$$\pi_i(\mathrm{Top}_n/O_n) \cong \pi_i(\mathrm{Top}/O), \qquad i \leqq n, n \geqq 5.$$

REFERENCES

1. M. Brown, *Locally flat imbeddings of topological manifolds*, Ann. of Math. (2) **75** (1962), 331–341. MR **24** #A3637.

2. M. W. Hirsch, *On combinatorial submanifolds of differentiable manifolds*, Comment. Math. Helv. **36** (1961), 103–111. MR **24** #A3658.

3. W. C. Hsiang and J. Shaneson, *Fake tori, the annulus conjecture, and the conjectures of Kirby*, Proc. Nat. Acad. Sci. U.S.A. **62** (1969), 687–691.

———, *Fake tori*, Proc. Georgia Conference on Topology of Manifolds, 1969.

4. R. C. Kirby and L. C. Siebenmann, *On the triangulation of manifolds and the Hauptvermutung*, Bull. Amer. Math. Soc. **75** (1969), 742–749. MR **39** #3500.

5. R. Lashof, *Lees' immersion theorem and the triangulation of manifolds*, Bull. Amer. Math. Soc. **75** (1969), 535–538. MR **39** #960.

6. R. Lashof and M. Rothenberg, *Triangulation of manifolds*. I, II, Bull. Amer. Math. Soc. **75** (1969), 750–754; ibid. **75** (1969), 755–757. MR **40** #895.

7. J. Milnor, *Microbundles*, Proc. Internat. Congress Math. (Stockholm, 1962), Inst. Mittag-Leffler, Djursholm, 1963.

8. L. Siebenmann, *The obstruction to finding a boundary for an open manifold of dimension greater than five*. Thesis, Princeton University, Princeton, N.J., 1965.

9. C. T. C. Wall, *On homotopy tori and the annulus theorem*, Bull. London Math. Soc. **1** (1969), 95–97. MR **39** #3498.

10. R. Lashof, *The immersion approach to triangulation*, Proc. Georgia Conference on Topology of Manifolds, Markham, Chicago, Ill., 1970.

11. J. F. P. Hudson, *Piecewise linear topology*, Benjamin, New York, 1969.

UNIVERSITY OF CHICAGO

SOME REMARKS ON THE KERVAIRE INVARIANT PROBLEM FROM THE HOMOTOPY POINT OF VIEW

M. E. MAHOWALD[1]

The object of this note is to discuss some results which were obtained in an effort to settle the Kervaire invariant conjecture.

There is a secondary cohomology operation based on the Adem relation which expands $Sq^{2^j}Sq^{2^j}$. Call this $\varphi_{j,j}$, after Adams.

CONJECTURE A. There exists a two cell complex $S^n \cup e^{n+N+1}$ $(N = 2^{j+1} - 2)$ with $\varphi_{j,j}$ nonzero. We call any such element θ_j. (Note that this defines only a coset.)

Browder has shown that this conjecture is equivalent to the existence of a framed manifold with nonzero Kervaire invariant.

There are many statements which imply the conjecture. Most are conjectured to be equivalent. Let us begin with the simplest.

Suppose $X = S^0 \cup_{2^\iota} e^1$, that is, the space in the stable category which represents $\Sigma^{-1}RP^2$.

THEOREM 1. *An element of Hopf invariant* 1 *in* $\Pi_{2^{j+1}-1}(X)$ *implies* A *in* dim $2^{j+1} - 2$.

PROOF. An element of Hopf invariant one implies a three cell complex so that $Sq^{2^{j+1}} \neq 0$. Adams has shown that $Sq^{2^{j+1}} = \sum a_{i,k,j}\varphi_{i,k}$. We apply this to the Spanier-Whitehead dual of the complex and conclude $\varphi_{j,j} \neq 0$ and $a_{j,j,j} = Sq^1$.

COROLLARY 2. *If* $[\iota_N, \iota_N] = 2\alpha$, $N = 2^{j+1} - 1$, *then* $\alpha = \theta_j$.

We can use Theorem 1 to try and construct the θ_j's inductively. Indeed,

---

AMS 1970 *subject classifications.* Primary 57D15, 55E45; Secondary 55G20, 55E50.

[1] These notes are based on the joint work of M. G. Barratt and M. E. Mahowald.

suppose there is an element in dimension $2^j - 1$ of Hopf invariant 1. Let us call it $\{h_j\}$.

PROPOSITION 3. $2\{h_j\} = \eta\theta_{j-1}$.

This follows immediately from Theorem 1 and a standard relationship in the homotopy of X.

Thus $\{h_j\}$ admits an extension by 2 modulo η. This gives a map

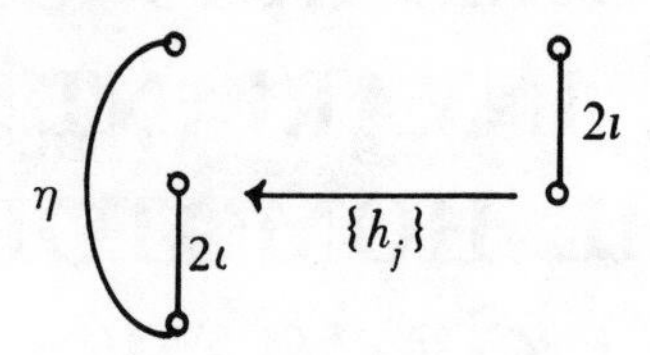

This clearly defines a map

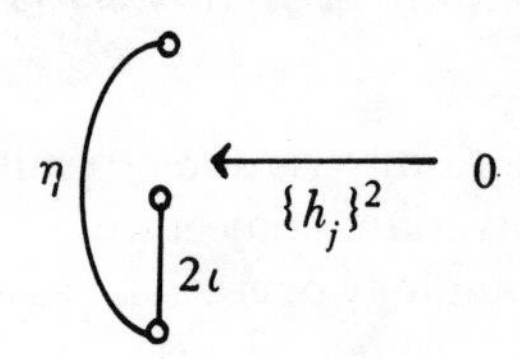

and a four cell space

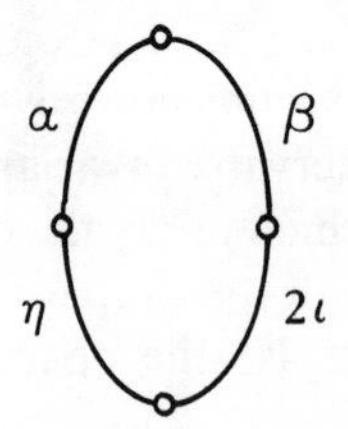

with attaching maps $\alpha = \theta_{j-1}^2$ and $\beta = \langle\theta_{j-1}, 2\iota, \theta_{j-1}\rangle$ such that $\varphi_{j,j}$ is nonzero from the bottom to the top cell. Summarizing, we have shown:

THEOREM 4. *There is a null-homotopy defined on S^{2^j} of*

$$\eta\theta_{j-1}^2 + 2\iota\langle\theta_{j-1}, 2\iota, \theta_{j-1}\rangle$$

which carries a $\varphi_{j,j}$.

COROLLARY 5. *If there is a non-$\varphi_{j,j}$ carrying null-homotopy of*

$$\eta\theta_{j-1}^2 + 2\iota\langle\theta_{j-1}, 2\iota, \theta_{j-1}\rangle$$

then θ_j exists.

Relevant to this is the following canonical relation.

PROPOSITION 6. *If $\alpha \in \Pi_k^S$, $2\alpha = 0$, $k \equiv 2 \pmod 4$, then $\langle\alpha, 2\iota, \alpha\rangle = 0$.*

Thus Corollary 5 has a weaker version.

COROLLARY 5'. *If there is a non-$\varphi_{j,j}$ carrying null-homotopy of $\eta\theta_{j-1}^2$, then θ_j exists.*

Thus we have the first part of the following theorem.

THEOREM 7. *If θ_{j-1} exists, $2\theta_{j-1} = 0 = \theta_{j-1}^2$ then θ_j exists and $2\theta_j = 0$.*

The second part follows by a similar analysis.

Another proof of Theorem 4 is based on the smash product. Let $S^{2^j-1} \to X$ represent $\{h_j\}$. Then $S^{2^j-1} \wedge S^{2^j-1} \to X \wedge X$ represents $\{h_j^2\}$ and this gives a map

$$S^{2^{j+1}-2} \longrightarrow \quad \eta \quad \longrightarrow$$

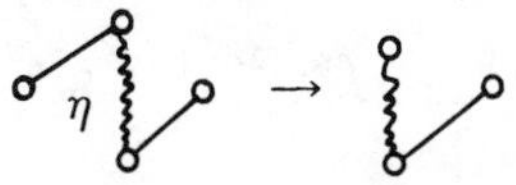

as before. This proof is a version of the central idea of our approach. The philosophy is to use the existence of $\theta_{j'}^{j'+j}$ and apply a functorial construction which hopefully gives θ_j. The Γ construction discussed earlier by Barratt is an example. It contains the "quadratic" construction and higher symmetries. In particular, explicit construction of the 30-manifold using $\mathscr{S}_4$, the symmetric group on four letters, has been given.

Milgram, using $\mathscr{S}_4$ symmetries, has proved the following:

THEOREM 8 [MILGRAM]. *With the hypothesis of Theorem 7, θ_{j+1} exists.*

REMARK. It can be shown that $\theta_4^2 = 0$ and thus Milgram's theorem implies θ_6 exists.

A more delicate argument about θ_{j-1} shows

PROPOSITION 9. *$\langle \theta_{j-1}, 2\iota, \theta_{j-1} \rangle_n = [\iota_n, \beta_{j-1}]$, where $n = 2^{j+1} - \varphi(j-1) - 1$, β_{j-1} is the generator of the* im J *in* stem *$\varphi(j-1) - 1$ and $\varphi(j-1)$ is the Adams function.*

COROLLARY 10. *$\Sigma^{-\varepsilon} 2 \langle \theta_{j-1}, 2\iota, \theta_{j-1} \rangle_n = [\iota_{2^{j+1}-\varphi(j)-1}, \beta_j]$ where $\varepsilon = \varphi(j) - \varphi(j-1)$.*

Thus

THEOREM 11. *If θ_{j-1} exists and has order 2 and θ_j exists, θ_j appears on the $2^{j+1} - \varphi(j)$ sphere with Hopf invariant β_j.*

There is a slightly less direct approach. First observe that there is a map $\lambda: RP \to S^0$ in the stable category. λ on each cell is the Whitehead product. To be precise, there is a map $\Sigma^n P^{n-1} \xrightarrow[\lambda_n]{} S^n$ and $S^{2n-1} \xrightarrow[\Sigma^n a_n]{} \Sigma^n P^{n-1}$ where a_n is the natural map $S^{n-1} \to P^{n-1}$. The composite $\lambda_n \Sigma^n a_n = [\iota_n, \iota_n]$. Thus

PROPOSITION 12. *If $\Sigma^n a_n$ can be halved for $n = 2^{j+1} - 1$ then θ_j exists.*

There is a similar statement for CP and QP.

Consider the situation with just a single suspension. We have the following fibration:

$$P * P \to \Sigma RP \to K(Z_2, 2).$$

PROPOSITION 13. *If there is a map* $f: S^{2^{j+1}-1} \to P * P$ *so that*

$$f^*(\alpha^{2^j-1} * \alpha^{2^j-1}) \neq 0$$

then θ_j *exists.*

A weaker version is also true.

PROPOSITION 13′. *If there is a stable map* $f: S^{2^{j+1}-2} \to P \wedge P$ *so that*

$$f^*(\alpha^{2^j-1} \wedge \alpha^{2^j-1}) \neq 0$$

then θ_j *exists.*

Even weaker versions than this are possible.

PROPOSITION 13″. *Let* v_j *be a cohomology class in* SO *which transgresses to* w_{2^j}. *Then* $v_j \otimes v_j$ *being spherical in* $SO * SO$ *or in* $S(SO \wedge SO)^*$ *implies Conjecture* A.

Another approach stems from the effort to construct large brackets.

THEOREM 14 (HOFFMAN). *If* $\langle \sigma, 2\sigma, 2\sigma, \ldots, 2\sigma, \sigma \rangle$ *can be defined then* θ_j *is in it.*

This is verified for $j = 4$. There is a family of spaces X_k which are defined by identifying particular subspaces of $\Lambda^k(S^8 \cup_\sigma e^{16})$. The cell structure looks like

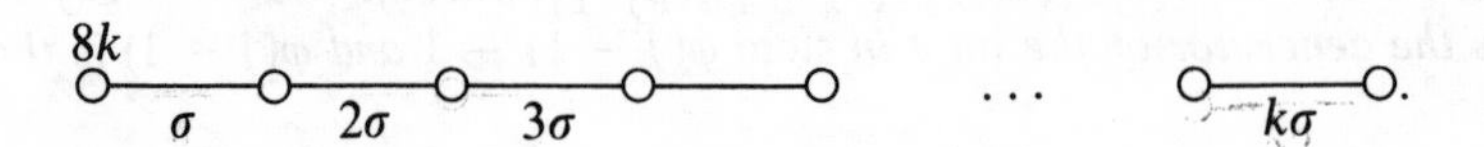

This shows $0 \in \langle \sigma, 2\sigma, \ldots, (k-1)\sigma, k\sigma \rangle$.

THEOREM 15. *If the bracket* $\langle \sigma, 2\sigma, \ldots, (2^{j-2}-1)\sigma, 2^{j-3}\sigma \rangle$ *can be formed then* θ_j *is in it.*

This requires a mild interpretation because $16\sigma = 0$, but it is not hard to see what should be done.

A paraphrase of Theorem 15 is the question of whether the attaching map in the construction X_k can be halved.

Another amusing approach:

THEOREM 16. *If* $2\theta_{j-1} = 0 = 2\theta_j$, *and* $\langle \theta_{j-1}, 2\iota, \theta_j \rangle = 0$ *then* θ_{j+1} *exists.*

PROOF. Take

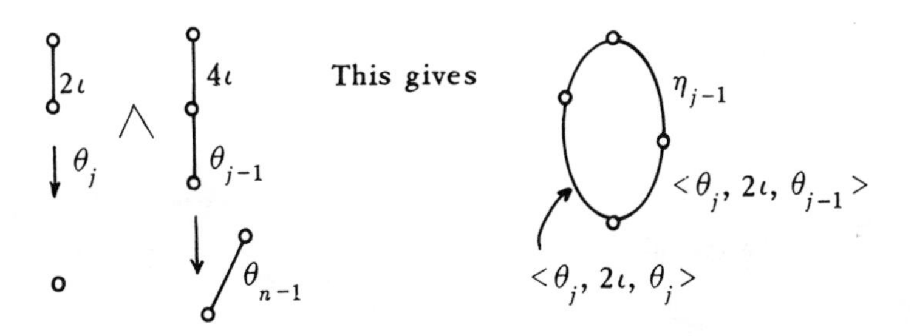

This construction is $\varphi_{j+1,j+1}$ carrying.

CONJECTURE. $\langle\theta_j, 2\iota, \theta_{j-1}\rangle = [\iota_n, \beta_{j-1}]$ where $n = 2^j + 2^{j-1} - \varphi(j-1)$. (Although $\langle\theta_4, 2\sigma, \sigma\rangle = 0$, it is not known whether $\langle\theta_4, 2\iota, \sigma\sigma\rangle = 0$.)

The strongest evidence for the existence of the θ_j's is

THEOREM 17. *There exists a sequence of integers n_i such that $2^i - 2 < n_i \leqq 2^{i+1} - 2$ and elements $\alpha_i \in \Pi^S_{n_i}$. If $n_i = 2^{i+1} - 2$ then the constructed elements are θ_i.*

PROOF. Simply stated, the α_i are the stable Hopf invariants of the β_i in stem $2^{i+1} - \varphi(i) - 1$. To be more precise consider the X_{2^j-2} constructing. The attaching map of the cell in dimension 2^{j+2} can be halved at least through the $2^{j+2} - 2^j$-skeleton. If it can be halved through the $8 + 2^{j+1}$-skeleton then we have θ_j, otherwise there is an obstruction. We call this obstruction α_j.

Another amusing result is the following.

Let $Y \to S^0 \to K(Z, 0)$ define Y. The map $\lambda: P \to S^0$ lifts to Y.

REMARK. Y/P has cohomology which is free over Sq^1 and Sq^2. Thus as far as *bo* homology is concerned, P and Y are equally interesting.

Consider the spectrum Im J defined by the fibration

$$\text{Im } J \to BO[8k, \ldots] \to BO[8k+4, \ldots].$$

THEOREM 18. $\Pi_i(P \wedge \text{Im } J) =$

$i =$ 0	1	2	3	4	5	−2	−1	0	1	2	3	4	5 (mod 8)
Z_2	Z_2	Z_8	Z_2	0	Z_2	Z_{λ_i}	Z_2^2	Z_2^2	Z_2	Z_8	Z_2	0	Z_2 (i)

where λ_i is the 2-primary order of the image of the J-homomorphism. (That is, if $i + 1 \equiv 2^{\rho(i)} \bmod 2^{\rho(i)+1}$ then $\lambda_i = 2^{\rho(i)+1}$.) If $\lambda: Y \to P \wedge \text{Im } J$, then λ_* on the elements in $\Pi_* Y$ given by the image of J, the μ's of Barratt and Adams, the elements η_j and θ_j generate the image of λ_*.

NORTHWESTERN UNIVERSITY

HOMOLOGY OPERATIONS ON INFINITE LOOP SPACES

J. PETER MAY

In the last few years, it has become clear that homology operations on infinite loop spaces play a very important role in algebraic topology. These operations are the fundamental tool in the study of the homology of such spaces as QX, BF, $B\,\mathrm{Top}$, etc. I recently observed that the language of topological PROP's developed by Boardman and Vogt in [2], allows extremely simple constructions of the operations and proofs of their properties. This approach to the operations was the subject of my talk at the summer institute, but is not yet ready for publication. Instead, I shall here summarize the basic algebraic results of the theory and shall briefly indicate the extent of our present information about the subject, including precise descriptions of the homology of F, F/O, and BF. The quoted results are due to various people. Only those results which are stated without historical references are due to the author; more complete statements and proofs will appear later. This paper is divided into three sections as follows:

(1) The Dyer-Lashof algebra and its dual,
(2) Allowable R-modules and $H_*(QX)$,
(3) The homology of F, F/O, BF, and B Top.

Much of this material has been previously circulated in preprint form, but some of the results in § 3 are new.

1. **The Dyer-Lashof algebra and its dual.** By an infinite loop sequence $B = \{B_i \mid i \geqq 0\}$, we understand a sequence of based spaces such that $B_i = \Omega B_{i+1}$; by a map $g: B \to C$ of infinite loop sequences, we understand a sequence of base-point preserving maps $g_i: B_i \to C_i$ such that $g_i = \Omega g_{i+1}$, $i \geqq 0$. B_0 and g_0 are then called (perfect) infinite loop spaces and maps. These notions are equivalent

AMS 1970 *subject classifications*. Primary 55G99, 55D35, 55F40, 57F25, 55-02; Secondary 55G10, 55F45, 57F35.

for the purposes of homotopy theory to the more usual ones in which equalities are replaced by homotopies [8]. We define $H_*(B) = H_*(B_0; Z_p)$ for some fixed prime p and regard H_* as a functor from the category of infinite loop sequences to that of graded Z_p-modules. $H_*(B)$ admits homology operations which are analogous to the Steenrod operations in cohomology. For $p = 2$, these operations were first introduced by Araki and Kudo [1]; their work was later simplified by Browder [4]. Dyer and Lashof [5] introduced the operations for odd primes and developed many of their algebraic properties; we shall therefore refer to the operations as Dyer-Lashof operations. The following theorem summarizes their properties. Parts (1) through (5) were proven by Dyer and Lashof, and part (6) was first observed by Milgram. The Adem relations (7) were implicit in the work of Dyer and Lashof and the Nishida relations (8) were proven by Nishida in [14]. We state the results for an arbitrary prime p; the modifications needed in the case $p = 2$ are indicated in square brackets.

THEOREM 1.1. *There exist natural homomorphisms* $Q^i: H_*(B) \to H_*(B)$, $i \geqq 0$, *of degree* $2i(p-1)$ [*of degree* i]. *They satisfy the following properties*:

(1) $Q^0(\phi) = \phi$ *and* $Q^i(\phi) = 0$ *for* $i > 0$, *where* $\phi \in H_0(B)$ *is the identity element for the loop product in* $H_*(B)$.

(2) $Q^i(x) = 0$ *if* $2i <$ degree (x) [*if* $i <$ degree (x)].

(3) $Q^i(x) = x^p$ *if* $2i =$ degree (x) [*if* $i =$ degree (x)].

(4) $\sigma_* Q^i = Q^i \sigma_*$, *where* $\sigma_*: IH_*(\Omega B) \to H_*(B)$ *is the homology suspension.*

(5) *Cartan formula*: $Q^s(xy) = \sum_{i=0}^{s} Q^i(x) Q^{s-i}(y)$ *and, if* $\psi(x) = \sum x' \otimes x''$, *then* $\psi Q^s(x) = \sum_{i=0}^{s} \sum Q^i(x') \otimes Q^{s-i}(x'')$.

(6) *If* $\chi: H_*(B) \to H_*(B)$ *is the conjugation* (*induced from the map* $\chi(l)(t) = l(1-t)$, $\chi: \Omega B_1 \to \Omega B_1$), *then* $\chi Q^i = Q^i \chi$.

(7) *Adem relations*: *If* $p \geqq 2$ *and* $a > pb$, *then*

$$Q^a Q^b = \sum_i (-1)^{a+i}(pi - a, a - (p-1)b - i - 1) Q^{a+b-i} Q^i;$$

if $p > 2$, $a \geqq pb$, *and* β *is the mod* p *Bockstein, then*

$$Q^a \beta Q^b = \sum_i (-1)^{a+i}(pi - a, a - (p-1)b - i) \beta Q^{a+b-i} Q^i$$

$$- \sum_i (-1)^{a+i}(pi - a - 1, a - (p-1)b - i) Q^{a+b-i} \beta Q^i.$$

(8) *Nishida relations*: *Let* $P^s_*: H_*(B) \to H_*(B)$, *of degree* $-2s(p-1)$, *be dual to* P^s (*i.e.* $P^s = \mathrm{Hom}_{Z_p}(P^s_*; 1)$ *with* $H^*(B) = \mathrm{Hom}_{Z_p}(H_*(B); Z_p)$) *if* $p > 2$, *and let* $P^s_* = Sq^s_*$, *of degree* $-s$, *if* $p = 2$; *then*

$$P^s_* Q^r = \sum_i (-1)^{i+s}(s - pi, r(p-1) - ps + pi) Q^{r-s+i} P^i_*;$$

if $p > 2$,

$$P^s_* \beta Q^r = \sum_i (-1)^{i+s}(s - pi, r(p-1) - ps + pi - 1) \beta Q^{r-s+i} P^i_*$$

$$+ \sum_i (-1)^{i+s}(s - pi - 1, r(p-1) - ps + pi) Q^{r-s+i} P^i_* \beta.$$

In (7) and (8), $(i, j) = (i + j)!/i!\,j!$ if $i > 0$ and $j > 0$, $(i, 0) = 1 = (0, i)$ if $i \geqq 0$, and $(i, j) = 0$ if $i < 0$ or $j < 0$; the sums are over the integers.

Define the Dyer-Lashof algebra R to be the quotient F/J of the free associative algebra F generated by $\{Q^s, \beta Q^{s+1} \mid s \geqq 0\}$ (not β itself) [by $\{Q^s \mid s \geqq 0\}$] modulo the two-sided ideal J consisting of all elements which annihilate every homology class of every infinite loop space. We shall explicitly describe R, and shall then describe its dual R^*. We need the following definition:

DEFINITION 1.2. (a) $p > 2$. Consider sequences $I = (\epsilon_1, s_1, \ldots, \epsilon_k, s_k)$, where $\epsilon_j = 0$ or 1 and $s_j \geqq \epsilon_j$. Define the degree, length, and excess of I by

$$d(I) = \sum_{j=1}^{k} [2s_j(p - 1) - \epsilon_j], \qquad l(I) = k,$$

$$e(I) = 2s_k - \epsilon_1 - \sum_{j=2}^{k} [2ps_j - \epsilon_j - 2s_{j-1}] = 2s_1 - \epsilon_1 - \sum_{j=2}^{k} [2s_j(p - 1) - \epsilon_j].$$

I is said to be admissible if $ps_j - \epsilon_j \geqq s_{j-1}$ for $2 \leqq j \leqq k$. Each I determines $Q^I = \beta^{\epsilon_1}Q^{s_1} \cdots \beta^{\epsilon_k}Q^{s_k} \in R$.

(b) $p = 2$. Consider sequences $I = (s_1, \ldots, s_k)$, $s_j \geqq 0$, and define $d(I) = \sum_{j=1}^{k} s_j$, $l(I) = k$, $e(I) = s_k - \sum_{j=2}^{k} (2s_j - s_{j-1}) = s_1 - \sum_{j=2}^{k} s_j$. I is said to be admissible if $2s_j \geqq s_{j-1}$ for $2 \leqq j \leqq k$. Each I determines the element $Q^I = Q^{s_1} \cdots Q^{s_k} \in R$.

(c) *Convention.* The empty sequence I is admissible and satisfies $d(I) = 0$, $l(I) = 0$, and $e(I) = \infty$ for all p; it determines $Q^I = 1 \in R$.

The structure of R is given by the following theorem:

THEOREM 1.3. *The ideal J is generated by the Adem relations* (7) *(and, if $p > 2$, the relations obtained by applying β to the Adem relations) and by the relations $Q^I = 0$ if $e(I) < 0$ (see* (2)*). R has the Z_p-basis $\{Q^I \mid I$ is admissible and $e(I) \geqq 0\}$. R_0 (degree zero) is the polynomial algebra generated by Q^0 and R is augmented via $\epsilon: R \to Z_p$ defined by $\epsilon((Q^0)^j) = 1$, $j \geqq 0$. R admits a structure of Hopf algebra with coproduct defined on generators by the formulas*

$$\psi(Q^s) = \sum_{i+j=s} Q^i \otimes Q^j; \qquad \psi(\beta Q^{s+1}) = \sum_{i+j=s} [\beta Q^{i+1} \otimes Q^j + Q^i \otimes \beta Q^{j+1}].$$

R admits a structure of left coalgebra over A^0, the opposite Hopf algebra of the Steenrod algebra; the operations P^i_* are determined by the Nishida relations (8) *and induction on the length of admissible monomials, starting with the formulas*

$$P^s_*Q^r = (-1)^s(s, r(p - 1) - ps)Q^{r-s};$$
$$P^s_*\beta Q^r = (-1)^s(s, r(p - 1) - ps - 1)\beta Q^{r-s}.$$

Here A^0 enters since we are writing Steenrod operations on the left in homology and $\mathrm{Hom}_{Z_p}(\ ; Z_p)$ is contravariant; for any space X, $H_*(X)$ is a left A^0-coalgebra. Observe that the inductive definition of the P^s_* on R could equally well be started with $P^s_*(1) = 0$ for $s > 0$ and $P^0_*(1) = 1$. The theorem is proven by showing that

R operates faithfully on a homology class of a certain infinite loop space [see Theorem 2.5].

If $i > 0$, then R_i is finite dimensional, although R_0 is not. Let $R[k] \subset R$ be the subspace spanned by $\{Q^I \mid I$ is admissible, $e(I) \geqq 0$, and $l(I) = k\}$ ($R[0]$ is spanned by 1). By (5), (7), and (8), each $R[k]$ is a sub A^0-coalgebra of R, and $R = \bigoplus_{k \geqq 0} R[k]$ as an A^0-coalgebra. $R[k]$ is connected and $R_0[k]$ is spanned by $(Q^0)^k$. The product takes $R[k] \otimes R[l]$ into $R[k+l]$ since, in marked contrast to the Steenrod algebra, the Adem relations for R are homogeneous with respect to length (because $Q^0 \neq 1$). As an A-algebra, $R^* = \prod_{k \geqq 0} R[k]^*$. The identity element of $R[k]^*$ is the dual ξ_{0k} of $(Q^0)^k$, and the identity of R^* is $\prod_{k \geqq 0} \xi_{0k}$. Strictly speaking, R^* is not a Hopf algebra, since R is not of finite type, but it is the inverse limit of its quotient A-algebras $\prod_{k=0}^n R[k]^*$, $n < \infty$, and each of these may be regarded as a sub Hopf algebra of R^* (dual to the quotient Hopf algebra $R/(\sum_{l>n} R[l])$ of R). $\prod_{k=0}^n R[k]^*$ is augmented by $\epsilon(\xi_{00}) = 1$ and $\epsilon(\xi_{0k}) = 0$, $k > 0$; clearly $\xi_{0k}^2 = \xi_{0k}$ and $\psi(\xi_{0k}) = \sum_{i=0}^k \xi_{0i} \otimes \xi_{0,k-i}$. We shall describe $R[k]^*$ as an A-algebra and shall give the coproduct on generators.

Define certain admissible sequences inductively as follows:

(a) I_{jk}, $1 \leqq j \leqq k$, $p \geqq 2$: $I_{11} = (0, 1)$; $I_{j,k+1} = (0, p^k - p^{k-j}, I_{jk})$ if $j \leqq k$; $I_{k+1,k+1} = (0, p^k, I_{kk})$ [if $p = 2$, omit the zeroes; if $p > 2$, the zeroes merely denote omission of Bockstein]. Then $d(I_{jk}) = 2(p^k - p^{k-j})$ if $p > 2$ and $d(I_{jk}) = 2^k - 2^{k-j}$ if $p = 2$.

(b) J_{jk}, $1 \leqq j \leqq k$, $p > 2$: $J_{1,1} = (1, 1)$; $J_{j,k+1} = (0, p^k - p^{k-j}, J_{jk})$ if $j \leqq k$; $J_{k+1,k+1} = (1, p^k, I_{kk})$. Then $d(J_{jk}) = 2(p^k - p^{k-j}) - 1$.

(c) K_{ijk}, $1 \leqq i < j \leqq k$, $p > 2$: $K_{ij,k+1} = (0, p^k - p^{k-i} - p^{k-j}, K_{ijk})$ if $j \leqq k$; $K_{i,k+1,k+1} = (1, p^k - p^{k-i}, J_{ik})$. Then $d(K_{ijk}) = 2(p^k - p^{k-i} - p^{k-j})$. Let $S_k = \{I_{jk}, J_{jk}, K_{ijk}\}$ if $p > 2$ and $S_k = \{I_{jk}\}$ if $p = 2$. Observe that the elements I_{jk} and K_{ijk} with $j < k$ all give pth powers when the corresponding Q^I is applied to a zero-dimensional class.

LEMMA 1.4. *$\{Q^I \mid I \in S_k\}$ is a basis for the primitive elements $PR[k]$ of $R[k]$. The Steenrod operations and Bockstein on $PR[k]$ are determined by the following formulas:*

(i) $P_*^{p^{k-j}} Q^{I_{jk}} = -Q^{I_{j-1,k}}$ *if* $2 \leqq j \leqq k$ ($P_*^{p^r} Q^{I_{jk}} = 0$ *otherwise*).

(ii) $P_*^{p^{k-j}} Q^{J_{jk}} = -Q^{J_{j-1,k}}$ *if* $2 \leqq j \leqq k$ ($P_*^{p^r} Q^{J_{jk}} = 0$ *otherwise*).

(iii) $P_*^{p^{k-j}} Q^{K_{ijk}} = -Q^{K_{i,j-1,k}}$ *if* $1 \leqq i < j-1 < k$;

$P_*^{p^{k-i}} Q^{K_{ijk}} = -Q^{K_{i-1,j,k}}$ *if* $2 \leqq i < j \leqq k$ ($P_*^{p^r} Q^{K_{ijk}} = 0$ *otherwise*).

(iv) $\beta Q^{I_{kk}} = Q^{J_{kk}}$; $\beta Q^{J_{jk}} = Q^{K_{jkk}}$ *if* $j < k$ ($\beta Q^I = 0$ *otherwise*, $I \in S_k$).

Let $\xi_{jk} = (Q^{I_{jk}})^*$, $\tau_{jk} = (Q^{J_{jk}})^*$, and $\sigma_{ijk} = (Q^{K_{ijk}})^*$ in $R[k]^*$ (in the dual basis to that of admissible monomials). Let $R^+[k] = R[k]$ if $p = 2$ and let $R^+[k]$ be the sub-coalgebra of $R[k]$ spanned by those Q^I such that $\epsilon_j = 0$ (that is, which do not involve β) if $p > 2$. Also, if $p > 2$, define elements $v_\rho \in R[k]^*$ for each set $\rho = \{r_1, \ldots, r_j\}$ such that $1 \leqq r_1 < \cdots < r_j \leqq k$ by the formulas:

(d) $v_\rho = \sigma_{r_1r_2k}\sigma_{r_3r_4k}\cdots\sigma_{r_{j-1}r_jk}$ if j is even; $v_\rho = \sigma_{r_1r_2k}\cdots\sigma_{r_{j-2}r_{j-1}}\tau_{r_jk}$ if j is odd.

Let $v_\rho = \xi_{0k}$ if ρ is the empty set, and let V_k be the subspace of $R[k]^*$ spanned by the v_ρ (the v_ρ are linearly independent). With these notations, the structure of R^* is determined by the following theorem.

THEOREM 1.5. *For $p \geqq 2$, $R^+[k]^*$ is the polynomial algebra generated by $\{\xi_{jk} \mid 1 \leqq j \leqq k\}$. If $p > 2$, the product defines an isomorphism $R^+[k]^* \otimes V_k \to R[k]^*$, and $R[k]^*$ is determined as an algebra by the relations ($\rho = \{r_1, \ldots, r_j\}$, $j = 2i - \epsilon$, $\epsilon = 0$ or 1):*

(i) $v_\rho\sigma_{stk} = 0$ if $s \in \rho$ or $t \in \rho$; $v_\rho\tau_{sk} = 0$ if $s \in \rho$.

(ii) $v_\rho\sigma_{stk} = (-1)^\alpha v_{\rho\cup\{s,t\}}$ if $s \notin \rho$ and $t \notin \rho$, where α is the number of indices l such that $s < r_l < t$.

(iii) $v_\rho\tau_{sk} = (-1)^{\beta+\epsilon}\xi_{kk}^\epsilon v_{\rho\cup\{s\}}$ if $s \notin \rho$, where β is the number of indices l such that $r_l < s$.

With sums taken over all integers which make sense ($\xi_{ij} = 0$ if $i < 0$ or $j \leqq i$, $\tau_{ij} = 0$ if $i < 1$ or $j \leqq i$, $\sigma_{hij} = 0$ if $h < 1$ or $i \leqq h$ or $j < i$), the coproduct is given on generators by the formulas:

(iv) $$\psi(\xi_{jk}) = \sum_{(h,i)} \xi_{k-i,k-i}^{p^i-p^{i-h}}\xi_{j-h,k-i}^{p^{i-h}} \otimes \xi_{hi},$$

(v) $$\psi(\tau_{jk}) = \sum_{(h,i)} \xi_{k-i,k-i}^{p^i-p^{i-h}}\xi_{j-h,k-i}^{p^{i-h}} \otimes \tau_{hi} + \sum_{i} \xi_{k-i,k-i}^{p^i-1}\tau_{j-i,k-i} \otimes \xi_{ii},$$

(vi) $$\begin{aligned}\psi(\sigma_{ijk}) = {} & \sum_{(f,g,h)}\left(\sum_{s=0}^{t} \xi_{k-h,k-h}^{p^h-p^{h-f}-p^{h-g}}\xi_{j-g+i-f-s,k-h}^{p^{h-g+i-f-s}}\xi_{i-f,k-h}^{p^{h-f}-p^{h-g+i-f-s}}\xi_{s,k-h}^{p^{h-g}}\right)\otimes \sigma_{fgh}\\ & -\sum_{(g,h)} \xi_{k-h,k-h}^{p^h-p^{h-g}-1}\xi_{i-g,k-h}^{p^{h-g}}\tau_{j-h,k-h} \otimes \tau_{gh}\\ & +\sum_{h} \xi_{k-h,k-h}^{p^h-1}\sigma_{i-h,j-h,k-h} \otimes \xi_{hh},\end{aligned}$$

where, in the first sum, $t = \text{minimum}(i - f, j - g, k - h)$.

The proof is by direct dualization, using the Adem relations. In the case $p = 2$, the structure of R^* was first discovered by I. Madsen [7]. While complicated, the A-algebra structure of R^* is not unmanageable. The Steenrod operations on the indecomposable elements $QR[k]^*$ are computed by Lemma 1.3 which implies the following useful corollary.

COROLLARY 1.6. *If $p = 2$, $R[k]^*$ is generated as an A-algebra by ξ_{1k}. If $p > 2$, $R[1]^*$ is generated as an A-algebra by τ_{11} and $R[k]^*$, $k > 1$, is generated as an A-algebra by ξ_{1k} and σ_{12k}.*

In other words, $R[k]^*$ is a quotient A-algebra either of $H^*(K(Z_p, n))$ or of $H^*(K(Z_p, n) \otimes K(Z_p, m))$, for appropriate integers n and m.

REMARK 1.7. R is very closely related to the E_1-term of the Curtis-Kan et al. [3] version of the Adams spectral sequence. Precisely, R is a quotient algebra of the opposite algebra E_1^0. This relationship, which was first noticed by I. Madsen, deserves further study.

2. **Allowable R-modules and $H_*(QX)$.** Having introduced the Dyer-Lashof algebra, we can now give analogs of the notions of unstable A-modules and A-algebras in the cohomology of spaces. These notions will yield a concise description of $H_*(QX)$ as a functor of $H_*(X)$, where $QX = \text{inj lim } \Omega^n S^n X$. Of course, $QS^i X = \Omega QS^{i+1}X$ and QX is therefore a (perfect) infinite loop space. The QX are, in a precise sense, the free infinite loop spaces (see the proof of Corollary 2.7); they play a role in the theory of infinite loop spaces which is roughly analogous to that played by $K(\pi, n)$'s in the cohomology of spaces. Clearly $\pi_n(QX)$ is the nth stable homotopy group of X.

The following lemma is required in order to obtain a sensible analog to the notion of unstable A-module.

LEMMA 2.1. *Let K^q, $q \geqq 0$, denote the subspace of R spanned by $\{Q^I \mid I$ is admissible and $0 \leqq e(I) < q\}$. Then K^q is a two-sided ideal (and sub A^0-module) of R, and K^q is precisely the set of all elements of R which annihilate every homology class of degree $\geqq q$ of every infinite loop space. The quotient algebra $R^q = R/K^q$ has basis $\{Q^I \mid I$ admissible, $e(I) \geqq q\}$.*

In order to deal with nonconnected spaces, which are crucial to the applications, we need a preliminary definition.

DEFINITION 2.2. By a homology coalgebra C, we mean a cocommutative, unital ($\eta : Z_p \to C$), augmented ($\epsilon : C \to Z_p$) coalgebra C such that C is a direct sum of connected coalgebras. We then define $GC = \{g \mid g \in C, \psi(g) = g \otimes g, g \neq 0\}$. GC is a basis for C_0, $\epsilon(g) = 1$ for $g \in GC$, and each $g \in GC$ determines a component C_g of C whose positive degree elements are $\{c \mid \psi(c) = c \otimes g + \sum c' \otimes c'' + g \otimes c\}$; clearly C is the direct sum of its components C_g for $g \in GC$.

If X is a based space, then $H_*(X)$ is a homology coalgebra; its base-point determines the unit and its components determine the direct sum decomposition.

DEFINITION 2.3. An R-module D is said to be allowable if $K^q D_q = 0$ for all $q \geqq 0$. The category of allowable R-modules is the full subcategory of that of R-modules whose objects are allowable; it is an Abelian subcategory which is closed under the tensor product. An allowable R-algebra is an allowable R-module and a commutative algebra such that the product and unit are morphisms of R-modules and such that $Q^i(x) = x^p$ if $p > 2$ and $2i = \deg(x)$ or if $p = 2$ and $i = \deg(x)$. (Here R operates on Z_p through its augmentation.) An allowable R-coalgebra is an allowable R-module and homology coalgebra whose coproduct, unit, and augmentation are morphisms of R-modules. An allowable R-Hopf algebra is an allowable R-module and Hopf algebra which is both an allowable R-algebra and an allowable R-coalgebra and which admits a conjugation χ [**13**, Definition 8.4] such that $\chi^2 = 1$ and χ is a morphism of R-modules (χ is necessarily a morphism of Hopf algebras and, by definition, $\phi(1 \otimes \chi)\psi = \eta\epsilon$). For any of these structures, an allowable AR-structure is an allowable R-structure and an unstable A^0-structure of the same type (in the sense of homology: its dual, if of finite type, is an unstable A-structure of the dual type) such that the A^0 and R operations satisfy the Nishida relations.

With these definitions, the entire content of Theorem 1.1 is that the mod p homology of an infinite loop space carries a natural structure of allowable AR-Hopf algebra. It should be observed that if B is connected and satisfies the definition of an allowable R-Hopf algebra, except for the condition about χ, then this last condition is automatically satisfied. This is false in the nonconnected case, where we have

LEMMA 2.4. *Let B be an allowable R-Hopf algebra. Then each $g \in GB$ is invertible and $\chi(g) = g^{-1}$. If $x \in B$, $\deg(x) > 0$, and $\psi(x) = x \otimes g + \sum x' \otimes x'' + g \otimes x$, then $\chi(x) = -x \cdot g^{-2} - \sum x' \cdot \chi(x'') \cdot g^{-1}$.*

To take advantage of our definitions, we require free functors taking values in our various categories of allowable R and AR-structures. These are obtained as follows.

(a) *Z_p-modules to allowable R-modules.* If M is a Z_p-module, define $D(M) = \bigoplus_{q \geqq 0} R^q \otimes M_q (D_n(M) = \bigoplus_{q \geqq 0} R^q_{n-q} \otimes M_q$ gives the grading). R operates on the left of $D(M)$ via the maps $R \to R^q$.

(b) *Homology coalgebras to allowable R-coalgebras.* If C is a homology coalgebra with unit $\eta : Z_p \to C$, and $JC = \operatorname{Coker} \eta$, define $E(C) = Z_p \oplus D(JC)$ as an R-module; $C \subset E(C)$ and, by induction on the length of admissible monomials, the coproduct on C together with the Cartan formula define a structure of allowable R-coalgebra on $E(C)$.

(c) *Homology unstable A^0-coalgebras to allowable AR-coalgebras.* Given C, the Nishida relations and A^0 operations on C define an allowable AR-coalgebra structure on $E(C)$ by induction on the length of admissible monomials.

(d) *Allowable R-modules to allowable R-algebras.* Given D, define $V(D) = A(D)/I$, where $A(D)$ is the free commutative algebra generated by D and I is the ideal generated by $\{d^p - Q^i(d) \mid 2i = \deg(d)\}$ if $p > 2$ or by $\{d^2 - Q^i(d) \mid i = \deg(d)\}$ if $p = 2$; the Cartan formula and the requirement that the unit be a morphism of R-modules define a structure of allowable R-algebra on $V(D)$.

(e) *Allowable R-coalgebras to allowable R-Hopf algebras.* Given E, define $W(E) = V(JE)$, $JE = \operatorname{Coker} \eta$; $E \subset W(E)$ and the coproduct on E induces a coproduct on $W(E)$ such that $W(E)$ becomes a Hopf algebra over R, not necessarily allowable unless E is connected. $GW(E)$ is a commutative monoid, $W_0(E)$ is its monoid ring, and $W(E) = V(\bar{E}) \otimes W_0(E)$ as an algebra, where $\bar{E}$ is the set of positive degree elements of E. Let $\tilde{G}W(E)$ be the (commutative) group generated by $GW(E)$, let $\tilde{W}_0(E)$ be its group ring, and define $\tilde{W}(E) = V(\bar{E}) \otimes \tilde{W}_0(E)$ as an algebra. The coproduct on $W(E)$ is determined by those on $W(E) \subset \tilde{W}(E)$ and on $\tilde{W}_0(E)$, and (as in Lemma 2.4) $\tilde{W}(E)$ admits a conjugation extending that on $\tilde{W}_0(E)$. The R-operations on $\tilde{W}(E)$ are determined by those on $W(E)$ and by $Q^s\chi = \chi Q^s$. With these structures, $\tilde{W}(E)$ is an allowable AR-Hopf algebra. Of course, if E is connected, then $\tilde{W}(E) = W(E)$.

(f) *Allowable AR-coalgebras to allowable AR-Hopf algebras.* Given E, the

Cartan formula (for the Steenrod operations) defines a structure of allowable AR-Hopf algebra on $\tilde{W}(E)$.

In each case, verifications are required to prove that these functors are well defined and are adjoint to the forgetful functor going in the other direction. The functors V and W occur in other contexts in algebraic topology and are discussed in [**9**].

Let X be a space. $H_*(X)$ is a homology unstable A^0-coalgebra, hence $\tilde{W}EH_*(X)$ is defined and is the free allowable AR-Hopf algebra generated by $H_*(X)$. The natural inclusion $X \to QX$ induces a monomorphism on homology; by the freeness of $\tilde{W}EH_*(X)$, there results a morphism of AR-Hopf algebras f: $\tilde{W}EH_*(X) \to H_*(QX)$ which defines a natural transformation of functors on the category $\mathcal{T}$ of based spaces. Dyer and Lashof [**5**] proved that f is an isomorphism of algebras if X is connected and computed a component of $H_*(QS^0)$ as an algebra. Simple proofs of their results are obtainable by use of the Eilenberg-Moore spectral sequence. By a reinterpretation and generalization of their methods, we can prove the following theorem; observations of I. Madsen were instrumental in obtaining this result.

THEOREM 2.5. *$f: \tilde{W}EH_*(X) \to H_*(QX)$ is an isomorphism of AR-Hopf algebras for every space X.*

REMARK 2.6. The mod p Bockstein spectral sequence of QX is determined by that of X and the following formula [**10**, Proposition 6.8], which is valid for any three-fold loop space B.

(1) Let $y \in H_{2q}(B)$ and let $\beta_{r-1}(y)$ be defined; then, modulo indeterminacy, $\beta_r(y^p) = \beta_{r-1}(y)y^{p-1}$ unless $r = 2$ and $p = 2$, when $\beta_2(y^2) = \beta(y)y + Q^{2q}\beta(y)$. (Of course, if $p = 2$, Theorem 1.1 (8) implies that $\beta Q^s = (s-1)Q^{s-1}$.) Thus the Bockstein spectral sequences of QX are functors of those of X, and the (additive) integral homology of QX is explicitly known as a functor of the integral homology of X.

The following corollary of Theorem 2.5 gives an analog to the statement that the cohomology of any space is a quotient of a free unstable A-algebra.

COROLLARY 2.7. *If B is an infinite loop space, then $H_*(B)$ is a quotient of the free allowable AR-Hopf algebra $H_*(QB_0)$.*

PROOF. By [**8**, Proposition 1], there is an adjunction isomorphism

$$\mathrm{Hom}_{\mathcal{T}}(X, B_0) \to \mathrm{Hom}_{\mathcal{L}}(\tilde{Q}(X), B),$$

where $\mathcal{L}$ is the category of infinite loop sequences and $\tilde{Q}(X) = \{QS^iX \mid i \geqq 0\}$. It follows that, for any infinite loop sequence $B = \{B_i\}$, there is a map $g: \tilde{Q}B_0 \to B$ in $\mathcal{L}$ such that the composite $B_0 \to QB_0 \xrightarrow{g_0} B_0$ is the identity in $\mathcal{T}$, the category of based spaces.

As an algebra, $H_*(QX) = VD(JH_*(X)) \otimes H_0(QX)$, where $VD(JH_*(X))$ is the free commutative algebra generated by the Z_p-module: $\{Q^I(x) \mid x \in JH_*(X),$ I admissible, $\deg(Q^Ix) > 0,\ e(I) + \epsilon_1 > \deg(x)\}$. (If $p = 2$, $e(I) > \deg(x)$ is required; $Q^I = 1$ is allowed if $\deg(x) > 0$.) The $Q^I(x)$ with $e(I) = \deg(x) > 0$ and, if $p > 2$, $\epsilon_1 = 0$ precisely account for the pth powers of positive degree elements. Note that:

$$\pi_0(QX) = \pi_0^s(X) = \text{inj lim}\ \pi_i(S^iX) = \text{inj lim}\ H_i(S^iX) = JH_0(X).$$

Therefore $H_0(QX)$, as a Hopf algebra with conjugation, is the group ring of the free commutative group with one generator $x \in JH_0(X)$ for each component of X other than that of the base-point, in agreement with (e). The coproduct and Steenrod operations on $H_*(QX)$ are induced from those on $H_*(X)$ by the Cartan formulas and Nishida relations; their explicit evaluation requires use of the Adem relations for the Dyer-Lashof operations.

3. **The homology of** F, F/O, BF, **and** $B\,\mathrm{Top}$. To illustrate the previous results and prepare for the discussion of F and BF, we consider QS^0 in detail. Let $\tilde{F}(n)$ denote the space of based maps $S^n \to S^n$, $S:\tilde{F}(n) \to \tilde{F}(n+1)$, and let $\tilde{F} = \text{inj lim}\ \tilde{F}(n)$. Then $\tilde{F}(n) = \Omega^n S^n$ and $\tilde{F} = QS^0$. Let $\tilde{F}_i$ denote the component of $\tilde{F}$ consisting of the maps of degree i, $i \in Z$. If $[i] \in H_0(\tilde{F})$ is represented by a map of degree i, then $\{[i] \mid i \in Z\}$ is a basis for $H_0(\tilde{F})$. Denote the loop product in $\tilde{F} = \Omega QS^1$ by $*$. Then $*:\tilde{F}_i \times \tilde{F}_j \to \tilde{F}_{i+j}$ and $[i] * [j] = [i+j]$. Each Q^s takes $H_*(\tilde{F}_i)$ to $H_*(\tilde{F}_{pi})$, and $Q^0[i] = [pi]$. $[0]$ is the identity for $*$ and $Q^s[0] = 0$ for $s > 0$. Let $\bar{R}$ denote the set of positive degree elements of R and let $H_*(S^0)$ have basis $[0]$ and $[1]$. Then $EH_*(S^0) = Z_p[0] \oplus R \cdot [1]$,

$$WEH_*(S^0) = V(R[1]) = V(\bar{R}[1]) \otimes P\{[1]\},$$

and

$$H_*(\tilde{F}) = \tilde{W}EH_*(S^0) = V(\bar{R}[1]) \otimes H_0(\tilde{F}).$$

Let $A(X)$ denote the free commutative algebra generated by the set X, then

$$V(\bar{R}[1]) = A\{Q^I[1] \mid I \text{ admissible}, d(I) > 0, e(I) + \epsilon_1 > 0\}.$$

$R \to H_*(\tilde{F})$ defined by $Q^I \to Q^I[1]$ is a monomorphism of A^0-coalgebras; by the Cartan formulas and Nishida and Adem relations, R determines the Steenrod operations and coproduct in $H_*(\tilde{F})$ and the Dyer-Lashof operations on $V(R[1])$; for $i > 0$, $Q^s[-i] = \mathscr{X}Q^s[i]$.

Of course, $\tilde{F}$ is a topological monoid under the product c defined by composition of maps, $c:\tilde{F}_i \times \tilde{F}_j \to \tilde{F}_{ij}$. This product is homotopic to that obtained from the smash product of maps $\tilde{F}(m) \times \tilde{F}(n) \to \tilde{F}(m+n)$ by passage to limits, and is therefore homotopy commutative.

With the product c, $\tilde{F}$ plays a special role in the theory of infinite loop spaces. Thus let $B = \{B_i \mid i \geqq 0\}$ be any infinite loop sequence. Then $B_0 = \Omega^n B_n$ is homeomorphic to the spaces of based maps $S^n \to B_n$, and composition of maps defines an operation $c_n:B_0 \times \tilde{F}(n) \to B_0$. Since $c_{n+1}(1 \times S) = c_n$, we obtain $c:B_0 \times \tilde{F} \to B_0$ by passage to limits. Clearly $c:QS^0 \times \tilde{F} \to QS^0$ coincides with the composition product on $\tilde{F}$. The basic properties of $c_*:H_*(B) \otimes H_*(\tilde{F}) \to H_*(B)$ are given in the following theorem.

THEOREM 3.1. *c_* gives $H_*(B)$ a structure of Hopf algebra over the Hopf algebra $H_*(\tilde{F})$ and, with $c_*(b \otimes f) = bf$,*

(i) *$\phi f = \epsilon(f)\phi$, $\epsilon: H_*(B) \to Z_p$, where $\phi \in H_0(B)$ is the identity.*

(ii) *$P^k_*(bf) = \sum P^i_*(b)P^{k-i}_*(f)$ and $\beta(bf) = \beta(b)f + (-1)^{\deg b} b\beta(f)$.*

(iii) *$Q^k(b) \cdot f = \sum_i Q^{k+i}(bP^i_*(f))$ and, if $p > 2$,*

$$\beta Q^k(b) \cdot f = \sum_i \beta Q^{k+i}(bP^i_*(f)) - \sum_j (-1)^{\deg b} Q^{k+j}(b \cdot P^j_*\beta(f)).$$

(iv) *$\sigma_*(b)f = \sigma_*(bf)$, where σ_* is the homology suspension.*

Thus $H_*(B)$ is a Hopf algebra over each of R, A^0, and $H_*(\tilde{F})$, all of these homology operations are stable, and we have precise commutation formulas relating these three types of operations. Applied to $B = \{QS^i \mid i \geq 0\}$, the theorem completely determines the structure of $H_*(\tilde{F})$ as an algebra under c_*. Explicitly, we have the following corollary.

COROLLARY 3.2. *If $x, y, z \in H_*(\tilde{F})$ and $\psi(z) = \sum z' \otimes z''$, then*

(a) *$(x * y)z = \sum (-1)^{\deg y \deg z'} xz' * yz''$, and*

(b) *$(x * [i])(y * [j]) = \sum (-1)^{\deg x'' \deg y'} x'y' * x''[j] * y''[i] * [ij]$.*

Thus c_* is determined by the products between the generators of $H_*(\tilde{F})$ under $*$ namely $[\pm 1]$ and the $Q^I[1]$. The products $Q^I[1] \cdot Q^J[1]$ are determined by (iii) of the theorem and induction on $l(I)$; in particular, all $Q^I[1]$ with $l(I) > 1$ are decomposable under c_* in terms of the $\beta^\epsilon Q^s[1]$. Finally, $x \cdot [1] = x$, $x \cdot [-1] = \mathscr{X}(x)$, and $x \cdot [0] = \epsilon(x)[0]$ for $x \in H_*(\tilde{F})$.

Now recall that $F = \tilde{F}_1 \cup \tilde{F}_{-1}$ is the space of based homotopy equivalences of spheres and that $SF = \tilde{F}_1$. It is easy to compute $H_*(F)$ as an algebra from the corollary. The following result was first proved by Milgram **[11]** in the case $p = 2$ and later by the author and Tsuchiya **[17]**, independently, in the case $p > 2$. The proofs of Milgram and Tsuchiya rely on similar, but less general, results than Theorem 3.1. It will be convenient to first fix notations for various elements and sets of elements in $H_*(SF)$. Thus define

(a) $x_s = Q^s[1] * [1 - p]$; $\deg x_s = s$ if $p = 2$, and $\deg x_s = 2s(p-1)$, if $p > 2$;

(b) $y_s = Q^{s(p-1)}Q^s[1] * [1 - p^2]$; $\deg y_s = 2s$ if $p = 2$ and $\deg y_s = 2sp(p-1)$ if $p > 2$;

(c) $z_s = Q^{s+1}Q^s[1] * [-3]$ if $p = 2$; $\deg z_s = 2s + 1$;

(d) $z_s = Q^{(p-1)s}\beta Q^s[1] * [1 - p^2]$ if $p > 2$; $\deg z_s = 2sp(p-1) - 1$;

(e) $I = Q^I[1] * [1 - p^{l(I)}]$, where I is admissible, $l(I) \geq 2$, and $e(I) + \epsilon_1 > 0$; let X denote the set of all such $I \in H_*(SF)$; observe that $z_s \in X$ but $y_s \notin X$ and, if $p > 2$, $\beta z_s \notin X$. Define

(f) $Y = \{y_s\} \cup X$ if $p = 2$ and $Y = \{\beta z_s\} \cup X$ if $p > 2$ (here $y_s \notin Y$).

COROLLARY 3.3. *As an algebra under c_*, $H_*(SF) = E\{x_s\} \otimes P(Y)$ if $p = 2$ and $H_*(SF) = E\{\beta x_s\} \otimes P\{x_s\} \otimes A(Y)$ if $p > 2$.*

Let $J:SO \to SF$ be the natural inclusion. We next give precise information on J_* and describe $H_*(F/O)$.

Let $p = 2$. Then $H_*(SO) = E\{\alpha_s \mid s \geqq 1\}$, where $\deg a_s = s$, $\psi(a_s) = \sum_{i=0}^{s} a_i \otimes a_{s-i}$, and $\langle w_{s+1}, \sigma_*(a_s)\rangle = 1$. We may define Stiefel-Whitney classes $W_s = \phi^{-1}Sq^s\phi(1) \in H^*(BSF)$, where ϕ is the Thom isomorphism, and then $(BJ)^*(w_s) = w_s$. Thus $(BJ)^*:H^*(BSF) \to H^*(BSO)$ is an epimorphism, hence so is J^*.

THEOREM 3.4. *Let $p = 2$. Then $J_*(a_s) = x_s$, hence $\mathrm{Im}\, J_* = E\{x_s\}$. $H_*(F/O) = H_*(SF)//J_* \cong P(Y)$, $H_*(SF) \to H_*(F/O)$ is the natural epimorphism, and $H_*(F/O) \to H_*(BO)$ is trivial.*

Now let $p > 2$. Then $H_*(SO) = E\{a_s \mid s \geqq 1\}$, where $\deg a_s = 4s - 1$, a_s is primitive, and $\langle P_s, \sigma_*(a_s)\rangle = 1$, with P_s the Pontrjagin class reduced mod p. We may define Wu classes $q_s = \phi^{-1}P^s\phi(1)$ in both $H^*(BF)$ and $H^*(BO)$. Then $(BJ)^*(q_s) = q_s$ and, in $H^*(BO)$, $q_s - k_sP_{ms}$ is decomposable, where $m = (1/2)(p-1)$ and $0 \neq k_s \in Z_p$ [**12**, p. 120]. By $\sigma^*(BJ)^* = J^*\sigma^*$ and dualization, $J_*(a_{ms}) \neq 0$ in $H_*(F)$.

We need the following lemma.

LEMMA 3.5. *Let $p > 2$. Then the sub Hopf algebra $E\{\beta x_s\} \otimes P\{x_s\}$ of $H_*(SF)$ contains unique primitive elements b_s such that $b_s - \beta x_s$ is decomposable, and $E\{\beta x_s\} \otimes P\{x_s\} = E\{b_s\} \otimes P\{x_s\}$.*

THEOREM 3.6. *Let $p > 2$. Then $J_*(a_{ms}) = lb_s$, $0 \neq l \in Z_p$, and $J_*(a_s) = 0$ for $s \not\equiv 0 \bmod m$, hence $\mathrm{Im}\, J_* = E\{b_s\}$. Further, $H_*(F/O) = [H^*(BO)/\!/P\{qs\}]^* \otimes H_*(SF)/\!/J_*$, where $H_*(SF)/\!/J_* \cong P\{x_s\} \otimes A(Y)$; $H_*(SF) \to H_*(F/O)$ is the natural epimorphism onto $H_*(SF)/\!/J_*$; and $H_*(F/O) \to H_*(BO)$ is the inclusion on $[H^*(BO)/\!/P\{q_s\}]^*$ and is trivial on $H_*(SF)/\!/J_*$.*

We next discuss the classifying space BF. If $p = 2$, $H_*(BF)$ is already determined, as a coalgebra, by $H_*(F)$ since $E^2 = E^\infty$ for dimensional reasons in the Eilenberg-Moore spectral sequence converging from $\mathrm{Tor}\, H_*(SF)(Z_2, Z_2)$ to $H_*(BSF)$. Thus Milgram [**11**] first computed $H_*(BF; Z_2)$ as a coalgebra. For $p > 2$, and to compute $H_*(BF; Z_2)$ as an algebra, more information is needed. This information for $p > 2$ was obtained first by Tsuchiya [**17**] and then by the author and for $p = 2$ by Madsen [**7**]; it will be discussed below.

THEOREM 3.7. *If $p = 2$, $H_*(BF) = H_*(BO) \otimes BC$ as a Hopf algebra, where BC is the primitively generated Hopf algebra*

$$E\{\sigma_*(y_s)\} \otimes P\{\sigma_*(z_s)\} \otimes P\{\sigma_*(I) \mid I \in X, e(I) > 1\}.$$

If $p > 2$, $H_*(BF) = [P\{q_s\} \otimes E\{\beta q_s\}]^* \otimes BC$ as a Hopf algebra, where BC is the primitively generated Hopf algebra

$$E\{\sigma_*(\beta z_s)\} \otimes P\{\sigma_*(z_s)\} \otimes A\{\sigma_*(I) \mid I \in X, e(I) + \epsilon_1 > 1\}.$$

In both cases, BC is closed under the Steenrod operations in $H_*(BF)$.

Of course, the pth powers in $(BC)^*$ are all zero. There are spaces $B\operatorname{Im} J$ (one for each p), constructed by Stasheff [**16**], such that $H^*(B\operatorname{Im} J) = P\{q_s\} \otimes E\{\beta q_s\}$ if $p > 2$. Peterson and Toda [**15**] first proved that $H^*(BF) \cong H^*(B\operatorname{Im} J) \otimes (BC)^*$, as a Hopf algebra over the Steenrod algebra, but without computing $(BC)^*$. In principle, the theorem determines the Steenrod operations in $H_*(BF)$ since the given generators of BC are suspensions of elements with known Steenrod operations in $H_*(SF)$. There is one practical difficulty, however. In applying the Nishida relations to the $Q^I[1] * [1 - p^{l(I)}]$, one sometimes reaches terms $Q^J[1] * [1 - p^{l(I)}]$, where $Q^J[1]$ is a pth power in the loop product $*$. It is not known that all such elements (other than the y_s if $p = 2$) are decomposable under c_*. Thus such elements could conceivably suspend nontrivially to $H_*(BF)$.

It is instructive to compare $H_*(BSF)$ to $H_*(QS^1)$. The latter is the free commutative primitively generated Hopf algebra

$$A\{Q^I i \mid I \text{ is admissible and } e(I) + \epsilon_1 > 1\},$$

where i is the fundamental class of $H_*(S^1)$. In the case $p = 2$, the presence of $H_*(BSO)$ in $H_*(BSF)$ forced the appearance of the additional generators $\sigma_*(y_s)$ and $\sigma_*(z_s)$. In the case $p > 2$, the presence of the Wu classes and their Bocksteins in $H^*(BF)$ forced the appearance of the additional generators $\sigma_*(\beta^\epsilon z_s)$.

The description of $H_*(BF)$ just given is clearly inappropriate for most applications. The interest in BF lies mainly in its relationship to BO, B Top, BPL, F/pl, etc. By the work of Boardman and Vogt [**2**], all of these spaces are infinite loop spaces and the maps between them are infinite loop maps. To study these maps and to compute $H_*(BBF)$, etc., one must describe $H_*(BF)$ in terms of its own Dyer-Lashof operations rather than in terms of the suspensions of the Q^s for the $*$ product in $H_*(\tilde{F})$.

To study this problem, it is again convenient to use all of $\tilde{F}$. Although $\tilde{F}$ is not an infinite loop space under c, since $\pi_0(\tilde{F})$ is not a group under c_*, it is easy to prove by use of topological PROP's that $H_*(\tilde{F})$ admits Dyer-Lashof operations $\tilde{Q}^s : H_*(\tilde{F}_i) \to H_*(\tilde{F}_{i^p})$ which coincide with the infinite loop operations for the composition product on $H_*(F)$ and which satisfy all of the properties stated in Theorem 1.1 except (6). The information required for the proof of Theorem 3.7 was, in the case $p = 2$, the determination of the first operation above the square, namely $\tilde{Q}^s(x)$ for $x \in H_{s-1}(SF)$ and, in the case $p > 2$, the determination of $\beta^\epsilon \tilde{Q}^s(x)$ for $x \in H_{2s-1}(SF)$. If $p > 2$, these $\beta\tilde{Q}^s(x)$ produced nontrivial differentials (d_{p-1}) in the Eilenberg-Moore spectral sequence and, for all p, these $\tilde{Q}^s(x)$ determined the algebra extensions from E_∞ to $H_*(BF)$. An earlier preprint of mine claimed a complete algebraic determination of the higher $\tilde{Q}^s$; the methods used did give considerable information, but less than was claimed. Tsuchiya [**17**] initiated a direct geometric study of the $\tilde{Q}^s$, but his results did not give an explicit hold on the higher operations. Recently Madsen [**7**] obtained precise formulas, in the case $p = 2$, for the evaluation of the $\tilde{Q}^s$ on $H_*(\tilde{F})$ in terms of the Q^s and the loop and composition products. Modulo one ambiguity, I have since obtained such an evaluation for all p. The key result is the following theorem, which evaluates the $\tilde{Q}^s$ on elements of $H_*(\tilde{F})$ which are $*$-decomposable.

THEOREM 3.8. *There exist operations* $\tilde{Q}^s_i: H_*(\tilde{F}) \otimes H_*(\tilde{F}) \to H_*(\tilde{F})$ *for* $0 \leqq i \leqq p$ *such that, for all* $x, y \in H_*(\tilde{F})$,

(1) $\tilde{Q}^s(x * y) = \sum \sum \tilde{Q}^{s_0}_0(x^{(0)} \otimes y^{(0)}) * \cdots * \tilde{Q}^{s_p}_p(x^{(p)} \otimes y^{(p)})$, $\sum s_i = s$, *where* $\psi(x \otimes y) = \sum x^{(0)} \otimes y^{(0)} \otimes \cdots \otimes x^{(p)} \otimes y^{(p)}$ *gives the iterated coproduct in* $H_*(\tilde{F}) \otimes H_*(\tilde{F})$.

The $\tilde{Q}^s_0$ *and* $\tilde{Q}^s_p$ *are determined from the* $\tilde{Q}^s$ *by*

(2) $\tilde{Q}^s_0(x \otimes y) = \tilde{Q}^s(\epsilon(y)x)$ *and* $\tilde{Q}^s_p(x \otimes y) = \tilde{Q}^s(\epsilon(x)y)$, $\epsilon: H_*(\tilde{F}) \to Z_p$.

The $\tilde{Q}^s_i$, $0 < i < p$, *are determined from the* Q^s *by the formulas*

(3) $\sum_{k \geqq 0} \tilde{Q}^{s+k}_i P^k_*(x \otimes y) = \sum \tilde{Q}^s_i([1] \otimes [1]) x^{(1)} \cdots x^{(p-i)} y^{(1)} \cdots y^{(i)}$, *where* $\psi(x) = \sum x^{(1)} \otimes \cdots \otimes x^{(p-i)}$ *and* $\psi(y) = \sum y^{(1)} \otimes \cdots \otimes y^{(i)}$ *give the iterated coproducts: this formula is inductively solvable for* $\tilde{Q}^s_i(x \otimes y)$ *in terms of the Steenrod operations, the composition product, and the elements* $\tilde{Q}^t_i([1] \otimes [1])$;

(4) $\tilde{Q}^s_i([1] \otimes [1]) = \sum \tilde{Q}^{s_1}_1([1] \times [1]) * \cdots * \tilde{Q}^{s_{r_i}}_1([1] \otimes [1])$, $\sum s_j = s$, *for* $1 < i < p$, *where* $r_i = (1/p)(i, p - i)$ *and*

(5) $\tilde{Q}^s_1([1] \otimes [1]) = Q^s[1]$.

In particular, formulas (3) *and* (5) *imply*

(6) $\tilde{Q}^s_1([1] \otimes y) = Q^s(y)$ *for all* $y \in H_*(\tilde{F})$.

The crucial formula (5) is obtainable algebraically, from knowledge of $H_*(O)$, if $p = 2$, but requires a very explicit geometric hold on the operations if $p > 2$. The theorem reduces the problem of calculating the $\tilde{Q}^s$ on $H_*(\tilde{F})$ to their evaluation on elements which are indecomposable under both the loop and composition products. We have the following lemma.

LEMMA 3.9. *If* $s > 0$, *then* $\tilde{Q}^s[0] = 0$, $\tilde{Q}^s[1] = 0$, *and, if* $p > 2$, $\tilde{Q}^s[-1] = 0$; *if* $p = 2$, *then* $\tilde{Q}^s[-1] = x_s = Q^s[1] * [-1]$.

Thus, by Corollary 3.2, we are reduced to the evaluation of the $\tilde{Q}^s\beta^\epsilon Q^r[1]$, and, by Theorem 3.8, it suffices to evaluate the $\tilde{Q}^s\beta^\epsilon x_r$ instead. Now Kochman [6] has succeeded in completely determining all Dyer-Lashof operations in the homology of all spaces involved in Bott periodicity. By Theorems 3.4 and 3.6, his results on $H_*(SO)$ compute the $\tilde{Q}^s(x_r)$ if $p = 2$ and the $\tilde{Q}^s(b_r)$ if $p > 2$. With $H_*(SO) = E\{a_r\}$, as above, Kochman has proven the following theorem.

THEOREM 3.10. *If* $p = 2$, *then, with* $a_0 = 1$,

$$Q^s(a_r) = (r, s - r - 1)a_{r+s} + \sum_{0 \leqq i < j < k,\, i+j+k=r+s} [(r - i, i + j - 2r - 1) + (r - i, s - j - r - 1) + (r - j, s - i - r - 1)]a_i a_j a_k$$

If $p > 2$, then $Q^s(a_r) = (-1)^s(2r - 1, s - 2r)a_{r+ms}$, $m = (1/2)(p - 1)$. If $p = 2$, these formulas completely determine the $\tilde{Q}^s$; if $p > 2$, the $\tilde{Q}^s(x_r)$ are not yet determined, but this problem appears to be solvable geometrically. Of course, the algebraic complexity of these results is enormous. They evaluate the $\tilde{Q}^s$ in terms of our basis for $H_*(\tilde{F})$ defined by means of the loop product and its operations Q^s, but their purpose is to enable us to prove that $H_*(F)$ admits a reasonable basis described in terms of the $\tilde{Q}^s$ themselves. The important result is therefore the following, which we state provisionally as a conjecture, although a complete proof should be available shortly.

CONJECTURE 3.11. Theorem 3.7 remains true with $\sigma_*(I)$ replaced by $\tilde{Q}^J\sigma_*(K)$ for those $I = (J, K) \in X$ such that $l(K) = 2$ and $l(J) > 0$ (with $e(I) + \epsilon_1 > 1$).

The conjecture implies the weaker statement that BC is generated as an R-algebra by the elements $\sigma_*(K)$ with $l(K) = 2$ (and the $\sigma_*(y_s)$ and $\sigma_*(\beta^\epsilon z_s)$), and so reduces the number of R-algebra generators of $H_*(BF)$ to manageable proportions. For $p = 2$, this weaker statement has been proven by Madsen [7].

Finally, we shall very briefly discuss $H_*(B\,\mathrm{Top})$. Let $p > 2$. Sullivan (unpublished) has shown that BPL, which is mod p homotopy equivalent to $B\,\mathrm{Top}$ by the triangulation theorem, splits as a space, after localization at p, into a product $BO \times B\,\mathrm{Coker}\,J$. (Here $BO \overset{(p)}{\cong} Y \times Y'$, where $H^*(Y) = P\{q_s\}$, and the relevant map $BO \to BPL$ is the natural inclusion on $Y \subset BO$ and is the composite $Y' \to BO \overset{(p)}{\cong} F/PL \to BPL$ on Y'.) By the Adams conjecture, F is mod p homotopy equivalent to $\mathrm{Im}\,J \times \mathrm{Coker}\,J$ (the analogous statement for BF is an exceedingly difficult open question). Using these facts and Theorem 3.8, I have proven the following result.

THEOREM 3.12. *If $p > 2$, then $H_*(BPL)$ is isomorphic as a Hopf algebra to $H_*(BO) \otimes BC$, and the following composite is an isomorphism of A^0-coalgebras*:

$$H_*(B\,\mathrm{Coker}\,J) \to H_*(BPL) \to H_*(BF) \to H_*(BF)//(E\{\beta q_s\} \otimes P\{q_s\})^* \cong BC.$$

I do not claim that $H_*(B\,\mathrm{Coker}\,J)$ maps into $BC \subset H_*(BF)$, and I thus do not have complete information on the map $H_*(BPL) \to H_*(BF)$. The previous result had been conjectured, and proven in low dimensions, by Peterson.

For $p = 2$, the problem is considerably more difficult Madsen [7] has obtained very useful information on the Dyer-Lashof operations in $H_*(F/\mathrm{Top}; Z_2)$ and his work may well lead to a computation of $H_*(B\,\mathrm{Top}; Z_2)$.

ADDED IN PROOF. A number of changes have occurred, since the summer conference, in the state of our knowledge.

(1) For $p = 2$, Madsen has proven Conjecture 3.11 and has shown that BF does not split as $BJ \times B\,\mathrm{Coker}\,J$; however, such a splitting has always appeared far less likely for $p = 2$ than for $p > 2$, and the question for $p > 2$ is still open.

(2) Madsen, Brumfiel, and Milgram have succeeded in computing $H_*(B\,\mathrm{Top}; Z_2)$, and Tsuchiya has given an independent proof of Theorem 3.12, in "Characteristic classes for PL-microbundles" (mimeographed notes).

(3) I have obtained very simple proofs of the results of Boardman and Vogt, and this work has greatly streamlined the construction of the operations and the proofs of the results of this paper.

BIBLIOGRAPHY

1. T. Kudo and S. Araki, *Topology of H_n-spaces and H-squaring operations*, Mem. Fac. Sci. Kyūsyū Univ. Ser A. **10** (1956), 85–120. MR **19,** 442.

2. J. M. Boardman and R. M. Vogt, *Homotopy-everything H-spaces*, Bull. Amer. Math. Soc. **74** (1968), 1117–1122. MR **38** #5215.

3. A. Bousfield, E. B. Curtis, et al., *The* mod *p lower central series and the Adams spectral sequence*, Topology **5** (1966), 331–342. MR **33** #8002.

4. W. Browder, *Homology operations and loop spaces*, Illinois J. Math. **4** (1960), 347–357. MR **22** #11395.

5. E. Dyer and R. K. Lashof, *Homology of iterated loop spaces*, Amer. J. Math. **84** (1962), 35–88. MR **25** #4523.

6. S. Kochman, Ph.D. Thesis, University of Chicago, Chicago, Ill., 1970.

7. J. Madsen, Ph.D. Thesis, University of Chicago, Chicago, Ill., 1970.

8. J. P. May, *Categories of spectra and infinite loop spaces*, Category Theory, Homology Theory, and Their Applications (Battelle Inst. Conference, Seattle, Wash., 1968), vol. 3, Lecture Notes in Math., no. 99, Springer-Verlag, Berlin, 1969, pp. 448–479. MR **40** #2073.

9. ———, *Some remarks on the structure of Hopf algebras*, Proc. Amer. Math. Soc. **23** (1969), 708–713.

10. ———, *A general algebraic approach to Steenrod operations*, Steenrod Algebra and its Applications: A Conference to Celebrate N. E. Steenrod's Sixtieth Birthday, Springer-Verlag, 1970.

11. R. J. Milgram, *The* mod 2 *spherical characteristic classes*, Ann. of Math. (2) **92** (1970), 238–261.

12. J. Milnor, *Lectures on characteristic classes*, Princeton University, Princeton, N.J., 1957 (mimeographed notes).

13. J. Milnor and J. C. Moore, *On the structure of Hopf algebras*, Ann. of Math. (2) **81** (1965), 211-264. MR **30** #4259.

14. G. Nishida, *Cohomology operations in iterated loop spaces*, Proc. Japan Acad. **44** (1968), 104–109.

15. F. P. Peterson and H. Toda, *On the structure of $H^*(BSF; Z_p)$*, J. Math. Kyoto Univ. **7** (1967), 113-121. MR **37** #5878.

16. J. D. Stasheff, *The image of J as a space* mod $p > 2$ (mimeographed notes).

17. A. Tsuchiya, *Spherical characteristic classes* mod *p*, Proc. Japan Acad. **44** (1968), 617–622.

UNIVERSITY OF CHICAGO

PROBLEMS PRESENTED TO THE 1970 AMS SUMMER COLLOQUIUM IN ALGEBRAIC TOPOLOGY

EDITED BY
R. JAMES MILGRAM

There were four problem sessions held at Madison, the first on embeddings and immersions by S. Gitler, the second on general problems by the editor, the third on H-spaces by J. Stasheff, and the fourth on homotopy theory by F. Peterson.

We present here the collected results of these sessions: 94 problems and conjectures. They are grouped into four sections:

(A) Classification and structures on manifolds,

(B) Generalized cohomology theories, cobordism and related topics,

(C) H-spaces and classifying spaces,

(D) Homotopy theory.

Where appropriate, or where the editor was capable of doing so, we have included with each problem some comments and introduction. [In the eight months since this article was originally written many of the problems have been solved and considerable progress has been made on others. Where appropriate these developments have been included in the text in brackets.]

I would like to take this opportunity to thank J. Stasheff for largely writing up Part C, and F. Peterson for his help on Part D. Also, P. Schweitzer and W. Browder offered very important aid which made the editor's job much easier than it might have been.

(A) **Classification and structures on manifolds.** (1) The work of Browder, Novikov, Sullivan, and Wall on the classification of manifolds with a given homotopy

---

AMS 1970 *subject classifications.* Primary 55B20, 55D15, 55D45, 55E45, 55F15, 55F25, 55F40.

type can be summarized in part by the exact sequence

$$L^{\varepsilon}_{n+1}(\pi, \pi', w) \xrightarrow{\omega} \mathscr{S}^{\varepsilon}_{H}(X) \xrightarrow{\mathscr{N}} [X, G/H] \xrightarrow{\sigma} L^{\varepsilon}_{n}(\pi, \pi', w_1),$$

where X is a manifold with boundary Y, $\pi_1(X) = \pi$, $\pi_1(Y) = \pi'$, $w = w_1 : \pi \to Z_2$ is the first Stiefel-Whitney class, $H = \boldsymbol{O}$, $\boldsymbol{PL}$ or $\mathbf{Top}$, $\varepsilon = h$ (homotopy) or s (simple homotopy), respectively, and $\mathscr{S}^{\varepsilon}_{H}$ are the manifold structures of the appropriate types.

The results of Sullivan and Kirby-Siebenmann have given the structure of $\boldsymbol{G/PL}$ and $\boldsymbol{G}/\mathbf{Top}$ (mod p for p-odd; they are equivalent to B_O, and mod 2 $\boldsymbol{G}/\mathbf{Top}$ looks like a product $\prod K(Z, 4i) \times K(Z_2, 4i + 2)$ while $\boldsymbol{G/PL}$ has a single non-trivial k-invariant $\{2\beta\iota_2\}$). Thus the set $[X, G/H]$ is effectively calculable in these cases.

Much less is known about $\boldsymbol{G/O}$. Its mod 2 cohomology structure has been determined by the editor, and J. P. May has given its mod (p) cohomology for p odd. The recent affirmative solution of the Adams conjecture by Quillen and Sullivan has determined the homotopy groups of $\boldsymbol{G/O}$, provided one knows the homotopy of spheres and also has shown there are maps (usually injective in homotopy) $\alpha_p : B^{(p)}_{SO} \to \boldsymbol{G/O}^{(p)}$, where the ^{p} signifies that we localize away from the prime p.

PROBLEM 1 (BRUMFIEL, MADSEN, STASHEFF). Let p be an odd prime and $\mathbf{Im}(\boldsymbol{J})_p$ the fiber in the map

$$(\psi^p - 1): B^{(p)}_{SO} \to B^{(p)}_{SO}.$$

Then there are liftings

$$\begin{array}{ccccc} \mathbf{Im}(\boldsymbol{J})^{(p)} & \longrightarrow & B^{(p)}_{SO} & \xrightarrow{\psi^p - 1} & B^{p}_{SO} \\ \downarrow{\scriptstyle \alpha_1} & & \downarrow{\scriptstyle \alpha_2} & & \downarrow \\ \boldsymbol{SG}^{(p)} & \longrightarrow & \boldsymbol{G/O}^{(p)} & \longrightarrow & B^{p}_{SG} \end{array}$$

Can α_1 and α_2 be chosen as infinite loop maps (optimist)? Are α_1, α_2 at least loop maps?

An affirmative answer implies that $\boldsymbol{G/O}^{(p)}$ splits as a product $B^{(p)}_{SO} \times \mathrm{coker}(J)^{(p)}$.

REMARK (MADSEN). For $p = 2$, α_2 cannot be delooped more than two times, as a corollary of Madsen's recent calculation of the action of the mod 2 loop homology operations in $\boldsymbol{G/PL}$.

Next we have

PROBLEM 2 (BROWDER). Find means to calculate the Wall groups $L^{\varepsilon}_{n}(\pi, \pi')$. There are some groups for which it has been possible using geometric techniques such as free groups and free abelian groups, and others such as finite groups where algebraic methods have been most effective. But nothing like a general picture has emerged.

Further, Browder asks

PROBLEM 3. Find general methods to calculate the three maps ω, μ, σ for $H = \boldsymbol{O}, \boldsymbol{PL}$ or **Top**.

Related to this is

PROBLEM 4 (F. QUINN). Calculate $[K(\pi, 1), G/\mathbf{Top}]$. Even better, let $\mathscr{S}$ be the spectrum obtained from $G/\mathbf{Top}$ by periodicity, and compute $H_*(K(\pi, 1); \mathscr{S})$.

Quinn points out that the map σ factors through a universal homomorphism A_π, so the diagram

$$\begin{array}{ccc} [M, G/\mathbf{Top}] & \xrightarrow{\sigma} & L_m(\pi) \\ \downarrow & & \uparrow \\ H^0(M, \mathscr{S}) & & A_\pi \\ \text{Poincaré duality} \quad \downarrow \cong & & \\ H_m(M, \mathscr{S}) & \xrightarrow[H_m(\pi_1, \mathscr{S})]{} & H_m(K(\pi, 1), \mathscr{S}) \end{array}$$

commutes. A_π is an isomorphism for π free or free abelian, and is essentially trivial for $\pi = Z_p$.

PROBLEM 5 (QUINN). What is $H_m(\pi, \mathscr{S})$ for the most familiar nonsimply connected manifolds?

Further special cases of Problem 3 have been suggested by Browder.

PROBLEM 6. When $H = \boldsymbol{O}$ and $X = S^{2^i - 2}$, so that $L_{4k+2}(1) = Z_2$, σ is the Kervaire invariant. Is σ nontrivial?

This is the same as asking when is the Kervaire manifold smoothable. These are the only remaining dimensions where it may be, and it is known for $i \leqq 7$.

By the result of Browder [1] this is equivalent to the question, "Does h_i^2 in the E^2 term of the mod 2 Adams spectral sequence survive to E^∞?".

Browder next optimistically asks

PROBLEM 7. Is there an alternate geometrical construction of almost-framed manifolds of Kervaire invariant one which can be directly shown to be smoothable?

Related to this, one wonders if there are systematic ways of constructing framed manifolds which directly can be shown to represent the homotopy of spheres.

PROBLEM 8 (BROWDER). Find new ways of constructing surgery obstructions. Is it possible that all the invariants of surgery can be obtained by simple geometric constructions from the known invariants of quadratic forms (signature, Arf invariant, G-signature of Atiyah-Singer) and Whitehead or Reidemeister torsion?

[The editor has recently shown that the invariant I in $L_4(Z_-)$ is not of these types. It corresponds to a certain invariant of forms over arbitrary finite Abelian groups.]

PROBLEM 9 (BROWDER). Find simpler techniques than the general ones for determining when manifolds with extra structures (such as nonsingular algebraic varieties) are diffeomorphic.

PROBLEM 10 (BROWDER). Extend the theory to manifolds with singularities.

(2) *Triangulation and related topics.* Since the work of Sullivan, Kirby-Siebenmann and Lashof-Rothenberg, we have very complete information on the relation between ***PL*** and topological manifolds of dimensions greater than 4. In low dimensions, however, many problems remain. Browder presents the following list.

PROBLEM 11. The Poincaré conjecture in dimensions 3 and 4.

PROBLEM 12. In particular, is there a homotopy 3-sphere which bounds a parallelizable smooth manifold of index 8?

PROBLEM 13. The annulus conjecture in dimension 4.

PROBLEM 14. The classification of 1-connected 4-manifolds.

PROBLEM 15 (MILNOR). As a special case of 14, what are the quadratic forms of 4-manifolds?

PROBLEM 16. Is there a nontriangulable 4-manifold?

PROBLEM 17. How or when can one do surgery on a 4-manifold?

[Cappell and Shaneson [**13**] have largely solved this problem.]

Related to these is

PROBLEM 18. Is there a noncombinatorial triangulation of any manifold? In particular, is the double suspension of some non-simply-connected homology 3-sphere homeomorphic to S^5?

These problems are what is left of the more general problems in the 1963 AMS list (Milnor's Problems 23–29 in Lashof's list).

(3) *Immersions and embeddings.* In S. Gitler [**2**] the main problems in this area are presented. There was also a problem session, and we give here the most interesting problems which arose there. I thank Paul Schweitzer for his aid and comments in preparing this section.

PROBLEM 19 (D. ASIMOV). Let $e(M)$ be the least k so $M \subset R^k$. Let $M^{(n)} = M \times \cdots \times M$, n times. Is $\lim_{n\to\infty} e(M^{(n)})/n$ an integer?

REMARK (P. SCHWEITZER). This limit is defined. If $i(M)$ is the least k so $M \subseteq R^k$, then by the Sanderson-Schwarzenberger lemma, ($e(M \times N) \leqq e(M) + i(N)$ if $\dim N \leqq \dim M$), we have

$$\lim_{n\to\infty} \frac{e(M^{(n)})}{n} = \lim_{n\to\infty} \frac{i(M^{(n)})}{n} \leqq i(M).$$

This suggests a secondary question.

PROBLEM 20 (S. GITLER). Find M so that $2i(M) - i(M \times M)$ is: (1) nonzero, or (2) arbitrarily large.

PROBLEM 21 (J. HARPER). Let M^n be k-parallelizable and $M^n \subset R^{n+r+1}$, $r \geqq \frac{1}{2}(n + 1)$. Give necessary and sufficient conditions in terms of "characteristic classes" for $M^n \subseteq R^{n+r}$.

REMARK. There was some clarification that "characteristic classes" should be taken in a very broad sense, involving generalized cohomology theories as well. For some results in this direction see [**7**].

PROBLEM 22 (E. REES). If $M^n \subset R^{n+k}$ and $\tilde{M}$ is a double cover of M, does $\tilde{M} \subset R^{n+k+1}$?

Examples of manifolds which might be worth studying to gain further insight were brought up by the editor and C. Giffen.

PROBLEM 23. Let $N = S^{2q-1} \times_{Z_p} M^{(p)}$ or $N = S^q \times_T M^{(2)}$ (free actions on the spheres and permutation of factors on the products). What are $e(N)$ and $i(N)$? What is the least k so that $N \subset RP^k$ in the case $p = 2$ so as to induce the line bundle over N from the nontrivial bundle over RP^k?

The "Hirsch conjecture" still remains as a major problem in this area.

PROBLEM 24 (HIRSCH). Let M^n be s-parallelizable; then does M embed in R^k, where $k = n + [(n + 1)/2]$?

Another problem which seems important in view of the editor's work on embedding projective spaces (with M. Mahowald [**8**]) and recent work with E. Rees [**9**] is

PROBLEM 25. Classify those bundles which occur as the normal bundle to an embedding $M^n \subset R^{n+k}$.

REMARK. For n odd, $n \neq 3, 5, 9$; then $S^2 \times S^{n-2}$ embeds in R^{2n-2} with two distinct (and fiber homotopically distinct) normal bundles [**9**]. See also [**10**].

Finally, we have

PROBLEM 26 (BROWDER). For an exotic sphere Σ^n, is there a relation between the embedding dimension of Σ^n and the largest dimension of a compact Lie group G which acts effectively on Σ (degree of symmetry of Hsiang)?

(4) *Miscellaneous problems on manifolds.* Browder has asked two questions about normal maps.

PROBLEM 27. What Kervaire invariants K are possible for normal maps into smooth manifolds? It is seldom possible for $K \neq 0$ for the sphere S^{2q} (at most it can happen when $q = 2^i - 1$) but if it is possible, then it is also possible for any manifold of dimension $2q$. There are smooth normal maps of Kervaire invariant 1 into RP^{2q} for every q.

[Recent work of Brumfiel, Madsen, and the editor shows that $K \equiv 0$ for any π-manifold M^n if $n \neq 2^k - 2$. Partial results on the general problem are contained in [**14**].]

PROBLEM 28. Find sufficiently homotopical definitions of the surgery obstructions in $L_{2n+1}(\pi)$ so that they are defined for normal maps of Poincaré spaces with $\pi_1 = \pi$.

PROBLEM 29 (BROWDER). Consider the Brieskorn sphere

$$\Sigma_d : \{z_0^d + z_1^2 + \cdots + z_n^2 = 0, \|z\| = 1\},$$

d odd, n odd, and the involution $T:\Sigma_d \to \Sigma_d$, $T(z_0) = z_0$, $T(z_i) = -z_i$ for $i > 0$. Let $d = 2l + 1$, where $l = 2n$. Then $\Sigma_{d/T}$ is diffeomorphic to either RP^{2n-1} or $RP^{2n-1} \# \Sigma_0$, where Σ_0 is the Kervaire sphere which generates bP_{2n}. Which is it?

PROBLEM 30 (AGOSTON). What types of closed manifolds admit maps of degree d for an arbitrary integer d?

Are they all essentially of the form $M^n \times S^n$?

PROBLEM 31 (AGOSTON). What conditions on a manifold will insure that it admits a fixed point free map (not necessarily homotopic to the identity)?

PROBLEM 32 (D. ASIMOV). Given a closed connected C^∞ manifold M, call it *prime* if there is *no* fiber bundle whose base and fiber are connected manifolds and whose total space is M. Is there a Jordan-Hölder type theorem for manifolds? That is, can each M be associated to a sequence $M_1, M_2, \ldots, M_K$ of prime manifolds so that M is the total space of a bundle $F_1 \to M \to M_1$, F_1 is the total space of a fiber bundle $F_2 \to F_1 \to M_2 \cdots F_K \to F_{K-1} \to M_K$, with $\{F_j\}$ all connected and F_K prime?

[A. Hatcher points out that a counterexample occurs in [**11**].]

(B) **Generalized cohomology theories, cobordism, and related topics.** Several basic problems in cobordism theory were presented by Peter Landweber and appear in the write-up of his talk [**3**]. We do not repeat them here.

(1) We start with several problems relating to the slowly emerging picture of the PL-cobordism groups.

Sullivan asserts a splitting $B_{PL_p} \cong B_O \times B_{\text{coker}(J)}$ for p any odd prime. Calculations of the editor and J. P. May have determined $H^*(B_{\text{coker}(J)}; Z_p)$ for all p. Thus the main problem, for odd primes, in studying $\pi_*(MSPL)$ would seem to be evaluating the homomorphism

$$\varphi : \mathscr{A}(p) \to H^*(\boldsymbol{MSPL}; Z_p)$$

determined by $\varphi(\alpha) = \alpha(U)$.

[This has been completed by A. Tsuchiya [**12**]. In particular, he shows the answer to the first question in Problem 33 is yes.]

PROBLEM 33 (FRANK PETERSON). Is $\ker \varphi = \mathscr{A}(p)[Q_0, Q_1]$ for p odd? Also, is it possible to find an $\mathscr{A}(p)$-module $\mathscr{N}$ and an $\mathscr{A}(p)$ map $\theta : \mathscr{N} \to H^*(MSPL)$ so that $\theta_* H(\mathscr{N}, Q_i) \to H(H^*(MSPL), Q_i)$ is an isomorphism for $i = 0, 1$? This would show, under mild technical hypotheses, that $H^*(MSPL) \approx \mathscr{N} \oplus$ free $\mathscr{A}(p)$-module and would help in determining Ω_*^{PL}.

In the mod (2) case, the situation is quite different. A moment's reflection is enough to show that $\boldsymbol{MSPL}$ is isomorphic (for the prime 2) to a product of Eilenberg-Mac Lane spaces. However, the determination of $H^*(B_{SPL}; Z_2)$ is made difficult by the PL-Kervaire invariant one manifolds in dimensions $4i + 2$, as well as the problem involved in defining an index for a Z_n-manifold. The current situation is this: There is a fibering

$$\boldsymbol{SG} \xrightarrow{\pi} \boldsymbol{G/PL} \xrightarrow{j} B_{SPL}$$

and the work of Sullivan gives the structure of $\boldsymbol{G/PL}$ as a product

$$E_2 \times \prod_{i=2}^{\infty} K(Z, 4i) \times K(Z_2, 4i - 2).$$

In order to give the mod 2 structure of $H^*(B_{SPL})$ the map π^* must be evaluated on the primitives over $\mathscr{A}(2)$ in $H^*(\boldsymbol{G/PL}, Z_2)$. In this connection Browder and Brumfiel recalled the question first asked by M. Hirsch at the Seattle conference:

PROBLEM 34. Is the Kervaire manifold K^{4k+2} a PL-boundary?

Through dimensions approximately 20, it is known that only K^2, K^6, and K^{14} are boundaries (Brumfiel and the editor), and Ib Madsen has recently shown that K^{2^i-2} is PL cobordant to zero for all i. [The exact answer is K^{4k+2} is a ***PL***-boundary if and only if $k = 2^i - 1$ (Brumfiel, Madsen, and the editor).]

More particularly, if we have answered Problem 34, then we should be able to solve:

PROBLEM 35 (SULLIVAN, BRUMFIEL). Compute $H^*(B_{PL}; Z_2)$ and $H^*(\boldsymbol{PL/O}; Z_2)$.

One further problem should be mentioned that should be in range, given the solutions of Problems 33–35.

PROBLEM 36 (BRUMFIEL). Determine the ring structure of Ω_*^{SPL}.

[Problem 35 too has been solved, though the solution was a great deal more involved than the editor had expected. In particular, the new surgery invariant described in the remark after Problem 8 played a key role. For Problem 36 the 2-local structure of Ω_*^{SPL} is largely completed.]

Finally, Browder and Brumfiel ask about the possibilities of further "Kervaire invariant" theories.

PROBLEM 37 (BROWDER). Find cobordism theories in which the absolute Kervaire invariant can be defined and calculated by (simple) formulas.

PROBLEM 38 (BRUMFIEL). Define an "Arf invariant" on the $(4n + 1)$-bordism of ***PL/O***. This should measure something about cobordism between a manifold M and a different smoothing M_α of M.

The recent developments clarifying the structure of B_{SG} imply, with the work of N. Levitt, that it is well within range to calculate the Poincaré duality cobordism groups.

In particular, as a first step, F. Peterson asks

PROBLEM 39. Compute $\pi_*(\boldsymbol{MSG})$ as a ring explicitly.

PROBLEM 40. Compute $H^*(B_{SG}; Z_p)$ as an algebra over $\mathscr{A}(p)$. For example, is it true that q_{p^s} and certain generalized Gitler-Stasheff elements $e_s \in H^{p^s(2p-2)-1}$ generate over products, Steenrod operations and higher order Bocksteins?

REMARK. The answer to the second part of Problem 40 would seem to be no. Perhaps one should look for a basic set of generators over all the above, and the algebra of loop-homology operations as well. Even then, it seems one would need a larger set of generators than those suggested above.

(2) *Some problems in generalized cohomology theories.* The work of Novikov and Don Anderson on the Adams spectral sequences for generalized cohomology theories has shown that there is much room for development in this area. Fundamental problems exist, however, when we ask to what degree the spectral sequences converge. The difficulty is that there are nontrivial spaces which may be acyclic with respect to the exotic theories under consideration. Thus Frank Adams asks

PROBLEM 41. Let E be a ring-spectrum which is not connected (e.g. the classical B_U-spectrum). Then there can exist CW-complexes (even finite CW-complexes) which are not contractible but are acyclic for E-theory:

$$E_*(X) = 0, \qquad E^*(X) = 0.$$

Find out something about the class of such complexes.

Novikov's work on the Adams spectral sequence for complex cobordism theory together with Quillen's recent determination of the structures of the Steenrod algebras for the Brown-Peterson spectra give a very good handle for cranking out calculations. However, so far they have proved very cumbersome in practice. More recently, D. Anderson has determined the (2 localized) torsion free part of the Steenrod algebra for stable complex K-theory. It is the completed algebra A on three generators π, θ, τ (π is the periodicity of degree -2) subject to relations

$$3\tau\theta = \theta\tau, \qquad \theta\pi = -\pi\theta - 2, \qquad \theta^2 = 0, \qquad \tau\pi - 3\pi\tau + 4\theta = 0.$$

Moreover, for calculations, it is useful to write in terms of the Adams operations $\pi\theta = \psi^{-1} - 1$ and $\pi^2\tau = \psi^3 + \psi^{-1} - 2$. This seems much more manageable than the preceding Steenrod algebras.

PROBLEM 42. Calculate $\operatorname{Ext}_A^{s,*}(b_U^*(RP_n^m), b_U^*(pt))$ for $s = 0, 1, 2$, and interpret the results in terms of the classical Adams spectral sequence. There are similar problems with $b_U^*(X)$ in place of $b_U^*(pt)$. It would also seem useful to study the relation between these groups and those given by the Brown-Peterson spectrum.

The next type of question that arises deals with the problem of constructing spectra which are distinct from those already known. H. Margolis has some results in this direction.

PROBLEM 43 (F. PETERSON). Does there exist an Ω-spectrum $\boldsymbol{X}_{(n)}$ with homotopy groups

$$\pi_*(X(n)) = \begin{cases} Z_2 & \text{if } * = 2^r n, \quad r \geqq 0, \\ 0 & \text{otherwise,} \end{cases}$$

and so that the rth stage Postnikov system has k-invariants coming from the relations $Sq^{n+1} = 0, (Sq^{2n+1}Sq^{n+1}) = 0, \ldots, (Sq^{2^{r-1}n+1}(\cdots Sq^{n+1}) = 0)$.

One would expect that the answers were yes. Also, the 0th term of the spectrum should be a product $K(Z_2, n) \times \cdots \times K(Z_2, 2^i n) \times \cdots$ with a twisted Hopf algebra structure so the homology generator in dimension n generates a polynomial algebra.

D. S. Kahn and D. Kraines have partial results.

CONJECTURE 44 (F. PETERSON). Let G/Top be the Ω-spectrum whose 0th-term is $\boldsymbol{G/\mathrm{Top}}$. Then up to mod 2 homotopy, type

$$G/\mathrm{Top} \sim \prod_{k \geqq 1} \boldsymbol{K}(Z, 4k) \times \prod_{k \geqq 0} \Omega^2 \boldsymbol{X}(8k + 4)$$

as infinite loop spaces.

Evidence for the conjecture is that the loop-homology operations agree with those computed by Ib Madsen so far on $\boldsymbol{G/\mathrm{Top}}$.

Another way of obtaining new cohomology theories is to take Thom spectra of fancy classifying spaces associated to B_O. For example, F. Peterson proposes

PROBLEM 45. Let $I_n = \bigcap_\gamma \ker \gamma^* \subset H^*(B_O; Z_2)$, the set of normal characteristic classes zero on all n-manifolds. Form the fibering

$$B_n \to B_O \to \prod_{j \in J} K(Z_2, j),$$

where you kill a minimal generating set for I_n. Compute $H^*(B_n)$ in a reasonable way. Also, compute $\pi_*(M_n)$, the Thom space of the canonical bundle over B_n.

Peterson remarks that this associated cobordism theory seems to be a good setting for studying the Arf invariant, especially its multiplicative properties.

[Mahowald points out that a map $X \to B_G$ can always be extended to a map $\Omega^n\Sigma^n X \to B_G$ since B_G is an infinite loop space. Taking the resulting Thom spaces of the associated bundles gives a wholesale method for constructing new spectra. For example, if $\eta: S^1 \to B_O$ is the nontrivial map, then $\boldsymbol{M}[\Omega^2 S^3] = \boldsymbol{K}(Z_2, 0)$, the stable Eilenberg-Mac Lane spectrum.]

Finally, we conclude this section with

PROBLEM 46 (BROWDER). Let $\mathscr{E}^k$ be a linear C^k bundle. Suppose $\boldsymbol{CP}(\mathscr{E}^k)$ (the associated $\boldsymbol{CP}^{k-1}$ bundle) is fiber homotopy trivial; that is, there is an $f: \boldsymbol{CP}(\mathscr{E}^k) \to \boldsymbol{CP}^{k-1}$ which is a homotopy equivalence on each fiber. Is $\mathscr{E}$ trivial? More generally, one can define a K-theory using fiber spaces with fiber $\boldsymbol{CP}^{k-1}$ into which ordinary K-theory maps (unpublished work of Browder and T. Petrie). How can one calculate the image of ordinary K-theory in this exotic fiber space theory?

PROBLEM 47 (F. ADAMS). Use the recent solution of the Adams conjecture to compute the groups $J(\boldsymbol{QP}^n)$ and determine which of the symplectic Stiefel fiberings have cross sections.

Adams remarks that this should be within range of the determined graduate student.

(C) ***H*-spaces and classifying spaces.**

(1) *Finite H-spaces.* The study of finite H-complexes is modelled on the study of compact Lie groups, and, so far as possible, the main object of most research is to see how similar to Lie groups they actually are.

Background for many of the following problems is given in Stasheff **[4]**. Also, most of the problems stated in terms of finite H-complexes have analogues in terms of finite group complexes.

DEFINITION. X is a finite group complex if X has the homotopy type of both a finite complex and a topological group (of course, the topological group need not be a finite complex). (Equivalently, X is an s.h.a. finite H-complex.)

PROBLEM 48. Classify finite H-complexes

(a) up to Hopf algebra isomorphism of $H^*(X; Q)$,

(b) up to Hopf algebra over $\mathscr{A}(p)$ isomorphism of $H^*(X; Z_p)$,

(c) up to underlying homotopy type,

(d) up to H-equivalence; i.e. (X, m) and (Y, n) are H-equivalent if there is an H-homotopy equivalence $h: X \to Y$.

REMARKS. Curjel and Douglas have recently shown that there are only a finite number of H-spaces up to H-equivalence in each dimension.

Hopf algebra considerations have long been known to imply $H^*(X, Q) \cong E(x_1, \ldots, x_r)$, so (a) reduces to determining the possible values of r (called the rank of X), the possible dimensions of the x_i and the possible coproducts $m^*(x_i)$. [Extensive results are due to Hubbuck and in the infinite group complex case to Ewing.]

The recent examples of non-Lie H-spaces suggest the answer to (c) may include

CONJECTURE 49. Given a simply connected finite H-complex X, for each prime p there is a Lie group G_X and a product S_X of odd dimensional spheres so that X is mod p equivalent to $G_X \times S_X$. [Stasheff has since shown this to be false for finite H-complexes; perhaps it is true for finite group complexes.]

PROBLEM 50 (S. Y. HUSSEINI). If $H^*(X; Z) \cong E(x_1, \ldots, x_r)$, $\dim x_i \leqq \dim x_{i+1}$, when does there exist a map $X \to S^{(\dim x_r)}$ so that $H_*(X; Z)$ maps onto $H_*(S; Z)$?

PROBLEM 51 (MISLIN). In Problem 50, does there exist $X \to S^{(\dim x_r)}$ with fiber an H-space of rank $r - 1$?

PROBLEM 52. For a finite H-complex X, does $H_*(X)$ have no p-torsion unless $p \leqq 5$ and/or unless $H_*(X)$ has no two-torsion? [Mislin has a counterexample which is not a group complex.]

PROBLEM 53. For a finite H-complex X, is $H^*(\Omega X)$ torsion free?

From a homotopy point of view, strongly homotopy multiplicative maps are the natural analogues of homomorphisms. Is the distinction significant for Lie groups? More precisely,

PROBLEM 54. For connected Lie groups $\boldsymbol{G}$ and $\boldsymbol{H}$, is $\mathrm{Hom}(\boldsymbol{H}, \boldsymbol{G})_\pi \cong [B_H, B_G]$? (The subscript π denotes homotopy classes.)

PROBLEM 55. For what degrees d is a map $S^3 \to S^3$ of degree d an s.h.m. map? Equivalently, for which d does there exist $f: \boldsymbol{QP}^\infty \to \boldsymbol{QP}^\infty$ of degree d in dimension 4?

REMARK. Several people have shown d must be zero or of the form $(2r + 1)^2$. (The question could also be asked in terms of the exotic multiplications on S^3 which admit classifying spaces. For H-maps, a complete solution has been given by Arkowitz and Curjel, as it has for S^7 by Mislin.)

PROBLEM 56 (ARKOWITZ, RECTOR). Classify the s.h.a. structures on S^3 up to s.h.a. equivalence. The work of Slifker shows there may be infinitely many.

[Rector has since completed the classification using the localization techniques of Sullivan, and some algebraic theory of quadratic forms. There are, indeed, infinitely many.]

PROBLEM 57 (RECTOR). Let $\alpha \in \pi_3(X)$, X a finite H-complex. When is there a multiplication on S^3 such that α is represented by an H-map, or by an s.h.m. map if X is of the homotopy type of a group?

PROBLEM 58 (HUSSEINI). For simply connected simple Lie groups, a non-trivial homomorphism $\boldsymbol{G} \to \boldsymbol{G}$ is a rational homotopy equivalence. Is the same true for simply connected finite H-complexes, or for s.h.m. maps if $\boldsymbol{G}$ is a finite group complex?

(2) *H-maps and sub-H-spaces.* Turning to the more general problem of classifying H-maps, we have the following concepts:

DEFINITION (RECTOR). In the category of finite group complexes, a subgroup H of a group $\boldsymbol{G}$ is an s.h.m. map $\boldsymbol{H} \xrightarrow{i} \boldsymbol{G}$ such that the fiber of $B_H \xrightarrow{B(i)} B_G$ has the homotopy type of a finite complex. We call the fiber $\boldsymbol{G}/\boldsymbol{H}$.

DEFINITION (STASHEFF). In the category of finite H-complexes, a sub-H-complex H of an H-complex G is an H-map $H \xrightarrow{i} G$ of the homotopy type of the inclusion of the fiber F into the total space E of a fibration over a finite complex which we call "G/H".

PROBLEM 59 (RECTOR). In the above sense, $\boldsymbol{G}/\boldsymbol{H}$ or "G/H" satisfy Poincaré duality. When do these complexes have the homotopy type of manifolds?

Now consider the problem of maximal tori. For a Lie group $\boldsymbol{G}$, a maximal torus has the same rank as does $\boldsymbol{G}$. This need not be true for finite H-complexes however, as Rector has found s.h.a. multiplications on S^3 for which the only subtori are trivial. Rector would like to preserve the term "maximal torus" of X for one having the same rank as X.

PROBLEM 60 (RECTOR). If X is a connected group complex and $t_i : T_i \to X$, $i = 1, 2$ maximal tori, is B_{t_1} homotopic to B_{t_2}?

We now come to the analogue of the Weyl group.

DEFINITION (RECTOR). Given a maximal torus $t : T \to X$, let $\tau = B_t$. Define $\tilde{W}(X)$ as the set of homotopy equivalences $\{\alpha \in [B_T, B_T] \mid \tau\alpha \simeq \tau\}$.

For Lie groups, the classical Weyl group $W(\boldsymbol{G})$ maps monomorphically into $\tilde{W}(\boldsymbol{G})$.

PROBLEM 61 (RECTOR). For Lie groups, is $W(\boldsymbol{G}) \cong \tilde{W}(\boldsymbol{G})$? [Yes, Rector.]

PROBLEM 62 (RECTOR). For a finite group complex X, is $H^*(B_X; Q) \cong H^*(B_T; Q)^{\tilde{W}(X)}$, the $\tilde{W}$-invariant subspace?

PROBLEM 63 (RECTOR). Is $\tilde{W}(X)$ independent of the choice of t?

In addition to the tori, abelian subgroups of classical interest are the finite central subgroups.

PROBLEM 64 (MISLIN-RECTOR). Given a finite group complex G, when can one find a finite abelian subgroup π so that G/π is a finite group complex (i.e. when can one find a finite group complex H and a fibration $K(\pi, 1) \to B_G \to B_H$)? If π is a finite subgroup of a maximal torus of G and π is invariant under the action of $\tilde{W}(G)$, when is G/π a finite group complex?

(3) *Extensions.* In the category of compact simply connected Lie groups, if $\boldsymbol{H}$ is a normal subgroup of $\boldsymbol{G}$, then $\boldsymbol{G} \cong \boldsymbol{H} \times \boldsymbol{G}/\boldsymbol{H}$ as topological groups.

PROBLEM 65 (RECTOR). Let $H \to G \to K$ be an extension of finite group complexes. Is $G \cong H \times K$ as spaces? Is $B_G \cong B_H \times B_K$?

More generally, one can consider the problem of extensions $H \to G \to K$ whether in terms of finite complexes or not. In the category of topological groups, with the assumption that extensions are principal H-bundles, H-abelian, a reasonable solution has recently become available. Heller has shown that such extensions can be described in terms of appropriate homological algebra. Independently, Dennis Johnson and Graeme Segal have defined complexes $\mathscr{C}$ so that $H^2(\mathscr{C})$ classifies such extensions. In particular, they both provide an exact sequence

$$0 \to H_c^2(K; H) \to \operatorname{Ext}(K; H) \to [B_K, B_H],$$

where $H_c^*(K; H)$ is the cohomology of K with coefficients in H defined entirely in terms of continuous cochains $K \times \cdots \times K \to H$. The map $\operatorname{Ext}(K; H) \to [B_K, B_H]$ classifies the bundle.

PROBLEM 66 (STASHEFF). (a) Classify extensions (suitably redefined perhaps) of s.h.a. H-spaces. (b) Classify extensions of H-spaces.

(4) *Multiplications on* X. For a single space X, it is natural to look at the set $M(X)$ of equivalence classes of multiplications on X; i.e. equivalent via an H-homotopy equivalence. In the study of 2-stage Postnikov systems and more generally of infinite loop spaces, the following question arises:

PROBLEM 67 (HARPER). Study $\Omega: M(X) \to M(\Omega X)$ given by looping the multiplication. In particular, what can be said about the kernel and image of Ω?

Similarly, one can ask how many different ways can a given space admit the structure of an infinite loop space.

PROBLEM 68. Consider $X = \prod K(Z_2, 2^i)$ with the nonstandard multiplication so that on the homology duals α_{2^i} of the fundamental classes, $(\alpha_{2^i})^2 = \alpha_{2^{i+1}}$. Is this multiplication that of an infinite loop space structure on X? (This is a special case of Problem 43.)

PROBLEM 69. How many infinite loop space structures on B_O or B_U are there?

(5) *Classifying spaces*. Along the lines of Problems 68–69, we notice that the work of Boardman has shown $\mathbf{G}$, $\mathbf{G}/\mathbf{Top}$, $\mathbf{G}/\mathbf{PL}$, $\mathbf{G}/\mathbf{O}$ are all infinite loop spaces.

PROBLEM 70 (BROWDER). Give a description of the homotopy type of the iterated classifying spaces of $\mathbf{G}$, $\mathbf{T}/\mathbf{Top}$, $\mathbf{G}/\mathbf{PL}$, $\mathbf{G}/\mathbf{O}$.

Madsen's results show the third classifying space $B^3(\mathbf{G}/\mathbf{Top}) \not\simeq_{(2)} \prod K(\pi, n)$. One should try to find geometrical interpretations for maps into these spaces so as to relate them closely to geometrical problems.

A further question related to these is

PROBLEM 71 (STASHEFF, MILGRAM). Interpret the exotic characteristic classes for spherical fibrations in terms analogous to the various interpretations of Stiefel-Whitney classes.

This is probably related to Problem 40 and the remark preceding it.

Finally, we have

PROBLEM 72 (BROWDER). Compute π_*, H_*, etc., for $\mathbf{Diff}(S^n)$ and its classifying space.

Recent work of Antonelli, Burgelea, and Peter Kahn has shown that $\mathbf{Diff}(S^n)$ is not the homotopy type of a finite complex, and they have found some nonzero homotopy not coming from $\mathbf{O}(n + 1)$.

(D) **Homotopy theory.**

(1) *Adams spectral sequence*. During the past five years, and based on the calculational results of May and Tangora on $\operatorname{Ext}^{**}_{\mathscr{A}(2)}(Z_2; Z_2)$, Barratt, Mahowald, and Tangora have essentially determined $\pi_*^S(S^0, Z_2)$ through the 62 stem. At this point, there are many, Mahowald among them, who suggest further stem by stem tabulation is pointless. What is now needed is some global understanding of the structures so far exposed.

CONJECTURE 73 (JOEL COHEN). $E_\infty^{s,*}(S^0, S^0)$ is finite for each s. It is further suggested that it might be worthwhile examining the relationship between this and

(1) X, Y finite CW-complexes implies $E_\infty^{s,*}(X, Y)$ is finite;

(2) X, Y finite CW-complexes implies $E_\infty^{1,*}(X, Y)$ is finite;

(3) Y a finite CW-complex implies $E_\infty^{s,*}(S^0, Y)$ or $E_\infty^{1,*}(S^0, Y)$ is finite.

REMARK. Barratt calls 73 "the doomsday conjecture". He and Mahowald believe all the h_i^2 represent infinite cycles, and if this were known, they believe they would then be in a position to fully explain the 2-adic homotopy of spheres. However, if 73 is correct, then only a finite number of the h_i^2 represent infinite cycles, and they have no idea of what $\pi_*^S(S^0, Z_2)$ could then be.

A. Liulevicius asks a number of explicit questions which should result in enlarging our knowledge of homotopy theory if and when they are answered.

PROBLEM 74 (LIULEVICIUS). Hopf invariant one allows Bott periodicity in $\pi_*(B_O)$. Find a similar relationship to nontrivial d_r's in the Adams-Novikov spectral sequence over M_U localized at 3.

PROBLEM 75 (LIULEVICIUS). Let A_n be the Hopf subalgebra of $\mathscr{A}(2)$ generated by $Sq^1, \ldots, Sq^{2^n}$. A_n satisfies Poincaré duality and is injective over itself. Exploit these facts to determine $\mathrm{Ext}_{A_n}(Z_2; Z_2)$.

PROBLEM 76 (LIULEVICIUS). Prove that $\mathrm{Ext}_{\mathscr{A}(2)}(Z_2; Z_2)$ tries to be a polynomial algebra with noise born in low s and propagated by periodicity.

See, for example, recent work of Anderson, Mahowald, Margolis, Priddy, Tangora, and Zachariou.

(2) *Realizing modules over the Steenrod algebra.* An $\mathscr{A}$-module $\mathscr{M}$ is called real if there is a spectrum $X_{\mathscr{M}}$ so that $H^*(X_{\mathscr{M}}) \cong \mathscr{M}$.

PROBLEM 77 (PETERSON). Give good conditions on $\mathscr{M}$ which imply it is or is not real.

Examples of large real $\mathscr{M}$ are $\mathscr{A}$, $\mathscr{A}/\mathscr{A}(Sq^1, Sq^2)$, $\mathscr{A}/\mathscr{A}(Sq^3)$, $\mathscr{A}/\mathscr{A}(Q_i)$, $\mathscr{A}/\mathscr{A}(Q_o, Q_i)$, $\mathscr{A}/\mathscr{A}(\chi(Sq^k), \chi(Sq^{k+1})\cdots)$, $\mathscr{A}/\mathscr{A}(Q_o)$. See, for example, work of Adams, Stong, Margolis, Brown-Gitler, and Brown-Peterson. Mahowald can show that $\mathscr{A}/\mathscr{A}(Sq^{01}, Sq^{02})$ is not real.

CONJECTURE 78 (J. COHEN). For a given s, there are only a finite number of indecomposable real modules $\mathscr{M}$ with $\sum_{i=0}^{\infty} \mathrm{rank}(M^i) = s$.

This is known to Peter Hoffman if one assumes $H^*(X_{\mathscr{M}}, Z)$ has no p-torsion.

PROBLEM 79 (TODA). Given p, find the maximum n such that $E(Q_0, \ldots, Q_n)$ is real. A similar question can be asked when one demands that $X_{\mathscr{M}}$ be a ring spectrum.

(See Toda's talk, this Colloquium, for further details.)

PROBLEM 80 (TODA). Classify ring spectra with a small number of cells.

(3) *The stable homotopy of spheres.* $Q(X)$ is $\lim_{n\to\infty} \Omega^n\Sigma^n(X)$. In particular, $Q(S^0) \cong G_*$, and there is a natural map $Q(\boldsymbol{RP}^\infty) \to Q(S^0)$ obtained from the inclusion $\boldsymbol{RP}^\infty \subset \boldsymbol{O} \subset \boldsymbol{G}_*$. Adjointing this, there is the stable Hopf invariant map $H^S : \boldsymbol{RP}^\infty \to S^0$.

CONJECTURE 81 (MAHOWALD). H_*^S is an epimorphism in homotopy.

This has been verified (Mahowald, [5]) in dimensions less than 29.

There are also several "classical" questions which should be repeated.

CONJECTURE 82 (BARRATT). $\pi_*^S(S^0)$ is nilpotent. Possibly $\xi^{p^2} = 0$ or $\xi^{p^2-p} = 0$ if p is odd, and $\xi^4 = 0$ if ξ has order a power of 2.

CONJECTURE 83 (FREYD). Let X, Y be finite CW-complexes, $f: X \to Y$. Then $f_*: \pi_*^S(X) \to \pi_*^S(Y)$ trivial implies f is stably trivial.

CONJECTURE 84 (G. WHITEHEAD).

$$\mathrm{Ker}[\pi_*(SP^{p^i}(X)) \to \pi_*(SP^\infty(X))] = \mathrm{Ker}[\pi_*(SP^{p^i}(X)) \to \pi_*(SP^{p^{i+1}}(X))]$$

on the p-components. Here X is assumed to be a spectrum or alternately the conjecture is made only in the stable range.

PROBLEM 85 (BROWDER). Can one find geometrical points of view about framed or other types of manifolds which will begin to give a systematic way of constructing the homotopy of spheres and give an understanding of their structures?

(4) *Questions in unstable homotopy theory*. Kan asks three questions which are unstable but related to Problem 41 of J. F. Adams.

PROBLEM 86. Discuss spaces X such that $\tilde{H}_*(X; Z) = 0$.

PROBLEM 87. What is needed besides $\tilde{H}_*(X; Z)$ and homology operations of all orders to describe the homotopy type of X?

Let $\boldsymbol{E}$ be a connected ring spectrum. Let $E(X)$ be the 0th space of the Ω-spectrum $\boldsymbol{E} \wedge X$. Then the analogue of the unstable Adams spectral sequence for X and $\boldsymbol{E}$ has an E_1-term which involves $\pi_*(E \cdots EX)$.

PROBLEM 88. What are these groups for well-known connected ring spectra and various X?

See, for example, work of Anderson and Mahowald when $\boldsymbol{E} = \boldsymbol{b_O}$.

CONJECTURE 89 (BARRATT-CURTIS). $\pi_*(S^n) \to \pi_*^S(S^0)/\mathbf{Im}(J)$ has finite image.

PROBLEM 90 (TODA). What is the maximal subcategory W_p of finite, 1-connected CW-complexes in which mod p homotopy equivalence is an equivalence relation? Does W_p depend on p?

(See work of Mimura and Toda.)

(5) *Miscellaneous problems*.

PROBLEM 91 (GANEA). Is there a nonsuspension X of dimension approximately four times its connectivity with a homotopy associative comultiplication? Related to this is

PROBLEM 92 (DRACHMAN-PORTER). Construct coclassifying spaces for co-H-spaces. What are the obstructions to desuspending a space? [Recently Drachman has succeeded in constructing the coclassifying space.]

PROBLEM 93 (BARRATT AND OTHERS). $Q(X)$ is a direct factor of $Q(Q(X))$, the other being $K = \mathrm{Ker}[Q(Q(X)) \to Q(X)]$. What is the nature of K? In particular, is it determined by suspensions of smash products of X with itself? The same question can also be asked for the category of ith loop spaces for finite i.

For results on Problem 92 in the metastable range (and perhaps for some idea of how such information may be used), see the editor's mimeograph article [**6**].

PROBLEM 94 (AGOSTON). Let

$$S_1 : \pi_{n-1}(S^{n-i}) \to \pi_{n+k}(S^{n+k-i+1}) \quad \text{and} \quad S_2 : \pi_{n+k}(S^{n+k-i+1}) \to \pi_{i-1}{}^S(S^0)$$

be the suspension homomorphisms. Define $G^{n-1}_{n+k,i-1} = \operatorname{Ker} S_2 \cap \operatorname{im} S_1$. Does $G^{n-1}_{n+k,i-1} = 0$ when $2k \geqq n + 3$ and $2 \leqq i \leqq n - 2$? If yes, then every 1-connected π-manifold imbeds in the metastable range!

REFERENCES

1. W. Browder, *The Kervaire invariant of framed manifolds and its generalizations*, Ann. of Math. (2) **90** (1969), 157–186. MR **40** #4963.

2. S. Gitler, *Immersion and embedding of manifolds*, Proc. Sympos. Pure Math., vol. 22, Amer. Math. Soc., Providence, R.I., 1971.

3. P. Landweber, *Cobordism and classifying spaces*, Proc. Sympos. Pure Math., vol. 22, Amer. Math. Soc., Providence, R.I., 1971.

4. J. D. Stasheff, *H-spaces and classifying spaces*, Proc. Sympos. Pure Math., vol. 22, Amer. Math. Soc., Providence, R.I., 1971.

5. M. Mahowald, *The metastable homotopy of S^n*, Mem. Amer. Math. Soc. No. 72 (1967). MR **38** #5216.

6. R. J. Milgram, *Unstable homotopy theory from the stable point of view* (mimeograph, Stanford University, 1971).

7. J. C. Becker, *Extensions of cohomology theories*, Illinois J. Math. (to appear).

8. M. Mahowald and R. J. Milgram, *Embedding real projective spaces*, Ann. of Math. (2) **87** (1968), 411–422.

9. R. J. Milgram and E. Rees, *On the normal bundle to an embedding*, Topology (to appear).

10. F. X. Connolly, *Toward a classification of embeddings* (to appear).

11. J. Tollefson, *3-manifolds fibering over S^1 with nonunique connected fiber*, Proc. Amer. Math. Soc. **21** (1969), 79–80.

12. A. Tsuchiya, *Characteristic classes for PL-micro bundles*, Trans. Amer. Math. Soc. (to appear).

13. S. E. Cappell and J. Shaneson, *On four dimensional surgery and applications* (to appear).

14. J. Morgan, *Stable homotopy equivalences* (to appear).

STANFORD UNIVERSITY

SYMMETRIES AND OPERATIONS IN HOMOTOPY THEORY

R. JAMES MILGRAM

In this talk I propose to describe recent work of J. F. Adams, M. Barratt, D. S. Kahn, M. Mahowald, myself, and H. Toda which introduces some new and potentially very powerful machinery into homotopy theory. After introducing these techniques and indicating some of our previous applications, we will apply them to the problem of finding framed manifolds of Kervaire invariant one. The result obtained represents joint work with Barratt and Mahowald. By the result of Browder [**17**], such a manifold exists if and only if h_i^2 represents an infinite cycle in the mod 2 Adams spectral sequence. Let θ_i be a corresponding element in the homotopy groups of spheres. Then we have

THEOREM 6.2. *Let θ_i exist and satisfy $2\theta_i = 0$, $\theta_i^2 = 0$; then θ_{i+1} exists, has order* 2, *and θ_{i+2} also exists.*

In particular, $\theta_3 = \sigma^2$ satisfies these hypotheses. The theorem then gives the existence (without properties) of θ_5. Similarly, using the result of Barratt-Mahowald that θ_4 satisfies these conditions, we find $2\theta_5 = 0$ and θ_6 exists. The existence of θ_5 and θ_6 has also been demonstrated by Barratt and Mahowald by other means.

1. **The constructions.** Let W_G be any acyclic CW-complex acted on freely and cellularly by the finite group G. Suppose G is given as a subgroup of the symmetric group on p-letters $\mathscr{S}_p$; then G acts on the Cartesian product X^p (the smash product $X^{(p)}$) for any space X (space with base point X). Hence we can form the associated spaces

$$(1.1) \qquad \Gamma_G(X) = W_G \times {}_G(X^p), \qquad [S\Gamma_G(X) = W_G \times {}_G(X^{(p)})].$$

(W_G is the total space of the universal principal G-bundle over B_G, and $\Gamma_G(X)$, $[S\Gamma_G(X)]$ are associated bundles.)

AMS 1970 *subject classifications.* Primary 55D99, 55E45, 55E20, 55E35, 55H15, 57D15, 55G10, 57D99, 55-02; Secondary 55F35, 55F25.

There is also a reduced construction. The fibering $\pi: S\Gamma_G(X) \to B_G$ has a cross-section $\rho\ [\{x\} \to \{x, *\}]$. Then $R\Gamma_G(X) = S\Gamma_G(X)/\rho(B_G) = *$.

We write $\Gamma_p(X)$, $S\Gamma_p(X)$, $R\Gamma_p(X)$ for these spaces when $G = \mathscr{S}_p$. The constructions are functorial in both X and G. Thus if $K \xrightarrow{i} G \xrightarrow{j} \mathscr{S}_p$ are inclusions, we have maps f, g, h so the diagram

$$\begin{array}{ccccc} \Gamma_K(X) & \xrightarrow{f(i)} & \Gamma_G(X) & \xrightarrow{f(j)} & \Gamma_p(X) \\ \downarrow & & \downarrow & & \downarrow \\ S\Gamma_K(X) & \xrightarrow{g(i)} & S\Gamma_G(X) & \xrightarrow{g(j)} & S\Gamma_p(X) \\ \downarrow & & \downarrow & & \downarrow \\ R\Gamma_K(X) & \xrightarrow{h(i)} & R\Gamma_G(X) & \xrightarrow{h(j)} & R\Gamma_p(X) \end{array} \tag{1.2}$$

commutes. Moreover, these f, g, h are unique up to homotopy. Similarly for maps $k: X \to Y$, there are associated maps $\Gamma_G(X) \xrightarrow{\Gamma_G(k)} \Gamma_G(Y)$, etc.

In order to study the structure of these spaces, we use

THEOREM 1.3. *$R\Gamma_p(S^n)$ is the Thom space of the np-plane bundle associated to $n\mathbf{r}: \mathscr{S}_p \to O_{np}$, where $\mathbf{r}$ is the regular representation $\mathscr{S}_p \to O_p$.*

COROLLARY 1.4. *$R\Gamma_2(S^n) = \Sigma^n RP_n^\infty$.*

For a further example, using results of **[4]**, **[10]**, we have $H^*(B_{\mathscr{S}_4}, Z_2) = P(A, B, C)/AC = 0$ with A of dimension 1, B of dimension 2, and C of dimension 3. $Sq^1B = AB + C$, $Sq^2C = CB$, and $\beta_4(AB) = -\beta_4(C) = B^2$. Also, $(1 + A + B + C)$ is the Stiefel-Whitney class of the regular representation **[12]**. Thus the mod (2) structure of $H^*(R\Gamma_4(S^n))$ is readily calculated.

To the author's best knowledge, these constructions first appeared in **[13]** and were used to construct the Steenrod powers. They were also used by Dyer and Lashof **[5]** to define homology operations in the categories of iterated loop spaces, and also occurred there (in an unpublished preprint version) as essential modules in constructing a small model for $Q(X) = \lim_{n\to\infty}(\Omega^n\Sigma^n X)$.

2. **Constructing the operations in homotopy theory.** By ${}_{(q)}R\Gamma_p(X)$, we mean the q-skeleton of $W_{\mathscr{S}_p}$ crossed with $X^{(p)}$ and then reduced. If $n = q(p!)$, then the bundle $n\mathbf{r}\,|_{(q)}\, B_{\mathscr{S}_p}$ is trivial. Hence by Theorem 1.3 there is a coreduction

$$\varphi_p: {}_{(q)}R\Gamma_p(S^n) \tag{2.1}$$

of degree 1 on the Thom class.

For example with $p = 2$, $q = 1$, this is just the statement that 2ξ is trivial on the one-skeleton of RP^∞ (where ξ is the canonical line bundle). Another way of looking at the construction in this case is to observe that the obstruction to extending the identity coreduction ${}_{(0)}R\Gamma_2(S^n) = S^{2n} \to S^{2n}$ is the obstruction to making the diagram

(2.2)

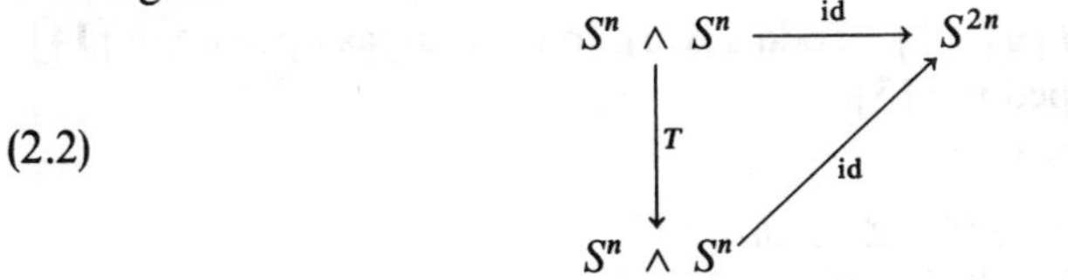

homotopy commute. In fact, this only happens for n even!

From the functoriality of the constructions, given a map $f: X \to S^n$, there is associated a map $\varphi(f): {}_{(q)}R\Gamma_p(X) \to S^{np}$. It is well defined up to homotopy, and on a suitably low skeleton of $B_{\mathscr{S}_p}$ turns out to be independent of the coreduction.

Define a map $\Delta_t: {}_{(q)}R\Gamma_p(S^{nt}) \to ({}_{(q)}R\Gamma_p(S^n))^{(t)}$ by

$$\Delta_t(x, y_{1,1} \cdots y_{1,t}, \ldots, y_{p,1} \cdots y_{p,t}) = \{(x, y_{1,1} \cdots y_{p,1}) \cdots (x, y_{1,t} \cdots y_{p,t})\}.$$

Then there are evident maps

$$(2.3) \qquad {}_{(q)}R\Gamma_p(S^{tn} \wedge X) \xrightarrow{\Delta_t} {}_{(q)}R\Gamma_p(S^n)^{(t)} \wedge {}_{(q)}R\Gamma_p(X) \xrightarrow{(\varphi^{(t)} \wedge \mathrm{id})} S^{tnp} \wedge {}_{(q)}R\Gamma_p(X),$$

and their composite is a homotopy equivalence. Now it is an easy exercise to check $(\mathrm{id} \wedge \varphi(f))(\varphi^{(t)} \wedge \mathrm{id})\Delta_t$ is homotopic to $\varphi(\Sigma^{tn}f)$. Hence $\varphi(f)$ passes to the stable category and we have

THEOREM 2.4. *There is a natural transformation in stable cohomotopy theory*

$$\varphi_\# : \pi^k(X) \to \pi^{kp}(R\Gamma_p(X))$$

associated to the maps $\varphi(f)$. Restricted to ${}_{(0)}R\Gamma_p(X)$, $\varphi_\#$ is the pth exterior product map.

Let $g: S^n \to R\Gamma_p(X)$ be given; then $g^*\varphi_\#(\alpha) \in \pi^{-n+pk}(S^n) \cong \pi_{n-pk}(S^0)$, and in this way one can construct operations in homotopy theory. For example, we have

THEOREM 2.5. *Let $n \equiv t + 1$ $(2^{h(t)})$, where $h(t) = 2t + 2^\varepsilon$, $0 \leqq \varepsilon < 4$; then there is a (nonlinear) operation*

$$Sq_t: \pi_n(S^0, Z_2) \to \pi_{2n+t}(S^0, Z_2)$$

and

$$Sq_t(a + b) = Sq_t(a) + Sq_t(b) + \theta ab + \tau a + \omega b + \gamma$$

with θ, τ, ω, γ in $\mathrm{im}(J)$. $Sq_0(a) = a^2$. Moreover, the framed manifold associated to $Sq_t(a)$ is $S^t \times {}_T M \times M = {}_{(t)}\Gamma_2(M)$ if M is associated to a. Finally, $2^{h(t)+\varepsilon}(Sq_t(a)) = 0$.

(Sq_t is defined by choosing an element in $\pi^s_{t+n}(RP_n)$ with nontrivial Hurewicz image.)

Special cases of these constructions have been analyzed by Barratt-Mahowald, F. Adams, and D. S. Kahn [6], [7]. Toda's work in these areas appears in [14], and is considerably developed in [15].

We close this section with

COROLLARY 2.6. *Let* $\alpha \in \pi_n(S^0, Z_2)$; *then*
(i) $n \equiv 3\ (4)$ *implies* $\eta\alpha^2 = 2\alpha^2 = 0$,
(ii) $n \equiv 2\ (4)$ *implies* $\eta\alpha^2 = 2Sq_1(\alpha)$,
(iii) $n \equiv 1\ (4)$ *implies* $2\alpha^2 = 0$.

(Indeed we have the diagram

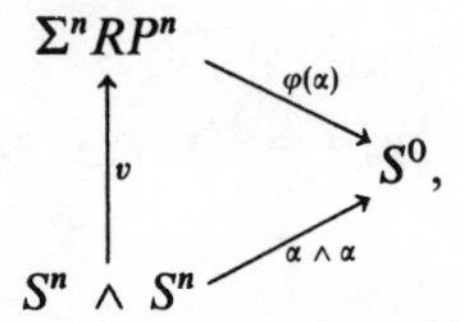

and Corollary 2.6 follows on examining the attaching maps of the first few cells.)

3. **An example, the Sq_1 series.**

LEMMA 3.1. *Let n be odd and* $X_i = S^n \cup_{2^i} e^{n+1}$; *then*

$$R\Gamma_2(X) = [\Sigma_n RP_n \cup_{2^i}(e^{2n+1})] \cup_{(W)} c\Sigma^n(RP_{n+1}),$$

and W attaches the bottom cell by
(i) $2^{i+1}(2^{i-1}\Sigma_n(e^{n+1}) - e^{2n+1})$ *for* $i > 1$,
(ii) $4(\Sigma^n e^{n+1} - e^{2n+1}) + \eta \circ (\Sigma^n(S^n))$ *for* $i = 1$.

In particular, for $i > 1$, there are Toda brackets $\{\eta, 2^{i+1}, (2^{i-1}\Sigma^n e^{n+1} - e^{2n+1})\}$, $\{\nu, 2^{i+1}, (2^{i-1}\Sigma^n e^{n+1} - e^{2n+1})\}$ which are nonzero in $\pi_*(R\Gamma_2(X))$. Similarly for $n \equiv 3\ (4)$, there is a homotopy class α_3 with nonzero Hurewicz image $e_1 \otimes (e^{n+1} \otimes e^{n+1})$ and α_5 with Hurewicz image $e_3 \otimes e^{n+1} \otimes e^{n+1}$. We have $4\alpha_5 = \eta^2\alpha_3$.

Using σ in $\pi_7(S^0)$ (σ is of order 16), we can map $S^7 \cup_{16} e^8$ into S^0. The construction above then gives homotopy classes $\beta_1(\sigma)$ corresponding to $(e^{2n+1} - 8\Sigma^n(e^{n+1}))$, $\mu_1(\sigma)$ corresponding to $\{\eta, 32, \beta_1\}$, $\gamma_1(\sigma)$ corresponding to $\{\nu, 32, \beta_1\}$. These being in odd dimensions, we can iterate the constructions.

On iterating, we find easily

LEMMA 3.2. (i) $\eta(\mu_1(\theta))^2 = 4\beta_1(\mu_1(\theta))$,
(ii) $\eta(\mu_1\beta_1(\theta)) = (\mu_1(\theta))^2$.

This gives rise, on iterating further, to infinite families of elements. (Instead of using X, we can use $S^{2n} \cup S^{2n+1} \cup_{\eta(\iota_{2n}) + 4(\iota_{2n+1})} e^{2n+2} = Y$ when iterating the construction on, say, $\mu_1(\theta)$. This leads to larger families.)

These families were first studied by M. Barratt in [3], and later were carefully explored in [2].

4. **Calculating with the construction.** Let $S^0 \supset E_1 \supset E_2 \supset \cdots \supset E_i \supset \cdots$ be an Adams Z_2-resolution of the sphere. By using the results in [**1**], D. S. Kahn and the author observed [**7**], [**12**] that it was possible to filter $R\Gamma_p(S^0)$,

$$(4.1) \quad \left\{ \mathscr{F}_r = \bigcup_{i_1+\cdots+i_p-q=r} {}_{(q)}R\Gamma_{(p)}(E_{i_1} \wedge \cdots \wedge E_{i_p} \cup \cdots \cup E_{\alpha(i_1)} \wedge \cdots \wedge E_{\alpha(i_p)}) \right\},$$

where $\alpha \in \mathscr{S}_p$, and φ_p is homotopic to a map $\tilde{\varphi}_p$ so $\tilde{\varphi}_p(\mathscr{F}_r) \subset E_r$. The spectral sequence associated to this filtration has $E^2 = \tilde{H}_*(W_{\mathscr{S}_p}, [\mathrm{Ext}^{**}_{\mathscr{A}(2)}(Z_2, Z_2)]^p)$, and $E^2(\tilde{\varphi})$ is seen to depend algebraically on $\mathscr{A}(2)$. Indeed, since $\mathscr{A}(2)$ is a cocommutative Hopf algebra, $\mathrm{Ext}_{\mathscr{A}(2)}(Z_2, Z_2)$ has operations analogous to the ordinary Steenrod squares ([**8**], [**12**]), and it is this structure which determines $E^2(\tilde{\varphi})$.

Motivated by (4.1), we set

DEFINITION 4.2. Let $f: X \to S^0$ be given; we say it is A.S. (Adams-skeletal) of degree i if f satisfies

$$f({}_{(0)}X) \subset E_i,\ f({}_{(1)}X) \subset E_{i-1}, \ldots, f({}_{(j)}X) \subset E_{i-j}.$$

Two such A.S. maps f, f' are A.S. homotopic if there is a homotopy H from f to f', and H is A.S. of degree i also.

There is a spectral sequence $\mathscr{E}$ converging to $\pi^s_*(X)$ with

$$E^2_{r,s} = \sum_k H_k(X, Z_2) \otimes \mathrm{Ext}^{k,t}_{\mathscr{A}(2)}(Z_2, Z_2),$$

and given an A.S. map $f: X \to S^0$, there is a well-defined map (varying f up to A.S. homotopies) $\mathscr{E} \to E^2(S^0)$.

THEOREM 4.3. *Let $f: X \to S^0$ be an A.S. map of degree i; then $\tilde{\varphi}_p \circ R\Gamma_p(f)$ is an A.S. map of degree 2i, and $\mathscr{E}(\tilde{\varphi}_p \circ R\Gamma_p(f))^2$ depends algebraically on $\mathscr{E}(f)^2$.*

(Again, the action of this modified Steenrod algebra on $\mathrm{im}(\mathscr{E}(f)^2)$ determines $\mathscr{E}(\tilde{\varphi}_p \circ R\Gamma_p(f))^2$.)

THEOREM 4.4 (D. S. KAHN). *Let $a \in \mathrm{Ext}^{s,n+s}_{\mathscr{A}(2)}(Z_2, Z_2)$ represent an infinite cycle in the Adams spectral sequence, and hence an element α in $\pi_n(S^0)$. Then $Sq_t(a)$ in $\mathrm{Ext}_{\mathscr{A}(2)}(Z_2, Z_2)$ is also an infinite cycle and represents $Sq_t(\alpha)$, provided $n \equiv t + 1\ (2^{h(t)})$ and $t < s$.*

Similarly, we can use Theorem 4.3 to show the existence of differentials in the Adams spectral sequence.

THEOREM 4.5. *Let $\alpha \in \mathrm{Ext}^{s,t}_{\mathscr{A}(2)}(Z_2, Z_2)$; then*

$$d_2(Sq_i(\alpha)) = \begin{cases} h_0 Sq_{i-1}(\alpha) & \text{for } t \equiv s - i\ (2), \\ 0 & \text{otherwise}. \end{cases}$$

There are, of course, analogous results for the odd primes. For further details, the reader is referred to [**7**], [**12**].

5. **Further details on the modified Steenrod algebra.** The operations referred to in §4 are indexed so that $Sq_i: \mathrm{Ext}^{r,s} \to \mathrm{Ext}^{2r-i,2s}$ are homomorphisms.

Alternatively, we define

$$Sq^i : \mathrm{Ext}^{r,s} \to \mathrm{Ext}^{r+i,2s} \quad \text{as} \quad Sq_{r-i}.$$

Then the Sq^i generate an algebra subject to relations

(5.1) $\qquad Sq^0 \neq 1,$

(5.2) $\qquad Sq^i(a) = a^2 \quad \text{for } a \in \mathrm{Ext}^{i,t},$

(5.3) $\qquad Sq^i(ab) = \sum_{r=0}^{i} Sq^r(a)Sq^{i-r}(b),$

(5.4) $\qquad Sq^a Sq^b = \sum_{j=0}^{[a/2]} \binom{b-1-j}{a-2j} Sq^{a+b-j}Sq^j \quad \text{for } a < 2b.$

In particular, $Sq^0Sq^i = Sq^iSq^0$, $Sq^1Sq^{2i} = Sq^{2i+1}Sq^0$.

Actually calculating in $\mathrm{Ext}^{**}_{\mathscr{A}(2)}(Z_2, Z_2)$, we have

THEOREM 5.5. (i) $Sq^0(h_i) = h_{i+1}$,
(ii) $Sq^3(c_0) = h_1^2 d_0$, $Sq^2 c_0 = h_0 e_0$, $Sq^1(c_0) = f_0$, $Sq^0(c_0) = c_1$,
(iii) $Sq^1(e_0) = x$, $Sq^2(e_0) = t$,
(iv) $Sq^0(c_i) = c_{i+1}$, $Sq^0(d_i) = d_{i+1}$, $Sq^0(e_i) = e_{i+1}$, $Sq^0(f_i) = f_{i+1}$.

(For notation, see [9].) Theorem 5.5 essentially determines the action of $\tilde{\mathscr{A}}(2)$ in $\mathrm{Ext}^{**}_{\mathscr{A}(2)}(Z_2, Z_2)$ for $t - s < 40$.

COROLLARY 5.6 (J. F. ADAMS). *$\alpha \in \pi_j(S^0)$ has mod 2 Hopf invariant one if and only if $j = 1, 3, 7$.*

(Since $h_0 h_i^2 \neq 0$ for $i \geq 3$, from Theorem 4.5, $d_2(h_{i+1}) = h_0 h_i^2 \neq 0$, and h_{i+1} is not an infinite cycle.)

Further work on the action and definition of $\tilde{\mathscr{A}}(2)$ and $\tilde{\mathscr{A}}(p)$ can be found in [8], [11], [12], [16].

6. **Application to constructing framed manifolds of Kervaire invariant one.** Since $Sq_0(h_i) = h_i^2$ and $Sq_2(h_i^2) = h_{i+1}^2$, it seems reasonable to hope the constructions studied above should give information on when h_i^2 can be an infinite cycle in the Adams spectral sequence. For example,

THEOREM 6.1 (BARRATT-MAHOWALD). *If θ_i exists, $\theta_i^2 = 0$ and $Sq_1(\theta_i) = 0$; then θ_{i+1} exists.*

(From the fact that in this case the map

$$\begin{array}{ccc} & \Sigma^n RP_n & \\ \nearrow & & \searrow \varphi \\ S^{2n} & \xrightarrow{\theta_i^2} & S^0 \end{array}$$

can be factored through $\Sigma^n RP_{n+2}$, and the bottom cell, by Theorem 4.3, goes to θ_{i+1}.)

More generally, Barratt and Mahowald showed that it is actually enough that θ_i have order two and $\theta_i^2 = 0$ in order to show θ_{i+1}, in effect by examining the first few cells of $R\Gamma_2(S^n \cup_2 e^{n+1})$.

We now apply the preceding theory and use the symmetry group $\mathscr{S}_4$ to show the promised

THEOREM 6.2. *Let θ_i exist and satisfy $2\theta_i = 0$, $\theta_i^2 = 0$; then θ_{i+1} exists, has order 2, and θ_{i+2} also exists.*

PROOF. Consider the extension

$$\begin{array}{ccc} S^n & \xrightarrow{\theta_i} & S^0 \\ & {\scriptstyle I}\searrow \quad \nearrow{\scriptstyle \lambda} & \\ & S^n \cup_2 e^{n+1} & \end{array} \quad . \tag{6.3}$$

Using λ, we have a map $\tilde{\varphi}_4 \circ R\Gamma_4(\lambda): R\Gamma_4(X_n) \to S^0$. Moreover, embedding $\mathscr{S}_2$ in $\mathscr{S}_4$ by sending the generator to the permutation $(1, 2)$, we find that the composition

$$R\Gamma_2(S^n) \wedge S^n \wedge S^n \xrightarrow{in} R\Gamma_4(X_n) \xrightarrow{\tau} S^0 \tag{6.4}$$

is homotopically trivial (since $\tau \circ in$ on $(S^n \wedge S^n)$ is θ^2).

Consider the cohomology structure of $R\Gamma_4(X_n)$. It has a basis of the form $H^*(\mathscr{S}_4, Z_2) \otimes (U, e^{n+1} \wedge \cdots \wedge e^{n+1})$ together with the three further elements

$$X_1 = \langle S^n, S^n, S^n, e^{n+1}\rangle, \qquad \langle S^n, S^n, e^{n+1}, e^{n+1}\rangle = X_2,$$
$$X_3 = \langle S^n, e^{n+1}, e^{n+1}, e^{n+1}\rangle.$$

Since n is congruent to 6 mod 8, the first six Stiefel-Whitney classes of $n\mathbf{r}$ are $1, 0, A^2, 0, B^2, 0, A^6 + C^2$. Also, the first three classes for $7\mathbf{r}$ are $1, A, B + A^2, A^3$. Finally, $Sq^1U = X_1$, $Sq^1X_2 = X_3$, $Sq^2X_1 = X_3$, while $Sq^4U = B^2U + (e^{n+1} \wedge \cdots \wedge e^{n+1})$. These remarks determine the action of $\mathscr{A}(2)$ in $R\Gamma_4(X_n)$.

λ and X satisfy the hypothesis of Theorem 4.3, and it is easy to check that the only classes mapping onto h_{i+2}^2 in the resulting map of spectral sequences are the classes dual to C^2U and $B(e^{n+1} \wedge \cdots \wedge e^{n+1})$. Now, construct a new map $\tilde{\tau}$ (of A.S. type 6) on the mapping cone of

$$_{(3)}R\Gamma(S^n) \wedge S^n \wedge S^n \xrightarrow{in} R\Gamma_4(X_n),$$

extending τ (by (6.4)). The resulting space Y is easily seen to have a homotopy class α (carried by its 4 skeleton), so

$$\langle h_*(\alpha), C^2U\rangle = 0, \qquad \langle h_*(\alpha), Be^{n+1} \wedge \cdots \wedge e^{n+1}\rangle = 1.$$

Then $\tilde{\tau}_*(\alpha)$ is θ_{i+2}, and Theorem 6.2 follows.

REFERENCES

1. J. F. Adams, *On the structure and applications of the Steenrod algebra*, Comment. Math. Helv. **32** (1958), 180–214. MR **20** #2711.

2. ———, *On the groups J(X)*. IV, Topology **5** (1966), 21–71. MR **33** #6628.

3. M. Barratt, *Lecture notes*, Amer. Math. Soc. Colloq. Algebraic Topology, 1963 (mimeograph).

4. H. Cardénas, *El algebra de cohomologia del groupo simétrico de grado* p^2, Bol. Soc. Mat. Mexicana (2) **10** (1965), 1–30.

5. E. Dyer and R. Lashof, *Homology of iterated loop spaces*, Amer. J. Math. **84** (1962), 35–88. MR **25** #4523.

6. D. S. Kahn, *A differential in the Adams spectral sequence*, Proc. Amer. Math. Soc. **20** (1969), 188–190. MR **38** #3866.

7. ———, *Cup-i products and the Adams spectral sequence*, Topology (to appear).

8. A. Liulevicius, *The factorization of cyclic reduced powers by secondary cohomology operations*, Mem. Amer. Math. Soc. No. 42 (1962). MR **31** #6226.

9. M. Mahowald and M. Tangora, *Some differentials in the Adams spectral sequence*, Topology **6** (1967), 349–369. MR **35** #4924.

10. R. J. Milgram, *The homology of symmetric products*, Trans. Amer. Math. Soc. **138** (1969), 251–265. MR **39** #3483.

11. ———, *Steenrod squares and higher Massey products*, Bol. Soc. Mat. Mexicana (2) **13** (1968), 32–57.

12. ———, *Group representations and the Adams spectral sequence* (preprint).

13. N. Steenrod and D. B. A. Epstein, *Cohomology operations*, Ann. of Math. Studies, no. 50, Princeton Univ. Press, Princeton, N.J., 1962. MR **26** #3056.

14. H. Toda, *Composition methods in homotopy groups of spheres*, Ann. of Math. Studies, no. 49, Princeton, Univ. Press, Princeton, N.J., 1962. MR **26** #777.

15. ———, *Extended pth powers of complexes and application to homotopy theory*, Proc. Japan Acad. Sci. **44** (1968), 198–203. MR **37** #5873.

16. A. Zachariou, Aarhus University, 1969/70 (mimeographed notes).

17. W. Browder, *The Kervaire invariant of framed manifolds and its generalizations*, Ann. of Math. (2) **90** (1969), 157–186. MR **40** #4963.

STANFORD UNIVERSITY

COBORDISM THEORY

F. P. PETERSON

1. **Introduction and background.** In these two lectures, I intend to discuss the development of cobordism theory since the Seattle conference on topology in 1963. I will concentrate on the results involving the description of the ground rings as we have already had lectures by Landweber, Smith, and Zahler giving results using the associated bordism theories.

Let me start with the basic definitions.[1] Let M^n be a closed, C^∞-manifold. Then M^n is cobordant to zero, $M^n \sim 0$, if $M^n = \partial W^{n+1}$, where W is a C^∞-manifold with boundary. More generally, $M_1^n \sim M_2^n$ if $M_1^n \cup M_2^n = \partial W^{n+1}$. This is an equivalence relation on n-manifolds and the set of equivalence classes is denoted by $\mathcal{N}_n$ and is a vector space over Z_2, the addition being given by disjoint union. Let $\mathcal{N}_* = \sum_n \mathcal{N}_n$; this is a graded algebra over Z_2, the multiplication given by Cartesian product. Let $\nu : M^n \to BO(k)$ be the normal bundle for M. Let $EO(k)$ be the universal k-plane bundle over $BO(k)$. Then the fundamental result of Thom is that $\mathcal{N}_n \approx \pi_{n+k}(MO(k))$, where $MO(k)$ is the one point compactification of $EO(k)$ and k is large. Using this, Thom proved that $\mathcal{N}_* = Z_2[x_i]$, where we have one generator x_i in each dimension which is not of the form $2^r - 1$. Let $u \in H^*(BO(k); Z_2)$. Then we can form the characteristic number $\langle \nu^*(u), [M] \rangle \in Z_2$. An important corollary of Thom's work is that $M^n \sim 0$ if and only if all Z_2 characteristic numbers are zero.

2. **More general cobordism.** Let $BG(k)$ be the classifying space for spherical fibrations, i.e. $G(k)$ is the "group" of homotopy equivalences of S^{k-1}. Let $\pi : BH(k) \to BG(k)$ corresponding to a "homomorphism" from the "group" $H(k) \to G(k)$. Then we can consider "manifolds" whose normal spherical fibration has an H-structure. Further, we can form cobordism groups of manifolds having an H-structure, Ω_*^H, and we can form the Thom space $MH(k)$. If H is small enough,

AMS 1970 *subject classifications*. Primary 57D75, 57D90, 57C20, 55G10; Secondary 16A24, 55F40.

[1] A good reference for the basic definitions and many of the results I will discuss is "Notes on Cobordism Theory" by R. E. Stong, Princeton Univ. Press, Princeton, N.J., 1968.

one has the theorem $\Omega_n^H \approx \pi_{n+k}(MH(k))$. For example, $H = SO, U$, Spin, Sp, SU, PL, SPL all work. Thus the computation of such cobordism theories reduces to a problem in homotopy theory.

Practice has taught us that the following is a good method for determining $\pi_*(MH)$. (1) Find $H^*(BH(k))$ and hence $H^*(MH(k))$ as a group. (2) Find $H^*(MH)$ as a module over $\mathscr{A}$, the Steenrod algebra. (3) Find $\operatorname{Ext}_{\mathscr{A}}(H^*(MH), Z_p)$, the E_2-term of the Adams spectral sequence. (4) Calculate E_∞ and read off $\pi_*(MH)$.

3. **Coalgebras over Hopf algebras, I.** In this section, we will show how to solve (2) above for many values of H. I will list various corollaries of a general theorem (to be stated and proved in the second lecture) and show how they apply to various cobordism theorems.

Let A be a connected Hopf algebra over Z_p. Let M be a connected coalgebra over A, i.e. M is a graded module over A and the diagonal $\psi: M \to M \otimes M$ is a map of A-modules. Let $\phi: A \to M$ be given by $\phi(a) = a(1)$, where 1 is the unit of M.

COROLLARY 3.1. *If* $\operatorname{Ker} \phi = 0$, *then M is a free A-module.*

EXAMPLES. 3.1(a). $H = O, p = 2$, and $A = \mathscr{A}_2$. Then $MO \sim \prod K(Z_2, n)$'s and we can read off Thom's results, in particular $M^n \sim 0$ if and only if all Stiefel-Whitney numbers vanish.

3.1(b). $H = PL$, $p = 2$, and $A = \mathscr{A}_2$. Then $MPL \sim \prod K(Z_2, n)$'s and PL-manifolds are detected by mod 2 characteristic numbers.

3.1(c). $H = SO$, p odd, $A = {}'\mathscr{A}_p$, the subalgebra of $\mathscr{A}_p$ consisting of the reduced powers. We read off that there is no p-torsion.

3.1(d). $H = U$, p arbitrary, $A = {}'\mathscr{A}_p$, and integral Chern classes detect U-cobordism classes.

COROLLARY 3.2. *If $A = \mathscr{A}_p\beta$, p arbitrary, then $M \approx \sum \mathscr{A}/\mathscr{A}\beta \oplus$ a free module as an $\mathscr{A}$-module.*

EXAMPLES. 3.2(a). $H = SO, p = 2$. Then

$$MSO \sim_2 \prod K(Z, n)\text{'s} \times \prod K(Z_2, m)\text{'s}$$

and Pontrjagin and Stiefel-Whitney numbers detect.

3.2(b). $H = SG$, p arbitrary. Then $MSG \sim \prod K(\pi, n)$'s and cohomology characteristic numbers detect.

3.2(c). $H = SU, p = 2$. Then one obtains the $\mathscr{A}$-structure and

$$\operatorname{Ext}_{\mathscr{A}}(H^*(MSU), Z_2).$$

COROLLARY 3.3. *Let $A = \mathscr{A}_2$. Let $Q_0 = Sq^1, Q_1 = Sq^3 + Sq^2Sq^1$. Assume $\operatorname{Ker} \phi = A/A(Sq^1, Sq^2) = A/A\bar{\mathscr{A}}_1$, where $\mathscr{A}_1$ is the subalgebra of A generated by Sq^1 and Sq^2. Assume there exists $\theta: N \to M$ such that $\theta_*: H(N, Q_i) \to H(M, Q_i)$ is an isomorphism for $i = 0, 1$. Finally assume that $N \approx A \otimes_{\mathscr{A}_1}$ (sum of cyclic $\mathscr{A}_1$-modules with none $\mathscr{A}_1$-free). Then* $\operatorname{Ker} \phi = 0$ *and* $\operatorname{Coker} \phi$ *is a free A-module.*

EXAMPLE 3.3. $H = \text{Spin}$, $N = \sum A/A(Sq^1, Sq^2) \oplus \sum A/A(Sq^3)$. $M\,\text{Spin} \sim_2 \prod$ connective BO's $\times \prod K(Z_2, n)$'s, and KO-characteristic numbers and Stiefel-Whitney numbers detect. This example will be elaborated on in the next lecture.

COROLLARY 3.4. *Similar to Corollary* 3.3 *with p odd and* $\text{Ker}\,\phi = A/A(Q_0, Q_1)$.

EXAMPLE 3.4. $H = SPL$, but we do not know N and θ. Also, we only know $\text{Ker}\,\phi \supset A/A(Q_0, Q_1)$. This example will also be discussed in the next lecture.

4. **The rest of the calculation.** For most of the above examples, steps (3) and (4) are not hard. $H = SU$ is some problem and $H = Sp$ seems very difficult. To calculate the spectral sequence for SU requires some outside geometric information given by Conner and Floyd. Determining $\text{Ext}_{\mathscr{A}}(H^*(MSp), Z_2)$ is hard and finding E_∞ is harder. Segal has partial results on this problem.

These methods are quite good at determining the image of one cobordism theory in another. For example

$$\begin{aligned} \text{Im}(\Omega_*^{fr} \to \Omega_*^{SU}) = \text{Im}(\Omega_*^{fr} \to \Omega_*^{\text{Spin}}) &= Z_2, \quad * \equiv 1, 2 \ (8), \\ &= 0, \quad \text{otherwise}. \end{aligned}$$

COROLLARY 4.1. *$K: \Omega_{8k+2}^{fr} \to Z_2$ is zero if $k \geqq 1$, where K is the Kervaire invariant.*

$\text{Im}(\Omega_*^{SU} \to \mathscr{N}_*)$ is known by work of Conner and Landweber and $\text{Im}(\Omega_*^{\text{Spin}} \to \mathscr{N}_*)$ is known.

Some more exotic groups have been studied such as Pin and Pin^c and Giambalvo has partial results on 7-connected cobordism.

Another problem one wants to answer is to find explicit manifolds which generate the groups. These are known for $H = O$, SO, and U, but are only partially known for SU and Spin, where many are given by P. G. Anderson, Conner, and Landweber, and one by Mann.

5. **Coalgebras over Hopf algebras,** II. I will start this lecture by stating and proving the theorem which has the results of §3 as corollaries.

Let B be a Hopf subalgebra of the Hopf algebra A. Assume B is a finite-dimensional vector space. Also, assume there is a collection of differentials $Q_j \in B, j \in J$, such that a B-module K is a free B-module if and only if $H(K, Q_j) = 0$ for all $j \in J$. For an A-module N, let $N^{(n)}$ be the sub A-module generated by all elements of dimension $\leqq n$.

THEOREM 5.1. *Let M be a connected coalgebra over A with* $\text{Ker}\,\phi = A\bar{B}$, *where B is as above. Assume there exists $\theta: N \to M$, an A-map, such that $\theta_*: H(N, Q_j) \to H(M, Q_j)$ is an isomorphism for all $j \in J$. Also, assume $N^{(0)} = A/A\bar{B}$ and that for every $x \in (N/N^{(n-1)})^n$, some element of B annihilates x. Then θ is a monomorphism and* Coker θ *is free, i.e. $M \approx N \oplus$ free A-module.*

PROOF. Let $\bar{M} = M/\bar{A}M$, and let $p: M \to \bar{M}$ be the natural map. Let $Z \subset M$ be a subvector space such that $p|Z$ is a monomorphism and $\bar{M} = p\theta(N) \oplus p(Z)$.

That is, Z is to be an A-base for the free module F. Let $\tilde{N} = N \oplus (A \otimes Z)$, and extend θ to $\tilde{\theta}: \tilde{N} \to M$, an A-map, by sending $1 \otimes Z \to Z \subset M$. Let $\tilde{\theta}^{(n)}: \tilde{N}^{(n)} \to M^{(n)}$. We shall prove, by induction on n, that $\tilde{\theta}^{(n)}$ is an isomorphism. By construction, $\tilde{\theta}^{(n)}$ is an epimorphism. $\tilde{\theta}^{(0)}$ is an isomorphism because $\tilde{N}^{(0)} = A/A\bar{B}$ by hypothesis and $M^{(0)} = A/A\bar{B}$ because Ker $\phi = A\bar{B}$. Assume $\tilde{\theta}^{(n-1)}$ is an isomorphism. Let $\lambda: \tilde{N}/\tilde{N}^{(n-1)} \to M/M^{(n-1)}$ be defined by $\tilde{\theta}$. Let $P =$ the B-submodule of $\tilde{N}/\tilde{N}^{(n-1)}$ generated by n-dimensional elements. We now prove two lemmas.

LEMMA 5.2. *$\lambda | P$ is a monomorphism.*

LEMMA 5.3. *Let $\{v_i\}$ be a vector space basis for P. Let $v \in \tilde{N}^{(n)}/\tilde{N}^{(n-1)}$. Then $v = \sum a_i v_i$ with $a_i = 0$ or $a_i \notin A\bar{B}$.*

PROOF OF LEMMA 5.2. A is a free B-module, by Milnor-Moore, hence $H(A, Q_j) = 0$. Thus $H(N, Q_j) \approx H(\tilde{N}, Q_j)$, and $\tilde{\theta}_*: H(\tilde{N}, Q_j) \to H(M, Q_j)$ is an isomorphism. Since $\tilde{\theta}^{(n-1)}$ is an isomorphism, by the 5-lemma we have $\lambda_*: H(\tilde{N}/\tilde{N}^{(n-1)}, Q_j) \to H(M/M^{(n-1)}, Q_j)$ is an isomorphism. λ is also an epimorphism. Let $K =$ Ker λ. Then $H(K, Q_j) = 0$ and K is a free B-module. Let $q =$ the top dimension of B.

Case 1. $K^n \cap P^n = 0$, $K \cap P \neq 0$. Let $x \in K \cap P$, $x \neq 0$. Then $bx \neq 0$, $|bx| \geqq n + q + 1$, because K is a free B-module with no elements in dimension $\leqq n$. However, $bx \in P$ and the top dimension of P is $\leqq n + q$. Contradiction.

Case 2. $K^n \cap P^n \neq 0$. Let $x \in (N/N^{(n-1)})^n$, $z \in Z^n$, such that $\lambda(x + z) = 0$. Then $\tilde{\theta}(x + z) \in M^{(n-1)}$ and $z = 0$ by construction of Z. Hence $x \in K^n$, but $bx = 0$ for some nonzero $b \in B$ by hypothesis. Since K is a free B-module, this is a contradiction.

Thus $K \cap P = 0$ and Lemma 5.2 is proved.

PROOF OF LEMMA 5.3. Let $v = \sum a_i v_i$. If $a_i \in A\bar{B}$, then $a_i = \sum a'b'$ with $a' \notin A\bar{B}$, and $b'v_i \in P$. Replace $a_i v_i$ in the sum by $\sum a'(b'v_i)$. This proves Lemma 5.3.

We now return to the proof of the theorem. To finish the induction we must show $\lambda | \tilde{N}^{(n)}/\tilde{N}^{(n-1)}$ is a monomorphism. Let $v \in \tilde{N}^{(n)}/\tilde{N}^{(n-1)}$ be such that $v \neq 0$, $\lambda(v) = 0$. Let $v = \sum a_i v_i$ as in Lemma 5.3. Let $k = \max |a_i|$, and let $a_{i_1}, \ldots, a_{i_r}$ have dimension k. Consider the image of v under the composite

$$\tilde{N}^{(n)}/\tilde{N}^{(n-1)} \xrightarrow{\lambda} M/M^{(n-1)} \xrightarrow{\psi} M \otimes M/M^{(n-1)}.$$

Using the fact that ψ is a map of A-modules, we obtain $0 = \sum_{j=1}^{r} a_{i_j}(1) \otimes \lambda(v_{i_j}) + \sum a''(1) \otimes v''$, where $|a''| < k$. This follows because $\psi\lambda(v_i) = 1 \otimes \lambda(v_i)$, $v_i \in P$ as $B(1) = 0$. By Lemma 5.2, $\lambda | P$ is a monomorphism, hence $\{\lambda(v_{i_j})\}$ are linearly independent, and $a_{i_j}(1) = 0$. Thus $a_{i_j} \in A\bar{B}$ which is a contradiction.

Corollary 3.1 follows immediately by taking $B = 1$. Corollary 3.2 follows from the fact that the map $\theta: \mathscr{A}/\mathscr{A}Sq^1 \otimes H(M, Q_0) \to M$ given by sending a Q_0-cycle $m \to m \in M$ and extending linearly induces an isomorphism on Q_0-homology. Corollary 3.3 is a direct corollary; however, to apply it to the case $H =$ Spin, we need to construct the map $\theta: N \to H^*(M \text{ Spin})$. In our paper, that was done by computing the filtration of various elements in $KO(M \text{ Spin})$ and involved a lot of hard work. I now give an elementary proof; unfortunately, this does not give the result that KO-characteristic classes detect, but it does lead to the determination of Ω_*^{Spin}.

Let $J = (j_1, \ldots, j_r)$ be a sequence of integers with $j_i > 1$. Let $n(J) = \sum j_i$. Let $P_J = P_{j_1} \cdot \ldots \cdot P_{j_r} \in H^{4n(J)}(B\,\mathrm{Spin})$ be the Pontrjagin class reduced mod 2.

THEOREM 5.4. *If $n(J)$ is even, there exists an element u_J such that $P_J + Q_0Q_1(u_J) \in \mathrm{Ker}\, Sq^1 \cap \mathrm{Ker}\, Sq^2$. If $n(J)$ is odd, there exists an element v_J such that $Sq^2(v_J) = P_J$.*

PROOF. By induction, it is enough to find u_J for $J = (2k)$ and $J = (2k + 1, 2l + 1)$. Let $Z_k = \sum_{i=0}^{\infty} W_{k-i} \cdot W_{k+2+i}$. Then easy computation shows that

$$W_{4k}^2 + Q_0Q_1(Z_{4k-3}) \in \mathrm{Ker}\, Sq^1 \cap \mathrm{Ker}\, Sq^2.$$

Also

$$W_{4k+2}^2 \cdot W_{4l+2}^2 + Q_0Q_1(Z_{4k} \cdot Z_{4l}) \in \mathrm{Ker}\, Sq^1 \cap \mathrm{Ker}\, Sq^2.$$

If $n(J)$ is odd, then $J = (j_1, J')$ with j_1 odd and $n(J')$ even. Then

$$P_J = Sq^2\{Z_{2j_1-2}(P_{J'} + Q_0Q_1(u_{J'})) + Sq^2(P_{j_1} \cdot u_{J'})\},$$

and the theorem is proven.

Set $N = \sum \mathscr{A}/\mathscr{A}(Sq^1, Sq^2) \otimes X \oplus \sum \mathscr{A}/\mathscr{A}(Sq^3) \otimes Y$, where X is a graded vector space on all J with $n(J)$ even and Y is a graded vector space on all J with $n(J)$ odd. The map θ is defined by sending $J \to U \cdot (P_J + Q_0Q_1(u_J)) \in H^*(M\,\mathrm{Spin})$ for $n(J)$ even and $J \to U \cdot v_J$ for $n(J)$ odd. Showing θ_* is an isomorphism is a computation.

6. **Super cobordism theories.** For groups H which are larger than O, we proved in Seattle that there is a Hopf algebra C^H over the mod 2 Steenrod algebra, such that $H^*(BH; Z_2) \approx H^*(BO; Z_2) \otimes C^H$ as Hopf algebras over $\mathscr{A}$. Furthermore, we proved that $\pi_*(MH) \approx \mathscr{N}_* \otimes C^{H^*}$ as algebras. C^G is now known (by May, Milgram and Tsuchiya) and C^{PL} and C^{Top} are about known. The main questions left involve Ω_*^{SPL} and $\Omega_*^{S\,\mathrm{Top}}$ and odd primes. $H^*(BSG; Z_p)$ is known and using the results, May has computed $H^*(BSPL; Z_p)$. I conjecture that Corollary 3.4 applies here but cannot as yet prove that $\mathrm{Ker}\, \phi = \mathscr{A}/\mathscr{A}(Q_0, Q_1)$ nor do I know the correct N. It seems to be true in dimensions $\leqq (p^2 + p + 1)(2p - 2)$ and I have computed the p-torsion in Ω_*^{SPL} for a range. An interesting phenomenon arises in Ω_{27}^{SPL}, where there is an element of order 9 and 3 times the generator is not detected by ordinary cohomology characteristic classes.

Finally, one can define cobordism for oriented Poincaré complexes, Ω_*^{SPD}. This is related to $\pi_*(MSG)$ by the following exact sequence of Levitt:

$$\cdots \to \pi_n(G/\mathrm{Top}) \to \Omega_n^{SPD} \to \pi_n(MSG) \to \pi_{n-1}(G/\mathrm{Top}) \to \cdots.$$

By Corollary 3.2, $MSG \sim \prod K(\pi, n)$'s so $\pi_*(MSG)$ is computable from the knowledge of $H^*(BSG)$.

MASSACHUSETTS INSTITUTE OF TECHNOLOGY

ON $\Omega^\infty S^\infty$ AND THE INFINITE SYMMETRIC GROUP

STEWART B. PRIDDY[1]

Introduction. By combining the work of Browder [3], Dyer, Lashof [5], and Nakaoka [8] we demonstrate a strong relationship between the infinite loop space $\Omega^\infty S^\infty$ and the infinite symmetric group $\mathscr{S}_\infty$. Recall that $\Omega^\infty S^\infty = \operatorname{inj}\lim \Omega^n S^n$, where $\Omega^n S^n$ is the n-fold iterated loop space on the n-sphere and that $\mathscr{S}_\infty = \operatorname{inj}\lim \mathscr{S}_n$, where $\mathscr{S}_n$ is the symmetric group on $\{1, 2, \ldots, n\}$. The homotopy groups $\pi_*(\Omega^\infty S^\infty) = \operatorname{inj}\lim \pi_{*+n}(S^n)$ are the stable homotopy groups of spheres, whereas the classifying space $B\mathscr{S}_\infty$ is an Eilenberg-Mac Lane space $K(\mathscr{S}_\infty, 1)$. Our main result is

THEOREM A. *Let $(\Omega^\infty S^\infty)_0$ be the connected component of the constant loop. Then*

1. *There is a canonical map*

$$\omega : B\mathscr{S}_\infty \to (\Omega^\infty S^\infty)_0$$

inducing isomorphisms of mod p and integral homology groups

$$\omega_* : H_*(B\mathscr{S}_\infty; Z_p) \xrightarrow{\approx} H_*((\Omega^\infty S^\infty)_0; Z_p), \qquad \omega_* : H_*(B\mathscr{S}_\infty) \xrightarrow{\approx} H_*((\Omega^\infty S^\infty)_0).$$

2. *Although $B\mathscr{S}_\infty$ is not an H-space the isomorphisms ω_* are algebra morphisms, where the monomorphisms $\mathscr{S}_m \times \mathscr{S}_n \to \mathscr{S}_{m+n}$ give $H_*(B\mathscr{S}_\infty)$ the structure of an algebra* [8] *and $H_*((\Omega^\infty S^\infty)_0)$ is an algebra under the Pontrjagin product. For mod p coefficients ω_* is a map of Hopf algebras.*

3. *The wreath products $\mathscr{S}_m \wr \mathscr{S}_n \to \mathscr{S}_{mn}$ induce operations on $H_*(B\mathscr{S}_\infty; Z_p)$ which are compatible with the Dyer-Lashof operations on $H_*((\Omega^\infty S^\infty)_0; Z_p)$* [5].

4. *In homotopy the induced map $\omega_i : \pi_i(B\mathscr{S}_\infty) \to \pi_i((\Omega^\infty S^\infty)_0)$ is zero for $i > 1$ while ω_1 is the abelianization $\mathscr{S}_\infty \to \mathscr{S}_\infty/\mathfrak{A}_\infty = Z_2$, where $\mathfrak{A}_\infty$ is the infinite alternating group.*

AMS 1970 *subject classifications.* Primary 55D35, 55D40, 55F35, 55G15; Secondary 55J10, 18H10.

[1] This talk is based on joint work with M. G. Barratt and D. S. Kahn.

While the theorem is stated in terms of topological spaces and a proof may be given in this context, a more explicit construction is obtained using simplicial sets: there is a simplicial map

$$p^{-\infty}: \overline{W}\mathcal{S}_\infty \to (\Gamma S^0)_0,$$

where $\overline{W}$ is Mac Lane's simplicial classifying functor [6] and Γ is Barratt's simplicial free group functor for the infinite loops on the infinite suspension [1]. This construction is discussed in §§1 and 2 and a proof of the theorem is given in §3.

We understand that D. Quillen has recently proved the same theorem. Substantial partial results have been obtained by Milgram [7].

1. **Barratt's functor** Γ **[1].** Let X, $*$ be a simplicial set with basepoint. Then the symmetric group $\mathcal{S}_n$ acts on the Cartesian product X^n and acts freely on the acyclic simplicial set $W\mathcal{S}_n$ [6]. Let $\mathcal{U}X = \bigcup_{n=1}^\infty (X^n \times_{\mathcal{S}_n} W\mathcal{S}_n)$ and define $\Gamma^+X = \mathcal{U}X/\mathcal{E}$, where $\mathcal{E}$ is the equivalence relation generated by $((x, *), w) \equiv (x, T_m w)$, where $x \in X^m$, $w \in W\mathcal{S}_{m+1}$ and $T_m: W\mathcal{S}_{m+1} \to W\mathcal{S}_m$ is the simplicial map defined extending the map on $\mathcal{S}_{m+1}$ defined by

(i) $T_m(\zeta\sigma) = \zeta T_m(\sigma)$ if $\zeta \in \mathcal{S}_m$,

(ii) $T_m((m+1, m, \ldots, k)) = 1$ for all k.

The formula $(x_m, w_m) \cdot (x_n, w_n) = ((x_m, x_n), J(w_m, w_n))$, where $x_m \in X^m$, $w_m \in W\mathcal{S}_m$, $x_n = X^n$, $w_n \in W\mathcal{S}_n$ and $J: W\mathcal{S}_m \times W\mathcal{S}_n \to W\mathcal{S}_{m+n}$ is the simplicial map induced by $\mathcal{S}_m \times \mathcal{S}_n \to \mathcal{S}_{m+n}$ gives Γ^+X the structure of a simplicial free monoid which contains James' monoid F^+X.

THEOREM B (BARRATT). *If X is connected then the geometric realization $|\Gamma^+X|$ is homotopy equivalent to $\Omega^\infty\Sigma^\infty |X|$.*

In case $X = S^0$ (the simplicial set on the two 0-simplices $*$, p) then

$$\Gamma^+S^0 = * \cup p \cup (p, p) \times \overline{W}\mathcal{S}_2 \cup \cdots \cup (\overbrace{p, p, \ldots, p}^{n}) \times \overline{W}\mathcal{S}_n \cup \cdots.$$

The monoid structure of Γ^+S^0 is given by $\overline{W}\mathcal{S}_n \times \overline{W}\mathcal{S}_m \to \overline{W}\mathcal{S}_{n+m}$ and $*$ is the unit. Now Γ^+S^0 has the wrong homology to be a model for $\Omega^\infty S^\infty$ (its components are not isomorphic). To correct this we convert Γ^+X into a simplicial group in a canonical way.

If M is a monoid then there is a *universal group* UM generated by M. Let FM be the free group generated by the elements of M and let N be the normal subgroup generated by xyz^{-1}, where $x, y, z \in M$ and $xy = z$ in M. Let $UM = FM/N$ and let $u: M \to UM$ be the composite $M \hookrightarrow FM \twoheadrightarrow UM$. Then u is a natural transformation and satisfies the obvious universal property (see [9]).

If M is a simplicial monoid then we may apply U in each dimension to obtain a simplicial group UM. Let $\Gamma X = U\Gamma^+X$. Let G denote Kan's simplicial loop functor.

THEOREM C (BARRATT). *If X is a simplicial set then ΓX is group homotopy equivalent to $G^\infty\Sigma^\infty X$.*

2. **The map** $p^{-\infty}:\overline{W}\mathcal{S}_\infty \to (\Gamma S^0)_0$. Given a free simplicial monoid M with $M_0 = Z^+ = \{p^i\}_{i \geqq 0}$ (such as $\Gamma^+ S^0$) let $M_\infty = \operatorname{inj\,lim}_{xp} M_i$ be the direct limit of the components M_i,

$$M_1 \xrightarrow{xp} M_2 \xrightarrow{xp} \cdots \xrightarrow{xp} M_i \xrightarrow{xp} \cdots,$$

where xp denotes right multiplication by p. The components of UM are specified as follows: for each $s \in Z$ let $(UM)_s \subset UM$ be the simplicial set consisting of all simplices $m_{a_1}m_{b_1}^{-1}m_{a_2}m_{b_2}^{-1}\cdots m_{a_k}m_{b_k}^{-1}$, where $m_{a_i} \in M_{a_i}$, $m_{b_i} \in M_{b_i}$ and $\sum a_i - \sum b_i = s$.

There is a simplicial map $p^{-\infty}: M_\infty \to (UM)_0$ given by $p^{-\infty} = \operatorname{inj\,lim}(xp^{-i})$

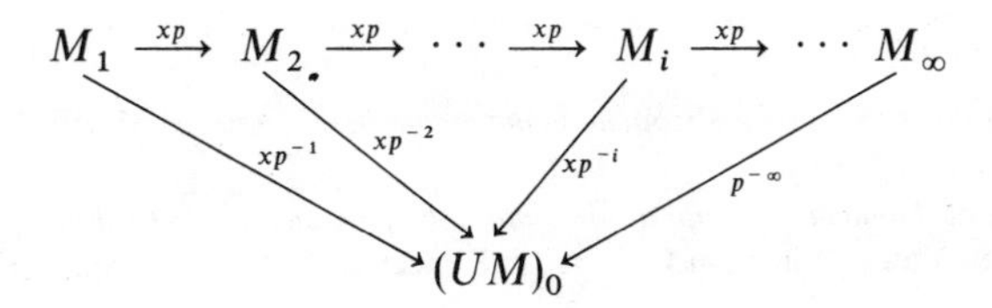

In case $M = \Gamma^+ S^0$, $M_\infty = \operatorname{inj\,lim} \overline{W}\mathcal{S}_n = \overline{W}\mathcal{S}_\infty$ and $p^{-\infty}:\overline{W}\mathcal{S}_\infty \to (\Gamma S^0)_0$.

3. **Proof of Theorem A.** Nakaoka [8] shows that the external product $H_*(\overline{W}\mathcal{S}_m) \otimes H_*(\overline{W}\mathcal{S}_n) \to H_*(\overline{W}\mathcal{S}_{m+n})$ induced by $\overline{W}\mathcal{S}_m \times \overline{W}\mathcal{S}_n \to \overline{W}\mathcal{S}_{m+n}$ is commutative. This is an immediate consequence of the fact that the operation of conjugation by an element of a group induces the identity homomorphism in homology. From this it follows that the noncommutative diagram

$$\begin{array}{ccc}
\overline{W}\mathcal{S}_n \times \overline{W}\mathcal{S}_n & \xrightarrow{(xp,xp)} & \overline{W}\mathcal{S}_{m+1} \times \overline{W}\mathcal{S}_{n+1} \\
\downarrow & & \downarrow \\
\overline{W}\mathcal{S}_{m+n} & \xrightarrow{xp^2} & \overline{W}\mathcal{S}_{m+n+2}
\end{array}$$

induces a commutative diagram of homology groups. Thus a well-defined product is defined in $H_*(\overline{W}\mathcal{S}_\infty) = \operatorname{inj\,lim} H_*(\overline{W}\mathcal{S}_n)$. Hence $p_*^{-\infty}: H_*(\overline{W}\mathcal{S}_\infty) \to H_*((\Gamma S^0)_0)$ is a map of graded algebras (for coefficients in Z or Z_p). This establishes part 2 of the theorem. Part 1 is now easy. Dyer and Lashof's computation of $H_*((\Omega^\infty S^\infty)_0; Z_p)$ shows that $p_*^{-\infty}$ is an epimorphism [5]. On the other hand Steenrod's reduced power correspondence $H_i(\overline{W}\mathcal{S}_m; Z_p) \to H^{mn-i}(SP^m(S^n); Z_p)$ given by $a \to (\iota_n)^m/a$ is an isomorphism for $i < n$, n even [9], [7]. The bijectivity of $p_*^{-\infty}$ for Z_p coefficients now follows from a simple counting argument using the known rank of $H^*(SP^m(S^n); Z_p)$ [4], [2]. Since the integral homology groups of $\overline{W}\mathcal{S}_\infty$ and $(\Gamma S^0)_0$ are finitely generated the integral isomorphism follows from the mod p case. Part 3 is a consequence of the definition of the Dyer-Lashof operations. (Another proof of part 1 can be obtained using parts 2 and 3 in conjunction with Nakaoka's calculation of $H_*(\overline{W}\mathcal{S}_\infty; Z_p)$.) Part 4 follows from the naturality of the Hurewicz homomorphism.

References

1. M. Barratt, *A free group functor for stable homotopy theory*, Proc. Sympos. Pure Math., vol. 22, Amer. Math. Soc., Providence, R.I., 1971.

2. A. Bousfield, *Operations on derived functors of non-additive functors* (to appear).

3. W. Browder, *Homology operations and loop spaces*, Illinois J. Math. **4** (1960), 347–357. MR **22** #11395.

4. A. Dold, *Homology of symmetric products and other functors of complexes*, Ann. of Math. (2) **68** (1958), 54–80. MR **20** #3537.

5. E. Dyer and R. K. Lashof, *Homology of iterated loop spaces*, Amer. J. Math. **84** (1962), 35–88. MR **25** #4523.

6. S. Mac Lane, *Constructions simpliciales acycliques*, Colloque Henri Poincaré, Paris, 1954.

7. R. Milgram, *The mod 2 spherical characteristic classes* (to appear).

8. M. Nakaoka, *Homology of the infinite symmetric group*, Ann. of Math. (2) **73** (1961), 229–257. MR **24** #A1721.

9. D. Puppe, *A theorem on semi-simplicial monoid complexes*, Ann. of Math. (2) **70** (1959), 379–394. MR **23** #A649.

10. N. Steenrod, *Cohomology operations and obstructions to extending continuous functions*, Colloquium Lecture Notes, Princeton University, Princeton, N.J., 1957 (mimeographed).

Northwestern University

ON NON-SIMPLY-CONNECTED MANIFOLDS

JULIUS L. SHANESON

Introduction. Let $h_i: K_i^n \to M^n$, n greater than four, $i = 1, 2$, be homotopy equivalences of closed piecewise linear (P.L.) manifolds. Let $\eta(h_i)$ denote their normal invariants (see §1 for definitions). One consequence of Browder-Novikov-Sullivan theory is the following:

THEOREM. *Assume that M is simply-connected. Then $\eta(h_1) = \eta(h_2)$ if and only if there is a P.L. homeomorphism φ of K_1 with K_2 such that $h_2\varphi$ is homotopic to h_1. Furthermore, the subset of $[M; G/PL]$ of elements of the form $\eta(h)$ can be calculated in terms of the homotopy and characteristic classes of M and the homotopy of G/PL.*

(As an example of the second statement, there are classes l_i in $H^{4i}(G/PL; Q)$ so that if $f: M^{4m} \to G/PL$, the homotopy class of f is of the form $\eta(h)$ if and only if the integer $(L(M)f^*(l_1 + l_2 + \cdots))[M]$ vanishes, where $L(M)$ is the total Hirzebruch class of M.)

In the non-simply-connected case, normal invariants do not in general distinguish between P.L. manifolds of the same homotopy type. One new type of invariant arises from the theorem of Rochlin. This theorem asserts that an almost parallelizable closed smooth four-manifold has index divisible by sixteen. By smoothing theory, this result is also valid for P.L. manifolds. On the other hand, for k at least two there is a closed almost parallelizable P.L. manifold of dimension $4k$ and index eight. This anomaly in dimension four leads to new invariants and to new P.L. (and smooth manifolds).

THEOREM 2.1. *Up to P.L. homeomorphism, there is a unique P.L. manifold K of the same homotopy type as $S^3 \times T^2$ but not P.L. homeomorphic to it. There is a homotopy equivalence from K to $S^3 \times T^2$ with vanishing normal invariant.*

In this paper we will discuss the above example and other examples of the

---

AMS 1970 subject classifications. Primary 57C25, 57D65; Secondary 57E25, 57E30.

effect of the theorem of Rochlin. We will also give some indication of when these new invariants do not arise despite the presence of a (free abelian) fundamental group.

1. **Surgery obstructions and normal invariants.** Let M^n, n at least five, be a closed, connected, oriented P.L. manifold. (Everything mentioned in this section generalizes to nonorientable manifolds and to manifolds with boundary.) By $ht(M)$ we denote a set of equivalence classes of homotopy equivalences $h:K \to M$ of closed P.L. manifolds. The homotopy equivalences h and h' are *equivalent* if there is a P.L. homeomorphism $\varphi:K \to K'$ with $h'\varphi$ homotopic to h.

Let $h:K \to M$ be a homotopy equivalence. Let ξ^k, k large, be the block bundle with $h^*\xi$ equivalent to the normal bundle of K, $\nu(K)$. Let $\bar{h}$ be a covering bundle for h. Then $(h, \bar{h})$ is a normal map.

In general a *normal map* (of degree one) is a map $f:P \to M$ of closed oriented manifolds of degree one, together with a bundle ξ and a bundle map $b:\nu(P) \to \xi$ over f. The normal map (f, b) determines a homotopy class $T(b)_*\alpha_P$ in $\pi_{n+k}(T(\xi))$. Here $T(\xi)$ is the Thom space of ξ and α_P the class in $\pi_{n+k}(T(\nu_P))$ of the natural collapsing map induced by an embedding of P in S^{n+k}. By the "enriched Spivak theorem" this gives a fibre-homotopy equivalence of ξ with $\nu(M)$, and so a fibre-homotopy trivialization of $\xi \oplus \nu(M)^{-1}$. The classifying space for isotopy classes of such is G/PL; in fact we have a bijection of cobordism classes of normal maps with $[M; G/PL]$, the set of homotopy classes of maps from M to G/PL. In particular $(h, \bar{h})$ determines the normal invariant $\eta(h)$ of h in $[M; G/PL]$.

Given a normal map (f, b), we can take its surgery obstruction $s(f, b)$, an element of the abelian group $L_n(\pi_1 M)$. This obstruction depends only upon the cobordism class of (f, b) and vanishes if and only if this class contains a homotopy equivalence. Taking surgery obstructions gives a well-defined map

$$s:[M; G/PL] \to L_n(\pi_1 M)$$

with Image $\eta = s^{-1}(0)$. (Surgery obstructions for normal maps were defined by Browder and Novikov in the simply-connected case and generalized by C. T. C. Wall. See [**1**] and [**9**].)

Let h and ξ be as above, and let β be an arbitrary element of $L_{n+1}(\pi_1 M)$. Then there is a normal map (F, B), $F: W \to M \times I$ and $B:\nu(W) \to \xi \times I$ satisfying the following: (I = interval $[0, 1]$)

(i) $\partial_- W = M$ and $F \mid \partial_- W = (h, 0)$;

(iii) $F \mid \partial_+ W:\partial_+ W \to M \times 1$ is a homotopy equivalence; and

(iii) $s(F, B) = \beta$.

Let $[h]$ denote the class of h in $ht(M)$. Put

$$\beta[h] = [F \mid \partial_+ W].$$

This defines a well-defined action of the Wall group $L_{n+1}(\pi_1 M)$ on $ht(M)$. It is not hard to see that two elements of $ht(M)$ have the same normal invariant if and only if they lie in the same orbit under this action.

This completes a brief review of Browder-Novikov-Sullivan theory. See [**1**], [**8**], and [**9**] for details.

To calculate Wall groups in the cases we need, we use the following special case of Theorem 5.1 of [**6**].

THEOREM 1.1. *Let M^{n-1} be a closed, orientable, connected P.L. manifold. Let $\pi = \pi_1 M$. Assume* $\mathrm{Wh}(\pi \times Z) = 0$, *and $n \gneq 5$. Then there is a split exact sequence*

$$0 \longrightarrow L_{n+1}(\pi) \xrightarrow{j_*} L_{n+1}(\pi \times Z) \xrightarrow{\alpha(M)} L_n(\pi) \longrightarrow 0.$$

REMARKS. 1. For π free abelian, $\mathrm{Wh}(\pi \times Z) = 0$.

2. The map j is the natural inclusion and j_* is the induced map.

3. The map $\alpha(M)$ is defined as follows: Let β be an element of $L_{n+1}(\pi \times Z)$. Let (F, B), $F: W \to M \times S^1 \times I$ and B a covering map on normal bundles, be a normal map satisfying:

(i) $F \mid \partial_- W: \partial_- W \to M \times S^1 \times 0$ is a P.L. homeomorphism;

(ii) $F \mid \partial_+ W: \partial_+ W \to M \times S^1 \times 1$ is a homotopy equivalence; and

(iii) $s(F, B) = \beta$.

We may take F to be transverse to $M \times I \subset M \times S^1 \times I$ and by [**2**] we may further assume, without losing (i)–(iii), the following:

(iv) If $P = F^{-1}(M \times I)$, the restriction of F is a homotopy equivalence of $\partial_+ P$ with $M \times 1$. (We say that $F \mid \partial_+ W$ is *split along* M.) Then we define

$$\alpha(M)\beta = s(F \mid P, B \mid P).$$

The homomorphism really depends not on M but only on how $\pi_1 M$ is identified with π.

A right inverse to $\alpha(M)$ is given by taking products with a circle.

4. In general Theorem 1.1 is valid modulo 2-torsion. The general formula is $L^s_{n+1}(\pi \times Z) = L_n(\pi) \oplus L^s_n(\pi)$. $L^s_m = L_m$ modulo 2-torsion, and are exactly equal when the Whitehead group of the fundamental group vanishes.

5. The same result holds in the smooth category. Since handle-theory is now available in the topological category in high enough dimensions [**4**], it follows (by folklore) that Theorem 1.1 holds in the topological category.

Finally, we recall the periodicity of surgery obstructions [**9**]. $L_n(\pi) = L_{n+4}(\pi)$ and $s(f, b) = s((f, b) \times CP^2)$, where CP^2 is complex projective 2-space. Also, $\alpha(M \times CP^2) = \alpha(M)$.

2. **An example.** Let $M = S^3 \times S^1 \times S^1$. We wish to study the P.L. manifolds of the homotopy type of M. (Recall that by smoothing theory every such manifold will have a unique smoothing.)

Now

$$[M; G/PL] = Z \oplus Z \oplus Z_2.$$

(To see this, note that G/PL is a loop space and the suspension of the disjoint union of M and a point is a one point union of spheres.) Furthermore, the map s vanishes on the element of order two and is monic on the two infinite cyclic summands. In fact $L_5(Z \oplus Z) = Z \oplus Z$ and the image of s is precisely $2Z \oplus 2Z$. For given a normal map (f, b) into $S^3 \times T^2$, $s(f, b)$ is the pair of integers which are the simply-connected surgery obstructions (index-type obstructions) of the following normal maps:

$$(f, b) \mid f^{-1}(S^3 \times S^1 \times pt), \quad \text{and} \quad (f, b) \mid f^{-1}(S^3 \times pt \times S^1).$$

(Of course, we first make f transverse along these submanifolds.) The divisibility by two follows from Rochlin's theorem and the fact that everything divisible by two is in the image follows from the existence of an almost parallelizable manifold of index sixteen and dimension four. Thus we have only one nontrivial normal invariant to consider.

Let $g: M \to M$ be a map such that $g \mid M - pt$ is homotopic to the identity and such that the obstruction to homotopy of g with the identity is the nonzero element of $H^5(M; \pi_5 M) = Z_2$. Then one can show that $\eta(g) \neq 0$. Thus normal invariants do not lead to anything new in the present example. (See §6 of **[6]** and §6 of [7] for more detailed discussions of similar situations.)

It remains to study the action of $L_6(Z \oplus Z)$. We have $L_6 = L_{10}$, and $L_{10}(Z \oplus Z) = Z \oplus Z_2$, by Theorem 1.1. The element of order two is in the image of the natural map of $Z_2 = L_6(e)$ into $L_6(Z \oplus Z)$ by Theorem 1.1. There is a normal map (f, b) into S^6, b mapping into the trivial bundle over S^6, with $s(f, b) \neq 0$. See **[5]**.

Let $h: K \to M$ be a homotopy equivalence. Let c be a covering bundle map of normal bundles. (See **[6]** or use the fact that the natural map of $[M; G/PL]$ to $[M; BPL]$ has the element of order two in its kernel.) It is not hard to see that

$$s((h \times \mathrm{id}) \# f, (c \times \mathrm{id}) \# b) \neq 0,$$

where "#" denotes connected sum in the interior and the identity map is that of the unit interval. (This is an example of the "local character" of surgery obstructions.) Thus the element of order two acts trivially.

Next we show that twice the infinite cyclic generator also acts trivially. Let P be a closed almost parallelizable manifold of index sixteen and dimension four. Let P_0 be the complement of an open P.L. 4-disk. Let

$$f: (P_0, \partial P_0) \to (D^4, S^3)$$

be a map of degree one, and let b be the bundle map over f of normal bundles. Then the surgery obstruction $s(f, b)$ is twice the generator of $L_4(e) = Z$.

Let $Q = (S^3 \times I) \amalg P_0$, where $\amalg$ denotes boundary connected sum taken, say, along the upper boundary. Let $G = \mathrm{id} \amalg f$ and let $B = \mathrm{id} \amalg b$. We want to calculate $s((G, B) \times T^2)$ using Theorem 1.1. To do this we must consider $\beta = s((G, B) \times T^2 \times CP^2)$, since Theorem 1.1 does not apply in dimension six. It is not hard to see from the definition that

$$\alpha(S^3 \times CP^2) \circ \alpha(S^3 \times CP^2 \times S^1)\beta = s((G, B) \times CP^2).$$

By periodicity for simply-connected surgery obstructions (due to Sullivan), the right side above is twice a generator of the infinite cyclic group $L_8(e)$. Hence β is twice an infinite cyclic generator and hence so is $s((G, B) \times T^2)$. Since connected sum with a sphere changes nothing, this shows that twice a generator acts trivially.

Let (F, C) be a normal map into $S^3 \times T^2 \times I$ with the following properties:

(i) $F \mid \partial_- W : \partial_- W \to S^3 \times T^2 \times 0$ is a P.L. homeomorphism, and

(ii) $F \mid \partial_+ W : \partial_+ W \to S^3 \times T^2 \times 1$ is a homotopy equivalence, and

(iii) $\mu = s(F, C)$ is an infinite cyclic generator of $L_6(Z \oplus Z)$.

Let $K = \partial_+ W$ and let $h : K \to S^3 \times T^2$ be the restriction of F. Then $[h] = \mu[\text{id}]$ and $[gh] = \mu[g]$. (Recall that g was a self equivalence of $S^3 \times T^2$ with non-vanishing normal invariant.) So K is the only candidate for a new P.L. manifold of the homotopy type of M.

Suppose that K were P.L. equivalent to M. Let π be the natural projection of $M = S^3 \times T^2$ on T^2. It is not hard to see that there is a P.L. (even smooth) homeomorphism U of T^2 with itself with πh homotopic to $U\pi$. By the covering homotopy property and the homotopy extension property we may assume, without disturbing (i)–(iii) above, that $\pi h = U\pi$. After (two) further homotopies of F, we can arrange to have the following nested sequence of transverse images and maps:

$$\begin{array}{ccc} W & \xrightarrow{F} & S^3 \times T^2 \times I \\ \cup & & \cup \\ F^{-1}(S^3 \times S^1 \times I) & \longrightarrow & S^3 \times S^1 \times I = \pi^{-1}(S^1) \times I \\ \cup & & \cup \\ Q = F^{-1}(S^3 \times I) & \xrightarrow{F|Q} & S^3 \times I = \pi^{-1}(pt) \times I. \end{array}$$

From the definition of α it is clear that

$$\alpha(S^3 \times CP^2)\alpha(S^3 \times S^1 \times CP^2)\mu = s((F \mid Q, C \mid Q) \times CP^2).$$

Hence the simply-connected obstruction $s(F \mid Q, C \mid Q)$ is a generator of $L_4(e)$, by Theorem 1.1 and periodicity. This obstruction is computed as one-eighth the index of bilinear form of intersection numbers on two-dimensional homology classes. Note that ∂Q is the disjoint union of two standard three-spheres. Let Z be obtained from Q by capping off the spheres. Then it is not hard to verify that Z has the following two contradictory properties:

(i) Z is an almost parallelizable closed P.L. manifold of dimension four, and

(ii) Z has index eight.

(Recall that the index is defined by intersection numbers on homology classes or cup products on cohomology classes.) Thus our hypothesis that K is P.L. equivalent to $S^3 \times T^2$ is incorrect. This completes an outline of the proof of the following result:

THEOREM 2.1. *Up to P.L. homeomorphism, there is a unique P.L. manifold homotopy equivalent to but not P.L. homeomorphic to $S^3 \times T^2$. There is a homotopy equivalence of this manifold with $S^3 \times T^2$ that has a vanishing normal invariant. The set $ht(S^3 \times T^2)$ has four elements.*

Let K be the unique manifold of the preceding theorem. Then one can show the following:

THEOREM 2.2. *Every finite covering space of K is P.L. homeomorphic to K.*

Using the fact that the natural map of $\pi_4(G/PL)$ into $\pi_4(G/\mathrm{TOP})$ send a generator to twice a generator [**4**] and using topological surgery, one can show that K is homeomorphic to $S^3 \times T^2$.

If L is a homotopy three-sphere, then its product with T^2 is either K or $S^3 \times T^2$. Recall that every homotopy three-sphere bounds a parallelizable manifold. Using techniques similar to the above, one can show the following:

PROPOSITION 2.3. *Let L be a homotopy three-sphere. Then $L \times T^2$ is P.L. homeomorphic to K if and only if L bounds a parallelizable manifold of index eight.*

3. **Fake tori.** Let T^n be the product of n circles. Then one can show (see [**3**]) that

$$\eta : ht(T^n) \to [T^n; G/PL]$$

is the trivial map. The proof that requires the fewest words goes as follows: The various standard subtori of T^n form a characteristic variety of T^n in the sense of [**8**]. The Farrell fibering theorem or the Farrell-Hsiang splitting theorem and periodicity of surgery obstructions imply that for a homotopy equivalence $h:\tau^n \to T^n$ the surgery obstructions along the various components of the characteristic variety all vanish. The characteristic variety theorem asserts that in this case, $\eta(h) = 0$. (There is a proof that does not use the characteristic variety theorem.)

We now define invariants for fake tori. Let J be a subset of $\{1, \ldots, n\}$. Let $T(J)$ be the subtorus of T^n determined by requiring that for i not in J, the ith coordinate remains fixed at a basepoint. Let J have *three* elements, and let

$$T(J) = M_0 \subset M_1 \subset \cdots \subset M_k = T^n \qquad (k = n - 3)$$

be a sequence of standard subtori, each of codimension one in the next. We assume from now on that *n is at least five*.

Let $h:\tau^n \to T^n$ be a homotopy equivalence of P.L. manifolds. Let (f, b) be a normal map, $f: W \to T^n \times I$, with $f \mid \partial_- W : \partial_- W \to T^n \times 0$ a P.L. homeomorphism and with $\partial_+ W = \tau^n$ and $f \mid \partial_+ W = (h, 1)$. This exists by the preceding paragraph. We consider $(f, b) \times CP^2$. Then $f \times \mathrm{id}$ is homotopic relative $\partial_- W$ to a map g that is transverse to $M_{k-1} \times CP^2 \times I$ and so that $g \mid \partial_+ W$ is *split* along $M_{k-1} \times CP^2 \times 1$. (See Remark 3 following Theorem 1.1 and in particular condition (iv) in that remark.) Let $(f_k, b_k) = (g, b \times CP^2)$. Let $W_k = W$. We define $W_{k-1} = f_k^{-1}(M_{k-1} \times CP^2 \times I)$. Let $(\bar{f}_{k-1}, b_{k-1})$ be the restriction of (f_k, b_k) to W_{k-1}. Let f_{k-1} be homotopic to $\bar{f}_{k-1}$ relative to $\partial_- W_{k-1}$, transverse to $M_{k-2} \times CP^2 \times I$, and split along $M_{k-2} \times CP^2 \times 1$. Continuing inductively, we finally arrive at a normal map (f_0, b_0), $f_0 : W_0 \to T(J) \times CP^2 \times I$, which is a homotopy equivalence on the boundary (in fact a P.L. homeomorphism on the lower boundary). So we can take the surgery obstruction $s(f_0, b_0)$ in $L_8(Z^3)$. Let $\alpha_J(h)$ be the reduction mod two of the image of this obstruction in $L_8(e) = Z$ under the natural map.

NOTE. One can split the above normal cobordisms along submanifolds with high codimension to obtain normal cobordisms which restrict to homotopy equivalences of boundaries. This procedure definitely does *not* lead to invariants in the present case; it is crucial to proceed, as above, one codimension at a time. The discussion on p. 224 of [**9**] greatly obscures this point.

THEOREM 3.1 ([**3**] and [**9**]). *The invariant*

$$\alpha_J : ht(T^n) \to Z_2$$

is well defined. The α_J *with* $J \subset \{1, \ldots, n\}$ *with three elements is a full set of invariants for* $ht(T^n)$, *and all possible invariants are realized.*

The proof of Theorem 3.1 is similar to the argument in §2. The main elements are Theorem 1.1 and the theorem of Rochlin, which is crucial in the proof that α_J is well defined.

Note that every homotopy equivalence of T^n with itself is homotopic to a P.L. homeomorphism. Hence Theorem 4.1 implies the existence of "fake tori".

Let x be in $ht(T^n)$. The correspondence $[T(J)] \to \alpha_J(x)$ is a map $H_3(T^n) \to Z_2$; i.e. a cohomology class $\alpha(x)$ in $H^3(T^n; Z_2)$. Clearly Theorem 3.1 is equivalent to the assertion that α is a bijection.

Let $p : T^n \to T^n$ be a covering map, in the P.L. category. Given $h : \tau^n \to T^n$, we get an induced covering $h^*(p) : M \to \tau^n$ and a bundle covering map $\bar{h} : M \to T^n$ over h. Let $p^t[h] = [\bar{h}]$. This defines $p^t : ht(T^n) \to ht(T^n)$.

THEOREM 3.2 ([**3**] and [**9**]). $\alpha p^t = p^*\alpha$.

COROLLARY 3.3. *Let* τ^n *be a P.L. manifold of the homotopy type of* T^n, n *at least five. Then every odd finite cover of* τ *is P.L. homeomorphic to* τ. *There are even finite covers of* τ *that are P.L. homeomorphic to* T^n.

The second part of this corollary was used by Kirby [**12**] to prove the generalized annulus conjecture. The first part was used in [**4**] to show that there is a fake torus homeomorphic to T^n. In fact, they all are.

For further details and generalizations, see [**3**] and [**9**]. For example, there is the set $ht(M, \partial M)$ of equivalence classes of homotopy equivalences which are P.L. homeomorphisms on the boundary. The equivalence relation is analogous to that for the absolute case, except that homotopies must all do nothing on the boundary. One has the following:

THEOREM 3.4. *There is a one-one correspondence between*

$$ht(D^k \times T^n, S^{k-1} \times T^n) \quad \text{and} \quad H^{3-k}(T^n; Z_2).$$

This correspondence is natural with respect to finite covering maps.

This result is also used in [**4**] and in [**11**].

4. **Avoiding Rochlin's theorem.**

THEOREM 4.1. *Let M^n be a closed, connected, orientable P.L. manifold of dimension at least five and with $\pi_1 M = Z^k$, the free abelian group on k generators. Assume the following*:

1. *$n - k$ is at least four;*
2. *There is a locally flat embedding $T^k \subset M$ with trivial normal bundle which induces an isomorphism of fundamental groups.*

Then if $h:K \to M$ is a homotopy equivalence, K a closed P.L. manifold, h is homotopic to a P.L. homeomorphism if and only if $\eta(h) = 0$.

An example of J. Morgan shows that hypothesis 2 cannot be eliminated. In any case, it is always satisfied for $\pi_1 M = Z$. Here are two consequences of Theorem 4.1 for this case.

COROLLARY 4.2. *Let h_1 and h_2 be homotopy equivalences of the closed P.L. manifolds K_1 and K_2 with M. Assume $\pi_1 M = Z$, $\dim M \geqq 5$, M is closed, and M is orientable. Then $[h_1] = [h_2]$ in $ht(M)$ if and only if $\eta(h_1) = \eta(h_2)$.*

COROLLARY 4.3. *If $h:K \to M$ is a homeomorphism of P.L. manifolds, M as in Corollary 4.2, then h is homotopic to a P.L. homeomorphism if and only if an obstruction $o(h)$ in the two torsion of $H^4(M; Z_2)$ vanishes.*

Corollary 4.3 follows because there is an obstruction $o(h)$ as above to the triviality of $\eta(h)$ in case h is a homeomorphism [**8**]. Y. Matsumoto has also proven Corollary 4.3 under more restrictive hypotheses.

To derive Corollary 4.2, let $f:K_1 \to K_2$ be a homotopy equivalence with $h_2 f$ homotopic to h_1. By Proposition 2.2 of [**7**],

$$\eta(h_1) = \eta(h_2) + (h_2^{-1})^* \eta(f).$$

Hence $\eta(f) = 0$. So by Theorem 4.1, f is homotopic to a P.L. homeomorphism, as desired.

To prove Theorem 4.1, let $T^k \times D^r \subset M$, $k + r = n$, be a neighborhood of T^k in M. Let β be an arbitrary element of $L_{n+1}(Z^k)$. Let (f, b) be a normal map, $f: W \to T^k \times D^r \times I$, satisfying the following:

(i) $\partial W = \partial_- W \cup \partial_+ W$, where $\partial_- W$ meets $\partial_+ W$ in their common boundary and where f restricts to a P.L. homeomorphism of $\partial_- W$ with $(T^k \times D^r \times 0) \cup (T^k \times S^{k-1} \times I)$ and to a homotopy equivalence of $\partial_+ W$ with $T^k \times D^r \times I$;

(ii) $s(f, b) = \beta$.

Let M_0 be the closure of the complement of $T^k \times D^r$. Let Q be obtained from the disjoint union of $M_0 \times I$ and W by identifying (x, t) in $\partial M_0 \times I$ with $f^{-1}(x, t)$. Let $F = f \cup \mathrm{id}$ and let $B = b \cup \mathrm{id}$. Then it is not hard to see that $s(F, B) = \beta$. (This is the local character of surgery obstructions again.)

Thus if $K = \partial_+ Q$, then $[F \mid \partial_+ Q] = \beta[\mathrm{id}]$. But by Theorem 3.4,

$$F \mid \partial_+ W : \partial_+ W \to T^k \times D^r \times 1$$

is homotopic relative to the boundary to a P.L. homeomorphism. Hence $F \mid \partial_+ Q$ represents the same class as the identity map in $ht(M)$. Thus $L_{n+1}(Z^k)$ acts trivially, which implies the desired result.

REFERENCES

1. W. Browder, *Surgery on simply-connected manifolds* (to appear).

2. F. T. Farrell and W.-C. Hsiang, *Manifolds with* $\pi_1 \times_\alpha G$ (to appear). See also: *A geometric interpretation of the Künneth formula for algebraic K-theory*, Bull. Amer. Math. Soc. **74** (1968), 548–553.

3. W. C. Hsiang and J. L. Shaneson, *Fake tori*, Topology of Manifolds, Proc. Univ. of Georgia Topology of Manifolds Inst. (1969), Markham, Chicago, Ill., 1970. See also: *Fake tori, the annulus conjecture, and the conjectures of Kirby*, Proc. Nat. Acad. Sci. U.S.A. **62** (1969), 687–691.

4. R. C. Kirby and L. C. Seibenmann, (to appear). See also: *On the triangulation of manifolds and the Hauptvermutung*, Bull. Amer. Math. Soc. **75** (1969), 742–749. MR **39** #3500.

5. S. P. Novikov, *Homotopically equivalent smooth manifolds*. I, Izv. Akad. Nauk SSSR Ser. Mat. **28** (1964), 365–474; English transl., Amer. Math. Soc. Transl. (2) **48** (1965), 271–396. MR **28** #5445.

6. J. L. Shaneson, *Wall's surgery obstruction groups for* $G \times Z$, Ann. of Math. (2) **90** (1969), 296–334. MR **39** #7614.

7. ———, *Non-simply-connected surgery and some results in low dimensional topology*, Comment. Math. Helv. **45** (1970), 333–352.

8. D. Sullivan, *Triangulating and smoothing homotopy equivalences*, Princeton University, Princeton, N.J., 1967. (mimeographed notes).

9. C. T. C. Wall, *Surgery on compact manifolds*, Academic Press, New York, 1970.

10. ———, *Homotopy tori and the annulus theorem*, Bull. London Math. Soc. **1** (1969),

11. R. K. Lashof and M. G. Rothenberg, *Triangulation of manifolds*. I, II, Bull. Amer. Math. Soc. **75** (1969), 750–754. MR **40** #895.

12. R. C. Kirby, *Stable homeomorphisms and the annulus conjecture*, Ann. of Math. (2) **89** (1969), 575–582. MR **39** #3499.

PRINCETON UNIVERSITY

ON THE EILENBERG-MOORE SPECTRAL SEQUENCE

LARRY SMITH

In these lectures we will attempt to survey some of the ideas revolving around the spectral sequence introduced by Eilenberg and Moore [6] to tie together the homology relations in fibre squares. This spectral sequence has proved to be a useful computational tool in many cases, particularly in the study of Postnikov type decompositions of H-spaces.

The general set-up may be described as follows. We suppose given a fibre square

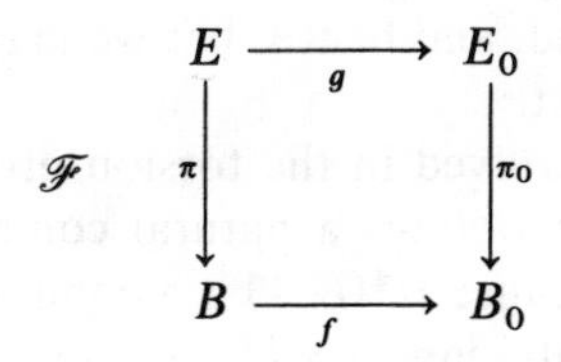

that is,

(1) $\pi_0 : E_0 \to B_0$ is a fibration.

(2) $\pi : E \to B$ is the fibration induced from π_0 by pullback along f.

Let us also assume that B_0 is 1-connected. Let k be a field, and denote by $H^*(\ ;k)$ the singular cohomology functor with coefficients in k. Then if suitable finiteness conditions are satisfied Eilenberg and Moore construct a spectral sequence $\{E_r(\mathscr{F}), d_r(\mathscr{F})\}$ such that

$$E_r(\mathscr{F}) \Rightarrow H^*(E;k),$$
$$E_2^{s,t}(\mathscr{F}) = \operatorname{Tor}^{s,t}_{H^*(B_0;k)}(H^*(B;k), H^*(E_0;k)),$$

which relates the various cohomology algebras of $\mathscr{F}$.

AMS 1970 *subject classifications*. Primary 55H20; Secondary 55F99.

REMARKS AND EXPLANATIONS. Let us first explain the term $E_2(\mathscr{F})$ in a little detail. From the diagram $\mathscr{F}$ we obtain the diagram

$$\begin{array}{ccc} H^*(E; k) & \xleftarrow{g^*} & H^*(E_0; k) \\ \uparrow{\scriptstyle \pi^*} & & \uparrow{\scriptstyle \pi_0^*} \\ H^*(B; k) & \xleftarrow{f^*} & H^*(B_0; k). \end{array}$$

With the aid of this diagram we may impose an $H^*(B_0; k)$-module structure on $H^*(B; k)$ by the composite

$$H^*(B; k) \otimes H^*(B_0; k) \xrightarrow{1 \otimes f^*} H^*(B; k) \otimes H^*(B; k) \xrightarrow{\Delta^*} H^*(B; k),$$

where Δ^* is the multiplication in the algebra $H^*(B; k)$. (The verification of the module identities uses the fact that f^* is a map of algebras.) In a similar manner we obtain an $H^*(B_0; k)$-module structure on $H^*(E_0; k)$. It is with respect to these module structures that

$$\operatorname{Tor}^{*,*}_{H^*(B_0;k)}(H^*(B; k), H^*(E_0; k))$$

is formed. Notice that this torsion product is *bi*graded. The first grading, called the external or resolution degree, is just that, the resolution degree. The second grading, called the internal grading, arises from the fact that all the modules and algebras involved are graded, and hence that we may form graded resolutions to compute the torsion product.

Now as the algebras involved in the torsion product are all commutative it follows that the above Tor obtains a natural commutative bigraded k algebra structure. This is good, because $H^*(E; k)$ is a commutative graded k algebra and we would like the spectral sequence $\{E_r(\mathscr{F}), d_r(\mathscr{F})\}$ to compute this algebra structure for us. Indeed it does.

A further word on gradings is now in order. The spectral sequence $\{E_r(\mathscr{F}), d_r(\mathscr{F})\}$ is a *second* quadrant *cohomology* spectral sequence. That is

$$E_2^{s,t}(\mathscr{F}) = 0\colon \text{ unless } s \leqq 0, t \geqq 0$$

(actually unless $2s + t \leqq 0$). Notice that as we are using upper index notation for the resolution degree of our torsion products this is consistent with the usual sign changing conventions for raising and lowering indices. The differentials $d_r(\mathscr{F})$ have bidegrees $(r, 1 - r)$, and are derivations of the algebra structure.

Next we note the map

$$\begin{array}{rcl} \pi^* \otimes g^*: H^*(B; k) \otimes H^*(E_0; k) & \longrightarrow & H^*(E; k) \otimes H^*(E; k) \\ & & \downarrow{\scriptstyle \text{multiply}} \\ & & H^*(E_0; k) \end{array}$$

factors through $H^*(B; k) \otimes_{H^*(B_0;k)} H^*(E_0; k)$ and yields a map

$$\pi^* \overline{\otimes} g^*: H^*(B; k) \otimes_{H^*(B_0;k)} H^*(E_0; k) \to H^*(E; k).$$

The spectral sequence has an edge homomorphism

$$e: E_2^{0,*}(\mathscr{F}) \to H^*(E; k)$$

and upon recalling that

$$E_2^{0,*}(\mathscr{F}) = H^*(B; k) \otimes_{H^*(B_0;k)} H^*(E_0; k)$$

it is not hard to show that the requirements of naturality entail the equality $e = \pi^* \overline{\otimes} g^*$.

To summarize (at least in part):

THEOREM (EILENBERG-MOORE). *Suppose that k is a field and*

$$\mathscr{F} \quad \begin{array}{ccc} E & \longrightarrow & E_0 \\ \downarrow & & \downarrow \\ B & \longrightarrow & B_0 \end{array}$$

is a fibre square with B_0 1-connected and all spaces of finite type. Then there exists a natural second quadrant cohomology spectral sequence of algebras $\{E_r(\mathscr{F}), d_r(\mathscr{F})\}$ such that

$$E_r(\mathscr{F}) \Rightarrow H^*(E; k),$$
$$E_2^{s,t}(\mathscr{F}) = \operatorname{Tor}^{s,t}_{H^*(B_0;k)}(H^*(B; k), H^*(E_0; k)).$$

A rather important special case of this spectral sequence is that where $B = *$, a one point space. One then has $E = \pi_0^{-1}(*) = F$, the fibre of the fibration π_0. Thus in this instance one has

$$E_r(\pi_0) \Rightarrow H^*(F; k),$$
$$E_2^{s,t}(\pi_0) = \operatorname{Tor}^{s,t}_{H^*(B_0;k)}(k, H^*(E_0; k)),$$

a spectral sequence that computes the cohomology of the fibre from the structure of the cohomology of the total space as a module over the cohomology of the base. Thus in particular if $\pi_0: E_0 \to B_0$ is the path space fibration over B_0 then one obtains

$$E_r(B_0) \Rightarrow H^*(\Omega B_0; k),$$
$$E_2^{s,t}(B_0) = \operatorname{Tor}^{s,t}_{H^*(B_0;k)}(k, k).$$

This latter spectral sequence was originally obtained by J. F. Adams **[1]** in slightly different guise.

Before turning to some elementary examples of computations with these spectral sequences we will examine an alternate formulation of the Eilenberg-Moore spectral sequence that has led to a recent new construction for it (see for example **[18]**, **[19]**, **[20]**).

1. **On generalized Künneth theorems.** We will here explain the point of view that makes it possible to regard the Eilenberg-Moore spectral sequence as a Künneth spectral sequence. We will fix throughout the remainder of this discussion a topological space B. We introduce the category Top/B, *of topological spaces over* B. An object of Top/B is a map $f: T(f) \to B$ the domain of f being denoted by $T(f)$ and often referred to as the total space of f. If $f, g \in \operatorname{obj}$ Top/B, a morphism $\phi: f \to g$ in Top/B from f to g is a commutative triangle

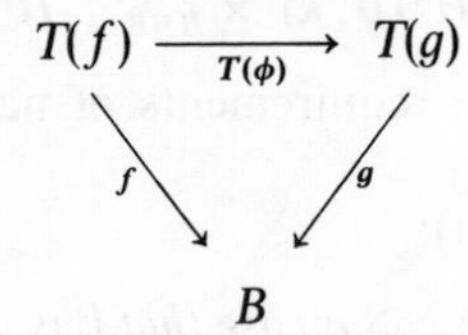

Note that if $B = *$ then the resulting category is just the usual category of spaces. The category Top/B and its pointed analog enjoys many of the properties of the category of all spaces. The usual constructions of homotopy theory may be made in these categories, such as cofibrations, suspensions, Puppe sequences, etc. (The reader is referred to [7], [8], [11] where some if not all of these things are done.) Note that the category Top/B has products, the familiar fibre product. More precisely, if $f, g \in \operatorname{obj}$ Top/B, define $f \times_B g \in \operatorname{obj}$ Top/B by

$$T(f \times_B g) = \{(x, y) \in Tf \times Tg \mid fx = gy\}$$

and

$$f \times_B g : T(f \times_B g) \to B \quad \Big| \quad fx = (f \times_B g)(x, y) = gy.$$

We may introduce the usual commutative diagram

$$\begin{array}{ccc} T(f \times_B g) & \xrightarrow{T(\pi_g)} & T(g) \\ {\scriptstyle T(\pi_f)}\downarrow & \searrow^{f \times_B g} & \downarrow{\scriptstyle g} \\ T(f) & \xrightarrow{f} & B \end{array}$$

defining the morphisms

$$f \xleftarrow[\pi_f]{} f \times_B g \xrightarrow[\pi_g]{} g$$

in Top/B. One readily checks that this is a product in Top/B.

Suppose that $\mathscr{H}^*(\ )$ is a cohomology theory on Top. We may prolong $\mathscr{H}^*(\ )$ to a sort of cohomology theory $\mathscr{H}^*_B(\ )$ on Top/B by setting $\mathscr{H}^*_B(f) = \mathscr{H}^*(Tf)$, and similarly for pairs and morphisms. The morphism $\mathrm{id}_B : B \to B$ is a point or terminal object in Top/B and so the coefficients of the prolonged theory ought to be $\mathscr{H}^*_B(\mathrm{id}_B)$.

With these facts in mind we may now ask what form the Künneth theorem *ought* to take for the cohomology theory $\mathscr{H}^*_B(\ )$ on the category Top/B. A moment's reflection suggests that under suitable regularity conditions we ought to have a spectral sequence $\{E_r(f, g), d_r(f, g)\}$ with

$$E_r(f, g) \Rightarrow \mathscr{H}_B^*(f \times_B g),$$

$$E_2(f, g) = \mathrm{Tor}_{\mathscr{H}_B^*(\mathrm{id}_B)}(\mathscr{H}_B^*(f), \mathscr{H}_B^*(g)).$$

Rewriting this in terms of our original theory $\mathscr{H}^*(\ )$ gives

$$E_r(f, g) \Rightarrow \mathscr{H}^*(Tf \times_B Tg),$$

$$E_2(f, g) = \mathrm{Tor}_{\mathscr{H}^*(B)}(\mathscr{H}^*(Tf), \mathscr{H}^*(Tg))$$

which is an Eilenberg-Moore type spectral sequence.

This reformulation as a Künneth theorem is actually quite useful, as it allows one to employ the geometric method Atiyah invented to handle the Künneth theorem in K-theory, to construct the Eilenberg-Moore spectral sequence and generalizations thereof. The details may be found in [**19**], [**20**].

2. **Elementary applications.** The key to applying the E-M spectral sequence to make practical computations in a specific problem rests on the ability to compute the term E_2 in closed form, that is to compute

$$\mathrm{Tor}_\Lambda(\Gamma', \Gamma''),$$

where Λ is a commutative algebra over a field k and Γ' and Γ'' are Λ algebras. Without the presence of some additional structures such computations constitute an unduly hard way to earn a living. We will therefore begin by considering some of the algebraic tools that ease the pain of computation. We can only indicate briefly some of these techniques which we will then apply to compute a few elementary examples. We will consider more sophisticated examples in the next section.

Some algebra. Most of the success or failure of a computation involving the E-M spectral sequence rests on the computation of $E_2(\ )$. For the various computational devices that have been successfully employed we refer to [**10**], [**14**], [**15**], [**16**], [**19**] of which we will only skim the surface.

Notations and conventions. Throughout this section k will denote a fixed field. All k modules will be positively graded and bigraded ones will live in the second quadrant. If V is a graded k module we denote by $S[V]$ the free graded commutative algebra generated by V. If the characteristic of k is 2 then

$$S[V] = k[V],$$

the graded polynomial algebra generated by V. If the characteristic of k is different from 2 then

$$S[V] = k[V^+] \otimes E[V^-],$$

where $V^+ \subset V$ is the submodule of elements of even degree, $V^- \subset V$ the submodule of elements of odd degree, and $E[W]$ denotes the exterior algebra generated by W.

DEFINITION. If Λ is a commutative k algebra, a sequence $\{\lambda_i \mid \lambda_i \in \Lambda\}$ is called an ESP-sequence in Λ iff λ_{i+1} is not a zero divisor in $\Lambda/(\lambda_1, \ldots, \lambda_i)$, for $i = 0, 1, \ldots$, where $\lambda_0 = 0$.

An ideal $I \subset \Lambda$ is called a Borel ideal iff $I = (\lambda_1, \lambda_2, \ldots)$ with $\{\lambda_i \mid \lambda_i \in \Lambda\}$ an ESP-sequence in Λ.

EXAMPLE. The sequence $x_1^{r_1}, \ldots, x_n^{r_n} \in k[x_1, \ldots, x_n]$ is always an ESP-sequence.

Notice that if $\{\lambda_i \mid \lambda_i \in \Lambda\}$ is an ESP-sequence in Λ then the elements $\lambda_1, \lambda_2, \ldots$ are algebraically independent. Hence $k[\lambda_1, \lambda_2, \ldots] \subset \Lambda$. (The converse is false, as xy, $yz \in k[x, y, z]$ are algebraically independent but not an ESP-sequence.) Moreover Λ is free as a module over $k[\lambda_1, \lambda_2, \ldots]$. (See for example [**15**, II§1-2].)

DEFINITION. An algebra Γ is called "nice" iff $\Gamma = S[V]/I$ for some V, where I is a Borel ideal.

If Γ is a nice k algebra then we may write

$$\Gamma \cong S[v_1, v_2, \ldots]/(w_1, w_2, \ldots),$$

where $w_1, w_2, \ldots \in S[v_1, v_2, \ldots]$ is an ESP-sequence with no w_i indecomposable. That is, there are no redundant generators in the collection $v_1, v_2, \ldots$.

The "nice" thing about nice k algebras is that one can compute $\operatorname{Tor}_\Gamma(k, k)$ in closed form in a fairly painless manner. The result is easily stated when k has characteristic zero, so we will confine our attention to that case for a moment.

PROPOSITION. *Suppose that k is a field of characteristic zero and Γ is a nice k algebra. Then*

$$\operatorname{Tor}_\Gamma(k, k) = S[s^{-1}V] \otimes k[s^{-2}W],$$

where

$$s^{-1}V = \operatorname{Tor}_\Gamma^{-1,*}(k, k) = Q\Gamma, \qquad s^{-2}W \subset \operatorname{Tor}_\Gamma^{-2,*}(k, k).$$

REMARKS AND EXPLANATIONS. The proof of the preceding result is not difficult and proceeds roughly as follows. Represent Γ as

$$\Gamma \cong S[v_1, v_2, \ldots]/(w_1, w_2, \ldots)$$

with $\{w_i \mid w_i \in S[v_1, v_2, \ldots]\}$ an ESP-sequence and no w_j indecomposable. Introduce the bigraded algebra

$$\mathcal{R} = \Gamma \otimes S[s^{-1}v_1, \ldots] \otimes k[s^{-2}w_1, \ldots],$$

where

$$\begin{aligned} \operatorname{bideg} \gamma &= (0, \deg \gamma) \colon \in \Gamma, \\ \operatorname{bideg} s^{-1}v_i &= (-1, \deg v_i) \colon i = 1, 2, \ldots, \\ \operatorname{bideg} s^{-2}w_j &= (-2, \deg w_j) \colon j = 1, 2, \ldots. \end{aligned}$$

It is then possible (see for example [**16**, §1], [**21**]) to introduce a differential ∂ in $\mathcal{R}$ compatible with the algebra structure so as to make it an acyclic complex. Then $(\mathcal{R}, \partial)$ is a free Γ resolution of k yielding the stated result.

If k has characteristic different from zero it is necessary to replace $k[s^{-2}W]$ by the algebra with divided powers generated by $s^{-2}W$. We refer to **[16]**, **[21]** for details.

We are now prepared to discuss some very elementary examples.

EXAMPLE 1. *The cohomology of ΩX when $H^*(X; k)$ is nice.* As a very special case of the E-M spectral sequence we have, as noted previously, a spectral sequence

$$E_r \Rightarrow H^*(\Omega X; k),$$
$$E_2 = \mathrm{Tor}_{H^*(X;k)}(k, k),$$

where X is a simply connected space with cohomology of finite type and k is a field. The preceding algebraic computations now lead easily to:

THEOREM. *Suppose that X is a simply connected cw-complex with cohomology of finite type. Let k be a field of characteristic 0 and suppose moreover that $H^*(X; k)$ is nice. Then the E-M spectral sequence of the pathspace fibration $\Omega X \to PX \to X$ has $E_2 = E_\infty$.*

PROOF. According to the computations in our algebraic prelims we have

$$E_2 \cong \mathrm{Tor}_{H^*(X;k)}(k, k) = S[s^{-1}V] \otimes k[s^{-2}W],$$

as k algebras. Hence the algebra E_2 is generated by $E_2^{-1,*}$ and $E_2^{-2,*}$ under multiplication. Now recall that bideg $d_r = (r, 1 - r)$. Therefore

$$d_r(E_2^{-1,*}) \subset E^{r-1,*} = 0 : r \geqq 2,$$

$$d_r(E_2^{-2,*}) \subset E_2^{r-2,*} = 0 : \begin{cases} r > 2, \\ r = 2, \quad * \neq 0. \end{cases}$$

As $E_2^{0,0} = k$ is the unit of E_2 it is never a boundary. Thus we see that each d_r must vanish on the algebra generators of E_2 for essentially degree reasons. As d_r is a derivation for each r we must conclude that $d_r = 0$; $r \geqq 2$. Hence $E_2 = E_\infty$ as claimed.

Notice that actually more is true. As

$$E_2 = S[s^{-1}V] \otimes k[s^{-2}W]$$

is a free commutative algebra there is actually no extension problem for the algebra structure, that is

$$H^*(\Omega X; k) = S[s^{-1}V] \otimes k[s^{-2}W]$$

as algebras, although not naturally.

Similar results may be obtained mod p in many cases. (But not always collapse!)

Examples of spaces to which the theorem applies are: $CP(n)$, $HP(n)$, S^k, G/H where G is a compact connected Lie group and $H \subset G$ is a closed connected subgroup of maximal rank, or finally an H-space X which is simply connected. Plugging these cases in, a whole host of familiar results fall out.

EXAMPLE 2. *Stiefel fiberings.* Let us denote by $V_{n,r}$ the homogeneous space $SO(n)/SO(n-r)$. Suppose that $\xi = (E, \pi, B, V_{n,r}, SO(n))$ is a fibre bundle with fibre $V_{n,r}$ and structural group $SO(n)$. Then we have a classifying diagram for ξ,

$$\begin{array}{ccc} E & \longrightarrow & BSO(n-r) \\ \pi \downarrow & & \downarrow \rho \\ B & \xrightarrow[f]{} & BSO(n) \end{array}$$

where $(BSO(n), \rho, BSO(n-r), V_{n,r}, SO(n))$ is a universal bundle and f is a classifying map for ξ.

For the sake of some numerical computations let us fix an odd prime p and suppose that n and r are even. Recall

$$H^*(BSO(2t); \mathbf{Z}_p) = \mathbf{Z}_p[\mathfrak{p}_1, \ldots, \mathfrak{p}_{t-1}, X_{2t}],$$

where $\mathfrak{p}_i$ are the reductions of the universal Pontrjagin classes and X_{2t} of the universal Euler class. Suppose that $n = 2m$ and $n - r = 2s$. Then

$$\rho^*: H^*(BSO(n); \mathbf{Z}_p) \to H^*(BSO(n-r); \mathbf{Z}_p)$$

is given by

$$\rho^* \begin{cases} \mathfrak{p}_i \to \mathfrak{p}_i & : \quad 1 \leqq i \leqq s-1, \\ \mathfrak{p}_s \to X_{2s}^2, & \\ \mathfrak{p}_i \to 0 & : \quad s+1 \leqq i \leqq m-1, \\ X_{2m} \to 0. & \end{cases}$$

Associated with the classifying diagram for ξ we have the E-M spectral sequence

$$E_r \Rightarrow H^*(E; \mathbf{Z}_p),$$

$$E_2 = \operatorname{Tor}_{H^*(BSO(2m); \mathbf{Z}_p)}(H^*(B; \mathbf{Z}_p), H^*(BSO(2s); \mathbf{Z}_p)).$$

The computation of E_2 may be accomplished in many cases with the aid of the two-sided Koszul complex [**3**]. By employing some change of rings results the following may be established.

THEOREM. *With the preceding notations suppose that the bundle ξ satisfies*

$$\mathfrak{p}_{s+1}(\xi) = \cdots = \mathfrak{p}_{m-1}(\xi) = X_{2m}(\xi) = 0.$$

Then there is a filtration on $H^*(E; \mathbf{Z}_p)$ such that

$$E_0H^*(E; \mathbf{Z}_p) = H^*(B; \mathbf{Z}_p)(\mathfrak{p}_s(\xi))^{1/2} \otimes E[u_{s+1}, \ldots, u_{m-1}, v_{2m}],$$

where bideg $v_{2m} = (-1, 2m)$, bideg $u_i = (-1, 4i)$, $i = s+1, \ldots, m-1$.

OUTLINE OF PROOF. First of all by $H^*(B; \mathbf{Z}_p)(\mathfrak{p}_s(\xi))^{1/2}$ we mean just that, the algebra obtained from $H^*(B; \mathbf{Z}_p)$ by adjoining a square root of $\mathfrak{p}_s(\xi)$ of grade $2s$.

The idea of the proof is to compute the E-M spectral sequence of the classifying diagram of ξ. We begin with E_2. Our hypotheses on ξ show by a simple change of rings argument that

$$E_2 \cong \operatorname{Tor}_{H^*(BSO(2m);\, \mathbf{Z}_p)}(H^*(B;\, \mathbf{Z}_p),\, H^*(BSO(2s);\, \mathbf{Z}_p))$$
$$\cong \operatorname{Tor}_{\mathbf{Z}_p[\mathfrak{p}_1,\dots,\mathfrak{p}_s]}(H^*(B;\, \mathbf{Z}_p),\, H^*(BSO(2s);\, \mathbf{Z}_p)) \otimes \operatorname{Tor}_{\mathbf{Z}_p[\mathfrak{p}_{s+1},\dots,\mathfrak{p}_{m-1}, X_{2m}]}(\mathbf{Z}_p, \mathbf{Z}_p).$$

Now $\rho^*(\mathfrak{p}_1), \dots, \rho^*(\mathfrak{p}_s) \in H^*(BSO(2s); \mathbf{Z}_p)$ is an ESP-sequence. Therefore since ρ^* is injective $H^*(BSO(2s); \mathbf{Z}_p)$ is free as a $\mathbf{Z}_p[\mathfrak{p}_1, \dots, \mathfrak{p}_s]$-module. Combining this with the fact that $\mathbf{Z}_p[\mathfrak{p}_{s+1}, \dots, \mathfrak{p}_{m-1}, X_{2m}]$ is super nice we see that

$$E_2 = H^*(B; \mathbf{Z}_p)(\mathfrak{p}_s(\xi))^{1/2} \otimes E[u_{s+1}, \dots, u_{m-1}, v_{2m}]$$

as algebras. Thus $E_2^{0,*}$ and $E_2^{-1,*}$ generate E_2 as an algebra. So since

$$\begin{aligned} d_r(E_2^{0,*}) &\subset E_2^{r,*} = 0, \\ d_r(E_2^{-1,*}) &\subset E_2^{r-1,*} = 0, \end{aligned} \qquad r \geqq 2,$$

we again obtain $E_2 = E_\infty$ for "placement" reasons. The result now follows from preceding computations.

EXAMPLE 3. *A theorem of Whitehead on the cohomology suspension.* Let X be a simply connected space and

$$\Omega X \to PX \xrightarrow[\pi]{} X$$

the path space fibering on X. One definition of the cohomology suspension σ^* is as the composite

$$\sigma^* : H^*(X, *) \xrightarrow[\pi^*]{} H^*(PX, \Omega X) \xleftarrow[\delta^*]{} H^*(\Omega X).$$

It is well known that σ^* vanishes on decomposable elements so we obtain

$$\sigma^* : QH^*(X; k) \to H^*(\Omega X; k),$$

where k is a field and $Q(\ \ )$ denotes the indecomposable functor. We now obtain:

THEOREM (G. W. WHITEHEAD). *If X is q-connected, $q \geqq 1$, then the cohomology suspension*

$$\sigma^* : QH^*(X; k)^{n+1} \to H^n(\Omega X; k)$$

is a monomorphism for $n \leqq 3q$.

OUTLINE OF PROOF. We consider the E-M spectral sequence for the path space fibration on X. We have seen before that in any E-M spectral sequence $E_2^{-1,*}$ consists entirely of infinite cycles. It is also easy to see that

$$s^{-1}QH^*(X; k) = \operatorname{Tor}^{-1,*}_{H^*(X;k)}(k, k).$$

As $E_2^{0,*} = 0$ for $* \neq 0$ and $E_2^{0,0} = k$, the unit of all the algebras $E_r^{*,*}$, we are able to define a k linear map

$$\phi^* : QH^*(X; k) \to F^{-1}H^*(\Omega X; k) \subset H^*(\Omega X; k)$$

of degree -1 by sending $u \in QH^*(X; k)$ to the class of u in $E_\infty^{-1,*}$ regarded as an element of $F^{-1}H^*(\Omega X; k)$. According to [15], $\phi^* = \sigma^*$. It will therefore suffice to show that no element of $E_2^{-1,n+1}: n \leqq 3q$ is a d_r boundary for any $r \geqq 2$.

To this end recall that $H^n(X; k) = 0: 0 < n \leqq q$. Therefore we see by use of the bar construction that

$$E_2^{-t,s} = \operatorname{Tor}_{H^*(X;k)}^{-t,s}(k, k) = 0: s \leqq tq.$$

Recall that

$$d_{t-1}: E_2^{-t,s} \to E_2^{-1,s+t-2}$$

and so we see that

$$\operatorname{Im}\{d_{t-1}: E^{-t,s} \to E^{-1,s+t-2}\} = 0: s \leqq tq.$$

Suppose now that $u \in E_2^{-1,n+1}$. Write $n + 1 = s + t - 2: t > 2$ and suppose that $n \leqq 3q$. Then

$$s + t - 2 \leqq 3q + 1 \leqq tq + t - 2$$

provided $t > 2$. Thus $s \leqq tq$ and hence u is not a d_{t-1} coboundary for any $t > 2$ and the result follows.

EXAMPLE 4. $H^*(\Omega\Sigma X; k)$. In this example we will begin to make use of the Hopf algebra structure present in the spectral sequence in many cases.

Suppose that B is a simply connected space. Then it may be shown that the spectral sequence

$$E_r \Rightarrow H^*(\Omega B; k), \qquad E_2 = \operatorname{Tor}_{H^*(B;k)}(k, k)$$

is a spectral sequence of Hopf algebras, that is, each E_r is a bigraded Hopf algebra and d_r preserves both the algebra and coalgebra structure.

For any space B one has the natural maps

$$\sigma: \Sigma\Omega B \to B, \qquad \eta: B \to \Omega\Sigma B$$

expressing the adjointness of the loop and suspension functors. The composite

$$H^*(B; k) \xrightarrow{\sigma^*} H^*(\Sigma\Omega B; k) \xrightarrow{\Sigma^{-1}} H^*(\Omega B; k)$$

is (up to a sign) the cohomology suspension. When $B = \Sigma X$ one also has the map

$$\Sigma\eta: \Sigma X \to \Sigma\Omega\Sigma X$$

and the composite

$$\Sigma X \xrightarrow{\Sigma\eta} \Sigma\Omega\Sigma X \xrightarrow{\sigma} \Sigma X$$

is known to be homotopic to the identity map. Therefore, we obtain that

$$\sigma^*: H^*(\Sigma X; k) \to H^*(\Omega\Sigma X; k)$$

is monic.

THEOREM. *Let X be a connected space and k a field. Then there is a filtration on $H^*(\Omega\Sigma X; k)$ such that*

$$E_0 H^*(\Omega\Sigma X; k) = \bar{B}(H^*(\Sigma X; k))$$

as Hopf algebras, where $\bar{B}(\ \ )$ denotes the reduced bar construction.

PROOF. Consider the E-M spectral sequence

$$E_r \Rightarrow H^*(\Omega\Sigma X; k), \qquad E_2 = \mathrm{Tor}_{H^*(\Sigma X;k)}(k, k).$$

Now of course $H^*(\Sigma X; k)$ is a trivial algebra, i.e. the product of any two elements of positive degree is zero. So the bar resolution shows

$$\mathrm{Tor}_{H^*(\Sigma X;k)}(k, k) = \bar{B}(H^*(\Sigma X; k))$$

as Hopf algebras. Recall that the coproduct in $\bar{B}(\ )$ is given by

$$\Delta[x_1|\cdots|x_n] = \sum_{r=1}^{n} [x_1|\cdots|x_r] \otimes [x_{r+1}|\cdots|x_n],$$

and so one easily sees that $H^*(\Sigma X; k) \subset \bar{B}(H^*(\Sigma X; k))$ is the submodule of primitives. Thus if $P(\ )$ denotes the primitive functor as applied to Hopf algebras $PE_2 = E_2^{-1,*}$.

We claim that the spectral sequence collapses. To see this we require a simple but potent lemma whose proof (left to the reader) is a tiny computation.

LEMMA. *Suppose that (E, d) is a differential Hopf algebra. Let $x \in E$ be an element of minimal degree such that $dx \neq 0$. Then dx is primitive.*

With the aid of this lemma we now see that $E_2 = E_\infty$ as follows. Suppose that $E_2 \neq E_\infty$. Let r be the smallest integer such that $d_r \neq 0$. Then $E_2 = E_r$. Let $x \in E_2$ have minimal degree such that $d_r x \neq 0$. Then by the lemma

$$d_r x \in PE_2^{*,*} = E_2^{-1,*}.$$

We have seen before that

$$H^*(\Sigma X; k) \to E_2^{-1,*} \to E_\infty^{-1,*} \hookrightarrow H^*(\Omega\Sigma X; k)$$

is the cohomology suspension. Therefore

$$d_r x \neq 0 \in \ker \sigma^*$$

which is contrary to the fact noted previously that σ^* is monic. Therefore our original assumption that $E_2 \neq E_\infty$ must be false and the result follows from the preceding computations.

3. **Stable Postnikov systems (2-primary).** In this section we will examine an application of the E-M spectral sequence to the study of Postnikov systems. It is from this example and its odd primary analogs that many of the results of [**10**], [**13**], [**17**], etc. find their origin.

DEFINITION. If π is an abelian group we denote by $K(\pi, n)$ the Eilenberg-Mac Lane space of type (π, n). If G is a graded abelian group we set

$$K(G) = \mathop{\mathsf{X}}_{i=0}^{\infty} K(G_i, i)$$

and refer to $K(G)$ as a GEM (= generalized Eilenberg-Mac Lane space).

A GEM $K(G)$ is an H-space in a natural way, with the H-structure derived from the addition in G. (Note in general $K(G)$ has lots of other H-structures; these are not of concern to us here.)

DEFINITION. A two stage Postnikov system is a fibre square

$$\mathcal{F}\quad \begin{array}{ccc} E & \longrightarrow & E_0 \\ \downarrow & & \downarrow \\ B & \xrightarrow{f} & B_0 \end{array}$$

where B, B_0 are GEM's and $E_0 \to B_0$ is the path space fibration.

The cohomology with field coefficients of GEM's was long ago computed by J. P. Serre [**12**] and H. Cartan [**4**]. These computations were important in the study of primary cohomology operations. In the study of cohomology operations of the second kind the two stage Postnikov systems play an analogous role. Their cohomology is therefore important. We will concentrate our attention on a particular class of two stage Postnikov systems associated with stable operations of the second kind.

DEFINITION. A two stage Postnikov system $\mathcal{F}$ is called stable iff f is an H-map.

In this section we will outline a proof of:

THEOREM. *Suppose that*

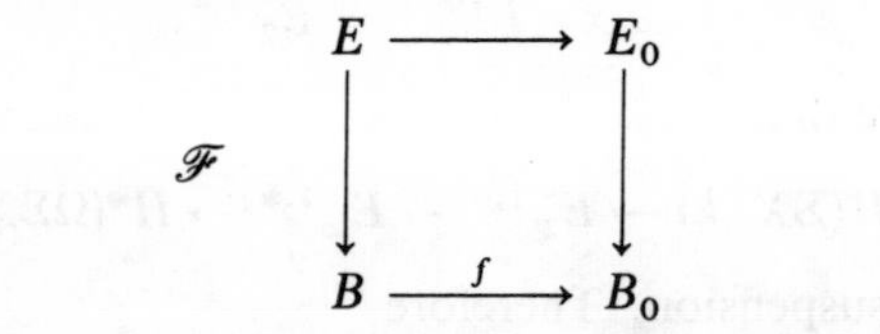

is a stable two stage Postnikov system of finite type with B_0 1-connected. Then the E-M spectral sequence with coefficients in $\mathbf{Z}_2$ of $\mathcal{F}$ collapses. Hence

$$E_0H^*(E; \mathbf{Z}_2) = \mathrm{Tor}_{H^*(B_0; \mathbf{Z}_2)}(H^*(B; \mathbf{Z}_2), \mathbf{Z}_2)$$

as bigraded Hopf algebras over $\mathcal{A}^*(2)$.

REMARKS. Actually as part of the proof we will compute the indicated torsion product as an algebra over $\mathcal{A}^*(2)$ in terms of f^* or, what is the same thing, the k-invariants of $\mathcal{F}$. It is also possible to obtain the algebra structure of $H^*(E; \mathbf{Z}_2)$, although we will not do this here, see for example [**14**]. The extension problem for the $\mathcal{A}^*(2)$ structure is much more delicate and has only been obtained in a few special cases. A recent example of C. Schochet has shown that if $\mathcal{F}$ is not stable then $E_2(\mathcal{F})$ need not equal $E_\infty(\mathcal{F})$.

We will present an outline of the proof of the preceding result as it will indicate (although not in detail) some of the tricks and gimmicks for making computations with the E-M spectral sequence.

Outline of Proof. The important "additional ingredient" of a *stable* two stage Postnikov system that will enable us to compute $E_2(\mathscr{F})$ in closed form is the presence of H-space and Hopf algebra structures. Note that if $\mathscr{F}$ is a stable two stage Postnikov system then all the spaces in $\mathscr{F}$ are H-spaces and all the maps are H-maps. Therefore

$$H^*(E; \mathbf{Z}_2), \qquad H^*(B; \mathbf{Z}_2), \qquad H^*(B_0; \mathbf{Z}_2)$$

are Hopf algebras over $\mathbf{Z}_2$ and f^* and π^* are maps of Hopf algebras. It is not too difficult to show that $\{E_r(\mathscr{F}), d_r(\mathscr{F})\}$ is a spectral sequence of Hopf algebras, that is, each $E_r(\mathscr{F})$ is a Hopf algebra and $d_r(\mathscr{F})$ is a derivation thereon. Let us turn our attention to the computation of

$$E_2(\mathscr{F}) = \mathrm{Tor}_{H^*(B_0; \mathbf{Z}_2)}(H^*(B; \mathbf{Z}_2), \mathbf{Z}_2).$$

It may be shown using Hopf algebra techniques **[10]**, **[14]** that

$$\ker\{f^* : H^*(B_0; \mathbf{Z}_2) \to H^*(B; \mathbf{Z}_2)\}$$

is a Borel ideal. Thus we may find an ESP-sequence $\{\lambda_i \mid \lambda_i \in H^*(B_0; \mathbf{Z}_2)\}$ such that

$$(\lambda_1, \ldots) = \ker f^*.$$

Let $\Lambda \subset H^*(B_0; \mathbf{Z}_2)$ be the subalgebra with 1 generated by $\{\lambda_i\}$. Since $\{\lambda_i\}$ are algebraically independent we have

$$\Lambda = \mathbf{Z}_2[\lambda_1, \ldots, \lambda_n, \ldots].$$

We are now in a position to apply the change of rings spectral sequence **[5**, XVI.6.1a**]** to conclude

$$\mathrm{Tor}_{H^*(B_0; \mathbf{Z}_2)}(H^*(B; \mathbf{Z}_2), \mathbf{Z}_2) = H^*(B, \mathbf{Z}_2)/\!/f^* \otimes \mathrm{Tor}_\Lambda(\mathbf{Z}_2, \mathbf{Z}_2),$$

where we have written (see for example **[10]**) $H^*(B; \mathbf{Z}_2)/\!/f^*$ for

$$H^*(B; \mathbf{Z}_2) \otimes_{H^*(B_0; \mathbf{Z}_2)} \mathbf{Z}_2.$$ [1]

Now $\Lambda = \mathbf{Z}_2[\lambda_1, \ldots, \lambda_n, \ldots]$ is a super nice algebra (see §2) and so we have $\mathrm{Tor}_\Lambda(\mathbf{Z}_2, \mathbf{Z}_2) = E[\mu_1, \mu_2, \ldots]$, where $\deg \mu_i = (-1, \deg \lambda_i)$. Thus we have shown

$$E_2(\mathscr{F}) = H^*(B; \mathbf{Z}_2)/\!/f^* \otimes E[\mu_1, \ldots]$$

as algebras (actually as Hopf algebras) over $\mathbf{Z}_2$. (This situation should be familiar, as it is somewhat like the situation of Example 2 of §2.) Namely, as an algebra

[1] Some of the details run as follows. The situation that we are in is this. Write Γ for $H^*(B_0; \mathbf{Z}_2)$. We have $\Lambda \subset \Gamma$ a normal subalgebra over which Γ is a free Λ-module. Setting $\Omega = \Gamma/\!/\Lambda$ we have a spectral sequence $E_r \Rightarrow \mathrm{Tor}_\Gamma^{*,*}(H^*(B; \mathbf{Z}_2), \mathbf{Z}_2)$, $E_2 = \mathrm{Tor}_\Omega^*(H^*(B; \mathbf{Z}_2), \mathbf{Z}_2) \otimes \mathrm{Tor}_\Lambda^*(\mathbf{Z}_2, \mathbf{Z}_2)$, there being no local coefficients as Λ is central in Γ. Now note that $\Omega \subset H^*(B; \mathbf{Z}_2)$ as a sub-Hopf algebra as $\Omega = \operatorname{Im} f^*$ and f^* is a morphism of Hopf algebras. Therefore **[9**, 4.4**]** $H^*(B; \mathbf{Z}_2)$ is a free Ω-module. Hence $\mathrm{Tor}_\Omega(H^*(B; \mathbf{Z}_2), \mathbf{Z}_2) = H^*(B; \mathbf{Z}_2) \otimes_\Omega \mathbf{Z}_2$ and the change of rings spectral sequence collapses to the edge isomorphism $H^*(B; \mathbf{Z}_2) \otimes_\Omega \mathbf{Z}_2 \otimes \mathrm{Tor}_\Lambda(\mathbf{Z}_2, \mathbf{Z}_2) \cong \mathrm{Tor}_\Gamma(H^*(B; \mathbf{Z}_2), \mathbf{Z}_2)$ as desired.

over $\mathbf{Z}_2$, $E_2(\mathscr{F})$ is generated by $E_2^{0,*}(\mathscr{F})$ and $E_2^{-1,*}(\mathscr{F})$. As

$$d_r(E_2^{0,*}) \subset E_2^{r,*} = 0, \qquad r \geqq 2,$$
$$d_r(E_2^{-1,*}) \subset E_2^{r-1,*} = 0,$$

and d_r is a derivation of the algebra structure we conclude that $d_r = 0$ for $r \geqq 2$ and hence that $E_2 = E_\infty$.

The assertion about the Steenrod algebra structure follows from an appeal to [**20**, 8.1].

REMARKS. The techniques used above can be easily extended to handle the case of stable Postnikov systems of arbitrary stage (see for example [**10**], [**19**]). Nor are they restricted to Postnikov systems, see for example [**13**].

4. **More applications.** Lest the reader get the impression that in all applications of the E-M spectral sequence it collapses we will begin this section by stating the odd primary analog of the results directly preceding.

THEOREM. *Suppose that*

$$\mathscr{F}\quad \begin{array}{ccc} E & \longrightarrow & E_0 \\ \downarrow & & \downarrow \\ B & \longrightarrow & B_0 \end{array}$$

is a stable two stage Postnikov system of finite type with B_0 *1-connected. Then for any prime p one has* $E_p(\mathscr{F}) = E_\infty(\mathscr{F})$.

OUTLINE OF PROOF. For $p = 2$ this is simply the result of the preceding section. For odd primes p this involves an analysis of the differentials in the spectral sequence. In outline the procedure runs as follows. We begin by computing $E_2(\mathscr{F})$. First we single out a certain $\mathbf{Z}_p$ subspace (see [**14**] or [**10**])

$$L \subset \ker Pf^*: PH^*(B_0; \mathbf{Z}_p) \to PH^*(B; \mathbf{Z}_p),$$

where $P(\ )$ denotes the primitive functor. Using various Hopf algebra techniques we find [**44**], [**10**],

$$E_2(\mathscr{F}) = H^*(B; \mathbf{Z}_p)/\!/f^* \otimes E[s^{-1}L^+] \otimes \Gamma[s^{-1}L^-],$$

where $\Gamma[W]$ denotes the divided power algebra generated by W. By use of the technique of universal example we find that $d_{p-1}(\mathscr{F})$ is the first possible nonzero differential and is given by the formula

$$d_{p-1}(\mathscr{F})(\gamma_p(s^{-1}\lambda)) = -s^{-1}\beta P^t\lambda$$

for $\lambda \in L^{2t+1}$, and the requirement that $d_{p-1}(\mathscr{F})$ be a derivation of Hopf algebras. (Here β denotes the Bockstein in $\mathscr{A}^*(p)$ and P^t is the usual Steenrod reduced pth power.) Brutish computation then shows

$$E_p(\mathscr{F}) = H^*(B; \mathbf{Z}_p)/\!/f^* \otimes E[s^{-1}\bar{L}^+] \otimes \mathbf{Z}_p[s^{-1}L^-]/((s^{-1}L^-)^p),$$

where $\bar{L}^+$ is the quotient of L^+ by the submodule generated by the elements βpt_λ, $\lambda \in L^{2t+1}$, $t = 1, 2, 3, \ldots$. Thus we are in the familiar situation where $E_p^{0,*}(\mathscr{F})$ and $E_p^{-1,*}(\mathscr{F})$ generate $E_p^{*,*}(\mathscr{F})$ as an algebra so we find that all further differentials are zero and hence $E_p(\mathscr{F}) = E_\infty(\mathscr{F})$.

Note again that the proof of the theorem contains an effective computation of $E_p(\mathscr{F})$ as a Hopf algebra over the Steenrod algebra. Again the algebra extension problem may be solved, but the Steenrod algebra structure remains unsolved except in a few special cases.

REMARKS. The techniques used above may be vastly extended and this is the subject of [**10**]. An alternate exposition appears in [**19**]. The situation exemplified by the previous theorem has proven to be the most tractable to computation with the E-M spectral sequence. In the case of the Hopf fibre squares of [**10**] formulas have been obtained for all the possible nonzero differentials by the method of universal example, and are analogous to the formula for $d_{p-1}(\mathscr{F})$ above. These results are too technical to discuss here, we refer to [**10**], [**17**] for some more details.

So far almost all the examples we have studied have been where the E-M spectral sequence collapses, or nearly does, for more or less trivial reasons concerning the placement of algebra generators in $E_2(\mathscr{F})$ (or $E_p(\mathscr{F})$). It is perhaps therefore necessary to close with an example of collapse where this is not known to be the case.

Let G be a compact connected Lie group and H a closed connected subgroup. Suppose that $\xi = (E, p, B, G/H, G)$ is a fibre bundle with fibre G/H and structural group G. There is then a classifying diagram

$$\mathscr{F}_\xi \qquad \begin{array}{ccc} E & \longrightarrow & BH \\ \downarrow & & \downarrow{\scriptstyle p} \\ B & \xrightarrow{\varphi_\xi} & BG \end{array}$$

for the bundle ξ. The following result, proved jointly with P. F. Baum [**3**], generalizes results of Borel, Cartan, and Hirsch.

THEOREM. *Suppose that $\xi = (E, p, B, G/H, G)$ is a differentiable fibre bundle over the closed Riemannian symmetric space*[2] *B, with fibre G/H and structural group G. Then with coefficients in the reals* **R** *we have*

$$E_2(\mathscr{F}_\xi) = E_\infty(\mathscr{F}_\xi).$$

Moreover there is an (unnatural) isomorphism of algebras

$$\operatorname{Tor}^{**}_{H^*(BG;\mathbf{R})}(H^*(B;\mathbf{R}), H^*(BH;\mathbf{R})) = H^*(E;\mathbf{R}).$$

[2] A manifold M is a Riemannian symmetric space iff each point $x \in M$ is an isolated fixed point of some involutive isometry $T_x : M \to M$.

It is to be emphasized that in the proof the term

$$E_2(\mathscr{F}_\xi) = \operatorname{Tor}^{**}_{H^*(BG;\mathbf{R})}(H^*(B;\mathbf{R}), H^*(BG;\mathbf{R}))$$

is by no means computed in closed form. By taking $B = *$ one obtains a theoretical computation of $H^*(G/H;\mathbf{R})$ equivalent to that originally obtained by Cartan.

References

1. J. F. Adams, *On the cobar construction*, Proc. Nat. Acad. Sci. U.S.A. **42** (1956), 409–412. MR **18,** 59.

2. M. F. Atiyah, *Vector bundles and the Künneth formula*, Topology **1** (1962), 245–248. MR **27** #767.

3. P. F. Baum and L. Smith, *Real cohomology of differentiable fibre bundles*, Comment. Math. Helv. **42** (1967), 171–179. MR **36** #4574.

4. H. Cartan, *Séminaire école normale supérieure* 1954/55, Secrétariat mathématique, Paris, 1955. MR **19,** 438.

5. H. Cartan and S. Eilenberg, *Homological algebra*, Princeton Univ. Press, Princeton, N.J., 1956. MR **17,** 1040.

6. S. Eilenberg and J. C. Moore, *Homology and fibrations*. I, *Coalgebras, cotensor product and its derivative functors*, Comment. Math. Helv. **40** (1966), 199–236. MR **34** #3579.

7. I. M. James, *Ex-homotopy theory*, Illinois J. Math. (to appear).

8. J. P. Meyer, *Relative stable homotopy theory* (to appear).

9. J. Milnor and J. C. Moore, *On the structure of Hopf algebras*, Ann. of Math. (2) **81** (1965), 211–264. MR **30** #4259.

10. J. C. Moore and L. Smith, *Hopf algebras and multiplicative fibrations*. I, II, Amer. J. Math. **90** (1968), 752–780, 1113–1150. MR **38** #2772.

11. J. F. McClendon (to appear).

12. J. P. Serre, *Cohomologie modulo 2 des complexes d'Eilenberg-MacLane*, Comment. Math. Helv. **27** (1953), 198–232. MR **15,** 643.

13. W. Singer, *Connective fibrings over BU and U*, Topology **7** (1968), 271–303. MR **38** #717.

14. L. Smith, *The cohomology of stable two stage Postnikov systems*, Illinois J. Math. **11** (1967). 310–329. MR **34** #8406.

15. ———, *Homological algebra and the Eilenberg-Moore spectral sequence*, Trans. Amer. Math. Soc. **129** (1967), 58–93. MR **35** #7337.

16. ———, *Cohomology of* $\Omega(G/U)$, Proc. Amer. Math. Soc. **19** (1968), 399–404. MR **37** #2259.

17. ———, *Primitive loop spaces*, Topology **7** (1968), 121–124. MR **37** #4816.

18. ———, *On the construction of the Eilenberg-Moore spectral sequence*, Bull. Amer. Math. Soc. **75** (1969), 873–878. MR **40** #3551.

19. ———, *Lectures on the Eilenberg-Moore spectral sequence*, Lecture Notes in Math., no. 134, Springer-Verlag, Berlin and New York, 1970.

20. ———, *On the Künneth theorem*. I, Math. Z. **116** (1970), 94–140.

21. J. Tate, *Homology of Noetherian rings and of local rings*, Illinois J. Math. **1** (1957), 14–27. MR **19,** 119.

University of Virginia

H-SPACES AND CLASSIFYING SPACES: FOUNDATIONS AND RECENT DEVELOPMENTS

JAMES D. STASHEFF[1]

I. HOW TO BUILD A CLASSIFYING SPACE

The original classifying space of Whitney **[37]**, **[31]**, **[22]** consisted of great k-spheres in the n-sphere. Later Steenrod **[32]**, Chern and Sun **[8]** called attention to the characterization of a classifying space in terms of acyclicity of the total space of the universal bundle. They and G. W. Whitehead exhibited classifying spaces for all the compact Lie groups. It was Milnor **[21]** who first gave a construction for an arbitrary topological group.

Although the uniqueness of a classifying space follows from the universal property, the existence is given by a specific construction. In this lecture, I would like to compare a few of the more common constructions and related results. What is it we want a classifying space to be? In my second lecture, I will discuss how (and what) classifying spaces do classify but here I would rather emphasize the fact that the classifying space of a topological group is an invariant (complete in an appropriate sense) of the combination of topology and algebra involved. I hope to clarify this point later in this talk and also to point out the relation of these constructions to certain well-known parts of homological algebra.

Prior to Milnor, there were many ways of constructing $K(\pi, n)$'s, the most algebraic being the W, $\overline{W}$-construction and the bar construction **[6]**, **[17]**. For any simplicial group $\mathscr{G}$, we have the bundle $\mathscr{G} \to W\mathscr{G} \to \overline{W}\mathscr{G}$ which is a simplicial

AMS 1970 *subject classifications*. Primary 55-02, 55-03, 55D35, 55D45, 55F05, 55F35, 55F15, 55F25, 55F30; Secondary 55F40, 55H20.

[1] Alfred P. Sloan Fellow. Supported in part by NSF grant 97836.

principal bundle. For G a topological group, we can consider $\mathscr{S}G$, the singular complex of G; technical difficulties apparently prevented the use of the realization $|\overline{W}\mathscr{S}G|$ as a classifying space for G, cf. **[39]**, **[40]**.

After Milnor was done, we saw that his construction (and most of those which followed) were related to the bar construction. Recall the following definition:

Let A be a D(ifferential) G(raded) A(ugmented) algebra with differential d. Let $[A]$ be the kernel of the augmentation. Define $\bar{B}A = \bigoplus_{n\geq 0} [A] \otimes \cdots \otimes [A]$ and $BA = A \otimes \bar{B}A$. Define

$$d_B(a[a_1 | \cdots | a_n])$$
$$= \sum \pm a[\cdots |a_ia_{i+1}| \cdots] + \sum \pm a[\cdots |da_i| \cdots] \pm aa_1[a_2 | \cdots | a_n] \pm da[a_1 | \cdots]$$

with signs adjusted so $dd = 0$. (This uses the associativity of A.) For $d_{\bar{B}}[a_1 | \cdots | a_n]$, omit the last two terms. $\bar{B}A$ is a differential coalgebra with respect to $\Delta[a_1 | \cdots | a_n] = \sum [a_1 | \cdots | a_p] \otimes [a_{p+1} | \cdots | a_n]$.

Prior to Milnor, the work of Cartan **[6]** pointed up the analogy between principal bundles and "constructions". Apparently there was no successful attempt to realize the bar construction geometrically and only after the fact did people become aware that Milnor's construction gave a geometric realization of a variant of the bar construction.

Let us look at Milnor's construction first. With some fine touches in the topology used, Milnor defines E_n to be the $n+1$-fold join $G * \cdots * G$ and B_n to be E_n/G, where G acts on $G * \cdots * G$ diagonally, i.e., $g \sum t_i g_i = \sum t_i g g_i$. He points out that in the spectral sequence derived from filtering $B_G = \lim B_n$ by B_n, the E^1 term with field coefficients is $E^1_{p,*} = \tilde{H}_*(G) \otimes \cdots \otimes \tilde{H}_*(G)$. The differential d^1 is not the bar construction differential but corresponds to it under a familiar algebraic automorphism.

To avoid the use of inverses, Dold and Lashof **[10]** gave a somewhat different construction. They proceed inductively: Given a principal X-fibration E, where X is an associative H-space, they define

$$\begin{array}{ccc} DL(E) = X \times CE \underset{\mu}{\cup} E \\ \downarrow \qquad\qquad \downarrow \quad \downarrow \\ DL(B) = \qquad CE \cup B \end{array}$$

where μ is the given action of X on E. An action of X on $DL(E)$ is given by

$$x(x', t, z) = (xx', t, z)$$

if the action of X on E is associative. The map $DL(E) \to DL(B)$ is not a bundle (as for Milnor) but at least a quasifibration (the homotopy groups behave as desired) and the construction can be iterated. Dold and Lashof define

$$\begin{array}{ccc} E_X & = & \lim DL^n(X) \\ \Big\downarrow & & \Big\downarrow \\ B_X & = & \lim DL^n(*) \end{array}$$

with some subtlety to the topology involved.

I subsequently pointed out **[28]** that if the construction is reworked using the reduced cone for C, then B_X is related to the bar construction as follows:

$$B_X = \bigcup_n \Delta^n \times X^n/\sim,$$

where

$$\begin{aligned} (t_0, \ldots, t_n, x_1, \ldots, x_n) &\sim (\ldots, \hat{t}_i, \ldots, x_i x_{i+1}, \ldots) && \text{if } t_i = 0, \\ &\sim (\ldots, t_{i-1} + t_i, \ldots, \hat{x}_i, \ldots) && \text{if } x_i = e. \end{aligned}$$

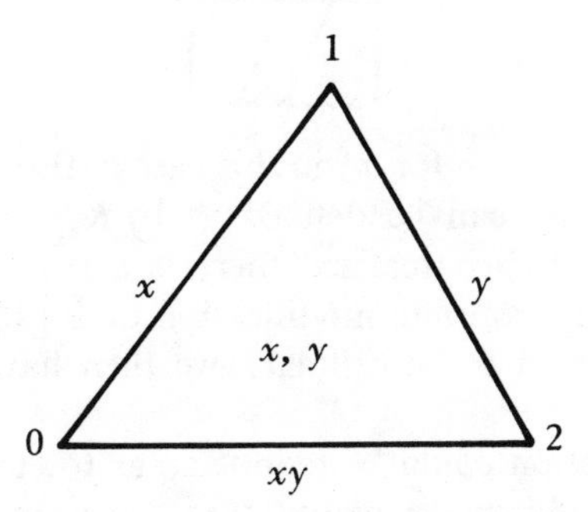

We then have $\bar{B}(\tilde{C}_*(X)) \subset C_*(B_X)$ as a deformation retract by

$$[u_1 | \cdots | u_n] \to 1 \times u_1 \times \cdots \times u_n,$$

where $1 : \Delta^n \to \Delta^n$, $u_i : \Delta^{n_i} \to X$.

Milgram **[20]** goes further and remarks that if X is CW and multiplication is cellular, then with cellular chains $\mathscr{C}_*$ we have $\bar{B}(\tilde{\mathscr{C}}_*(X)) \approx \mathscr{C}_*(B_X)$ as chain complexes. He was further able to show that if X is an abelian group so is B_X. Thus $K(\pi, n)$ can be constructed as

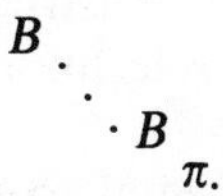

Recently Mac Lane **[18]** has pointed out that this construction of B_X as a quotient of $\bigcup \Delta^n \times X^n$ is precisely the "tensor product" of two functors—the standard simplex functor and the purely algebraic bar construction.

Independently of Dold and Lashof, Sugawara **[33]** showed that if X is an H-space with a multiplication which is homotopy associative up to suitable higher homotopies and with suitable homotopy inverses, then a modified Milnor construction can be carried through.

Somewhat independently, I was able to provide a simpler generalization [**28**]. For each n there is a complex K_n, homeomorphic to an $n-2$-cell with vertices indexed by the various ways of associating n variables, edges by applications of homotopy associativity, etc. For example,

$$K_2 = *, \qquad K_3 = \text{———}, \qquad K_4 =$$

An A_∞-space or s(trongly) h(omotopy) a(ssociative) space consists of a family of maps $m_n : K_n \times X^n \to X$ such that m_2 is a multiplication with unit and $m_n \mid \partial K_n \times X^n$ is suitably defined in terms of m_i for $i < n$. For example, $\Omega^1 X$, loops parametrized by $[0, 1]$, is well known to be homotopy associative. (This is usually depicted by

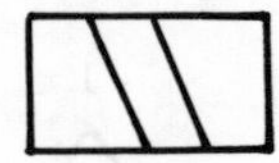

The corresponding description for m_4 in this case is also easy to depict.)

For an sha space, B_X can be defined as $\bigcup K_{n+2} \times X^n/\sim$ where $\sim$ now involves the maps m_i. I then realized there is a corresponding notion of sha DGA-algebra and an appropriate modification of $\bar{B}$ [**28**]. If X is CW, the sha structure can be deformed to be cellular; we then have $\mathscr{C}_*(B_X)$ isomorphic to $\bar{B}\mathscr{C}_*(X)$.

This entire presentation could be redone so as to emphasize the total spaces E_X which are acyclic. Attention would then focus on $BA = A \otimes \bar{B}A$, which is acyclic, rather than on $\bar{B}A$. The language would then be more that of homological algebra, regarding E_X as a "geometric resolution" of X [**30**].

Certain of the maps $E_X \to B_X$ will play more of a role in my second lecture where they appear as universal examples. One comment is in order, however, at a constructive level. Using inverses and just the right topology, Milnor obtains a principal G-bundle $E_G \to B_G$. Dold and Lashof settle for a principal quasi-fibration, I obtain a quasifibration and Sugawara something even weaker. Recently Fuchs [**11**] has shown that for an associative H-space with suitable homotopy inverses the Dold-Lashof construction can be modified slightly (using more connective tissue) to give a Dold fibration, i.e., having the weak covering homotopy property.

These fibrations play a fundamental role in establishing for connected CW X, Y the following homotopy equivalences:

$$(*) \qquad \Omega B_X \simeq X, \qquad B_{\Omega X} \simeq Y.$$

Actually the homotopy equivalence $X \simeq \Omega B_X$ respects the structure in a very strong sense.

A s(trongly) h(omotopy) m(ultiplicative) map [34] of an associative H-space X to one Y is a family of maps $f^n : I^{n-1} \times X^n \to Y$ such that

$$\begin{aligned} f^n(t_1, \ldots, t_{n-1}, x_1, \ldots, x_n) & \\ = f^{n-1}(\ldots, \hat{t}_i, \ldots, x_i x_{i+1}, \ldots) & \qquad \text{if } t_i = 0, \\ = f^i(t_1, \ldots, t_{i-1}, x_1, \ldots, x_i) f^{n-i}(t_{i+1}, \ldots, x_n) & \qquad \text{if } t_i = 1. \end{aligned}$$

Such maps induce maps $B_X \to B_Y$; indeed they are precisely the most general maps that do. Moreover, $X \cong \Omega B_X$ via shm maps.

It is in this sense that B_X is a complete invariant of the homotopical algebra of X. Further, consider $(\ )_\pi$ to denote "homotopy classes of". Given associative H-spaces X, Y we have

$$\begin{array}{c} \mathrm{Hom}(X, Y)_\pi \\ \downarrow \\ \mathrm{Shm}(X, Y)_\pi \\ \downarrow \\ \mathrm{Map}(B_X, B_Y)_\pi \\ \downarrow \\ \mathrm{Hom}(\Omega B_X, \Omega B_Y)_\pi \end{array}$$

where the maps should be clear except perhaps $\circlearrowleft$ which is induced by the natural shm equivalence $\Omega B_Z \simeq Z$.

Fuchs has studied this situation thoroughly.

THEOREM [12]. *The category of connected associative* CW *H-spaces and homotopy classes of shm maps is in 1-1 correspondence with the category of simply connected spaces and homotopy classes of maps.*

Further modifications of the above "realization of the bar construction" have been studied [15], [27], [29], corresponding to the algebraic gadgets $M \otimes \bar{B}A \otimes N$, where M, N are modules over A. These will play a role in studying the various ways of classifying fibrations. In particular, given a fibre space $F \to E \to X$, there is defined an action of ΩX on F [29]. The total space E then has the homotopy type of $\bigcup \Delta^n \times (\Omega X)^n \times F/\sim$ for a suitable relation $\sim$ determined by the action. In terms of chains we have

$$\mathscr{C}_*(E) \simeq \bar{B}(\mathscr{C}_*(\Omega X)) \tilde{\otimes} \mathscr{C}_*(F) \simeq \mathscr{C}_*(X) \tilde{\otimes} \mathscr{C}_*(F) \quad \text{(cf. [5])}.$$

A final direction for generalization is provided by groupoids and categories. The first such observation I am aware of is due to Segal [25] who works in terms of categories. We will denote by B_X the subset of $\bigcup \Delta^n \times X^n/\sim$ which makes sense, where X is the groupoid or the set of morphisms of the category. That is B_X is the quotient of $\bigcup \{(t_0, \ldots, t_n, x_1, \ldots, x_n) \mid x_i x_{i+1} \text{ is defined}\}$. It is far from clear what the properties of B_X in this generality are. It is certainly not the case that $X \simeq \Omega B_X$, even for reasonable X. As pointed out by Husseini, X^I with the usual composition of paths is a groupoid yet $X^I \simeq X$ which in general is not of the weak homotopy type of a loop space.

The following three applications do show that in certain cases at least this generality is worthwhile.

A. *Haefliger's foliation theory.* Haefliger works with foliations in terms of Γ-structures, where Γ is the groupoid of local diffeomorphisms of R^q and uses a classifying space B_Γ. My understanding (though I have not seen any details) is that B_Γ can be defined as above, or at least by the corresponding version of Milnor's construction.

B. *Lashof's classification of G-bundles by* $\mathrm{Hom}(\bar{\Omega}X, G)$ [**16**]. Here it is crucial that $\bar{\Omega}X$ is the groupoid $\{\lambda : I \to X \mid \lambda(0) = \lambda(1) = *\}$ with $\lambda * \mu$ defined iff $\lambda(t) = \mu(1 - t)$ for $\frac{1}{2} \leqq t \leqq 1$ by

$$(\lambda * \mu)(t) = \lambda(t), \qquad 0 \leqq t \leqq \tfrac{1}{2},$$
$$= \mu(t), \qquad \tfrac{1}{2} \leqq t \leqq 1.$$

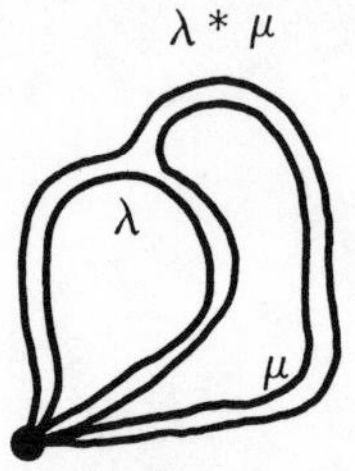

One might hope that $B_{\bar{\Omega}X} \simeq X$; the question has just arisen in the course of these lectures.

C. *The reconstruction of a space from a nice cover.* This procedure was apparently known to P. Conner in the late 50's, was discovered independently by J. Wirth [**38**] and G. Segal [**25**]. Given a cover $\{U_\alpha\}$ of a space Y, consider the groupoid $\mathscr{U}$ generated by the cover under the operation of $\cap$, nonempty intersection. Provided the covering is numerable, then we have $Y \simeq B_{\mathscr{U}}$; this is useful for constructing classifying maps in terms of transition functions as we shall see in our next lecture.

These are further generalizations of the realized bar construction due to McCord [**19**] and tom Dieck [**35**].

A somewhat different construction, due to G. W. Whitehead [**36**], [**13**], gives rise to a special case of $\mathrm{Tor}^{H_*(G)}(k, k) \Rightarrow H_*(B_G)$. Consider Y^{Δ^∞} to mean $\lim Y^{\Delta^n}$ with some suitable topology. In particular Y^{Δ^∞} like Y^{Δ^n} will have the homotopy type of Y. Thus we can define a filtration Y_i by $Y_i = \{f : \Delta^n \to Y \mid f(j) = *,\ j \leqq i \leqq n\}$. Notice that $Y_0 \simeq PY = \{\lambda : I \to Y \mid \lambda(0) = *\}$ and $Y_1 \simeq \Omega Y$. Further work reveals that the derived spectral sequence is that of Moore: $\mathrm{Tor}^{H_*(\Omega Y)} \Rightarrow H_*(Y)$. Notice in this approach we need to know Y as well as $X = \Omega Y$ before we start while all our other constructions have constructed something of the homotopy type of Y from structure maps on X. Significant computational results were obtained with this particular representation in terms of the category of Y [**13**].

Extensive computational results have been obtained using the bar construction spectral sequence by Moore and his students (Browder [**4**], Clark [**9**], Gitler [**14**], and myself [**28**]), Milgram, Rothenberg, and Steenrod [**23**], [**24**].

Recently there has been considerable interest in iterated classifying spaces or rather in infinite loop spaces. One way of showing a space X is an infinite loop space is

(1) exhibit sha structure on X so that B_X exists;

(2) exhibit additional homotopies on X so that B_X admits sha structure so that B_{B_X} exists;

(3) etc.

A major problem here is the description of precisely which homotopies are needed in (2) and even more so in (3). Part of the point of Boardman's machinery [2], [3] is to handle this without getting too explicit.

Other approaches used by Beck [1] and very recently Segal [26] have in common that they construct the family $B \cdots BX$ not by iteration of B but rather simultaneously in terms of some other family of structures.

The existence of a classifying space in the sense we have talked about most, particularly (*), has a long history of which we have tried to sketch the highlights. It has played and surely will continue to play a fundamental role in algebraic topology. It is conceivable that the viewpoint of resolutions, etc. (i.e., of homological algebra) may prove more useful, but I expect rather that the bar construction and its analogues will reign supreme.

REFERENCES

1. Jon Beck, *On H-spaces and infinite loop spaces*, Category Theory, Homology Theory and their Applications (Battelle Inst. Conference, Seattle, Wash., 1968), vol. 3, Springer-Verlag, New York, 1969, pp. 139–153. MR **40** #2079.

2. J. M. Boardman, *Monoids, H-spaces and tree surgery*, Math. Dept., Haverford College, Haverford, Pa., 1969 (mimeograph).

3. J. M. Boardman and R. M. Vogt, *Homotopy-everything H-spaces*, Bull. Amer. Math. Soc. **74** (1968), 1117–1122. MR **38** #5215.

4. William Browder, *On differential Hopf algebras*, Trans. Amer. Math. Soc. **107** (1963), 153–176. MR **26** #3061.

5. E. H. Brown, *Twisted tensor products*. I, Ann. of Math. (2) **69** (1959), 223–246. MR **21** #4423.

6. Henri Cartan, *Séminaire école normale supérieure* 1954/55, Secrétariat mathématique, Paris, 1955. MR **19,** 438.

7. ———, *Sur les groupes d'Eilenberg-MacLane $H(\pi, n)$*. I, II, Proc. Nat. Acad. Sci. U.S.A. **40** (1954), 467–471, 704–707. MR **16,** 390.

8. Shiing-Shen Chern and Yi-Fone Sun, *The imbedding theorem for fibre bundles*, Trans. Amer. Math. Soc. **67** (1949), 286–303.

9. Allan Clark, *Homotopy commutativity and the Moore spectral sequence*, Pacific J. Math. **15** (1965), 65–74. MR **31** #1679.

10. Albrecht Dold and R. Lashof, *Principal quasi-fibrations and fibre homotopy equivalence of bundles*, Illinois J. Math. **3** (1959), 285–305. MR **21** #331.

11. Martin Fuchs, *A modified Dold-Lashof construction that does classify H-principal fibrations* (preprint).

12. ———, *Verallgemeinerte Homotopie-Homomorphismen und klassifizierende Räume*, Math. Ann. **161** (1965), 197–230. MR **33** #3295.

13. Michael Ginsburg, *On the Lusternik-Schnirelmann category*, Ann. of Math. (2) **77** (1963), 538–551. MR **26** #6976.

14. Samuel Gitler, *Spaces fibered by H-spaces*, Bol. Soc. Mat. Mexicana (2) **7** (1962), 71–84.

15. Sufian Y. Husseini, *The topology of classical groups and related topics*, Gordon and Breach, New York, 1969.

16. R. K. Lashof, *Classification of fibre bundles by the loop space of the base*, Ann. of Math. (2) **64** (1956), 436–446. MR **18,** 497.

17. Saunders Mac Lane, *Homology*, Die Grundlehren der math. Wissenschaften, Band 114, Academic Press, New York; Springer-Verlag, Berlin, 1963. MR **28** #122.

18. ———, *Milgram's classifying space as a tensor product of functions*, Steenrod Conference, Lecture Notes in Math., no. 168, Springer-Verlag, Berlin and New York.

19. M. C. McCord, *Classifying spaces and infinite symmetric products*, Trans. Amer. Math. Soc. **146** (1969), 273–298.

20. R. James Milgram, *The bar construction and abelian H-spaces*, Illinois J. Math. **11** (1967), 242–250. MR **34** #8404.

21. John Milnor, *Construction of universal bundles*. II, Ann. of Math. (2) **63** (1956), 430–436. MR **17,** 1120.

22. L. S. Pontrjagin, *Classification of some skew products*, C. R. (Dokl.) Acad. Sci. USSR **47** (1945), 322–325. MR **7,** 138.

23. Melvin Rothenberg, Thesis, Berkeley, California.

24. Melvin Rothenberg and Norman Steenrod, *The cohomology of classifying spaces of H-spaces*, Bull. Amer. Math. Soc. **71** (1965), 872–875. MR **34** #8405.

25. Graeme Segal, *Classifying spaces and spectral sequences*, Inst. Hautes Études Sci. Publ. Math. No. 34 (1968), 105–112. MR **38** #718.

26. ———, *Homotopy-everything H-spaces* (preprint).

27. James D. Stasheff, *Associated fibre spaces*, Michigan Math. J. **15** (1968), 457–470. MR **40** # 2083.

28. ———, *Homotopy associativity of H-spaces*. I, II, Trans. Amer. Math. Soc. **108** (1963), 275–312. MR **28** #1623.

29. ———, *"Parallel" transport in fibre spaces*, Bol. Soc. Mat. Mexicana (2) **11** (1966), 68–84. MR **38** # 5219.

30. Norman E. Steenrod, *Milgram's classifying space of a topological group*, Topology **7** (1968), 349–368. MR **38** #1675.

31. ———, *The classification of sphere bundles*, Ann. of Math. (2) **45** (1944), 294–311. MR **5,** 214.

32. ———, *The topology of fibre bundles*, Princeton Math. Series, vol. 14, Princeton Univ. Press, Princeton, N.J., 1951. MR **12,** 522.

33. Masahiro Sugawara, *A condition that a space is group-like*, Math. J. Okayama Univ. **7** (1957), 123–149. MR **20** #3546.

34. ———, *On the homotopy-commutativity of groups and loop spaces*, Mem. Coll. Sci. Univ. Kyoto Ser. A Math. **33** (1960/61), 257–269. MR **22** #11394.

35. Tammo tom Dieck, *Faserbündel mit Gruppenoperation*, Arch. Math. **20** (1969), 136–143.

36. George W. Whitehead, *On the homology suspension*, Colloq. Topologie Algébrique, (Louvain, 1956), Georges Thone, Liège; Masson, Paris, 1957, pp. 89–95. MR **20** #1306.

37. Hassler Whitney, *Topological properties of differentiable manifolds*, Bull. Amer. Math. Soc. **43** (1937), 785–805.

38. James F. Wirth, *Fibre spaces and the higher homotopy cocycle relations*, Thesis, Notre Dame, Ind., 1965.

39. Alex Heller, *Homotopy resolutions of semi-simplicial complexes*, Trans. Amer. Math. Soc. **80** (1955), 299–344. MR **17,** 773.

40. ———, *Singular homology in fibre bundles*, Ann. of Math. (2) **55** (1952), 232–249. MR **13,** 967.

II. HOW TO CLASSIFY FIBRATIONS

Regarding B_G directly as an invariant of G and its multiplication is less common than regarding B_G as a classifying space. Originally this meant classifying G-bundles; more recent classification has been extended to analogues of G-bundles.

Let us review the heart of the classification of G-bundles, which can be done in one of two basic ways.

Let $LG(B)$ denote equivalence classes of G-bundles over B.

THEOREM 1 [**30**]. *$LG(B)$ is naturally equivalent to $H^1(B; \mathscr{G})$, where $\mathscr{G}$ is the sheaf associated to G, i.e., the group of local sections of $\mathscr{G}$ is G^U, $U \subset B$.*

The correspondence is given via transition functions as we shall recall further below.

THEOREM 2 [**5**], [**30**], [**9**]. *For paracompact B (e.g.* CW*), $LG(B)$ is naturally equivalent to $[B, B_G]$, the set of homotopy classes of maps $B \to B_G$.*

It is only fairly recently [**35**], [**32**] that a description of classifying maps in terms of transition functions has existed, and still more recently that the general relation between sheaf theoretic cohomology and homotopy classes have been worked out. The second method has often been used by passing first to the associated principal bundle, although the first classifying maps I am aware of, the Gauss map [**17**] and Whitney's generalization for sphere bundles [**33**], deal directly with the given bundle.

I would like to trace the development of classification from these early beginnings through to recent generalizations in terms of fibrations. Wherever possible, I will relate the various approaches, in a few instances reflecting the influences which were active historically, but much more often reflecting insight acquired after the fact. A chart summarizing this history appears on the next page.

Steenrod [**29**; 1944] and Pontrjagin [**23**; 1945] independently showed that Whitney's representing space [**33**; 1937] was in fact a classifying space. Chern and Sun [**9**; 1949] and Steenrod [**30**; 1951] carried out the classification over complexes in terms of a universal bundle with acyclic total space and constructed such bundles for compact Lie groups. Milnor [**21**; 1956] later provided a construction for arbitrary topological groups, which Dold [**5**; 1963] showed was universal for numerable bundles (e.g. over paracompact bases).

The classification proceeds via the associated principal bundle which is classified by a map

$$\begin{array}{ccc} P & \xrightarrow{\bar f} & E_G \\ \downarrow & & \downarrow \\ B & \xrightarrow{f} & B_G. \end{array}$$

For CW B with $\pi_i(E_G) = 0$ for $i > \dim B$, the map is constructed inductively. Consider $B = B' \cup_\alpha e^n$ so that $P = P' \cup G \times e^n$. Suppose we already have

$$\begin{array}{ccc} P' & \xrightarrow{\bar f} & E_G \\ \downarrow & & \downarrow \\ B' & \xrightarrow{f} & B_G \end{array}$$

then for $\bar{\alpha}: S^{n-1} \to P'$ such that $p\bar{\alpha} = \alpha$, extend $\bar{f}\bar{\alpha}$ to $\beta: e^n \to E_G$. Use the action of G to extend to $\beta: G \times e^n \to E_G$ which gives $\bar{f}: P \to E_G$ and induces $f: B \to B_G$.

C.S.S.	$B \to B_G$	$\Omega B \to G$	$g_{\alpha\beta}$
	Gauss 1828		
	Whitney 1937		
	Steenrod 1944		
	Pontrjagin 1945		
	Chern and Sun 1949		
	Steenrod 1951		Steenrod 1951
	Curtis 1954		
	Milnor 1956	Lashof 1956	
		Curtis-Lashof 1958	
Barratt, Gugenheim Moore 1959	Dold-Lashof 1959		
	Dold 1963		
	Stasheff 1963		Wirth 1964
			Wirth 1965
	Allaud 1966		tom Dieck 1966
	Dold 1966	Stasheff 1966	
		Drachman 1966	
			Segal 1968
	Porter 1969	Husseini 1969	
	Fuchs 1969		
	Dyer (to appear)		

Curtis [6] studied the applicability of the "universal bundle" approach to fibre spaces and in particular to the classification of certain principal ΩY-fibre spaces. A more general version of this result has recently been carried out by Fuchs [15].

The construction of Dold and Lashof [10; 1959] was aimed particularly at classification up to fibre homotopy equivalence. Let $H(F)$ denote the auto-homotopy-equivalences of F, then Dold and Lashof [10] showed that for a subgroup of Homeo(F) the image of $[B, B_G]$ in $[B, B_{H(F)}]$ is in 1-1 correspondence with fibre homotopy equivalence classes of G-bundles with fibre F.

Of course this suggested that $B_{H(F)}$ was the classifying space for fibre homotopy equivalence classes of fibrations (in some sense) with fibre F. I was in fact able to show [26; 1963]: Let $LF(B)$ denote f.h.e. classes of Hurewicz fibrations with fibres of the homotopy type of F.

THEOREM 3. *For a finite complex F over a* CW-*base B, $LF(B)$ is in* 1-1 *correspondence with* $[B, B_{H(F)}]$.

The essence of the proof consisted of two pieces:

(1) A modified Dold-Lashof construction applicable to any q.f. Namely given $F \to E \to B$, define Prin $E = \{\varphi: F \to E \mid \varphi: F \xrightarrow{\simeq} p^{-1}(b)$ for some $b\}$.

Define

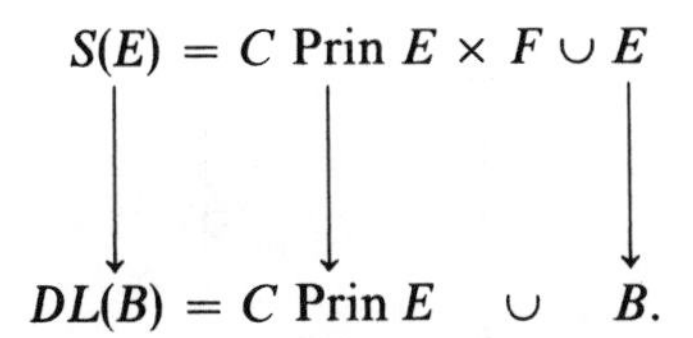

Iterating beginning with $F \to *$, we obtain in the limit a universal example $UE \to B_{H(F)}$.

(2) Iterating S on E and F we obtain

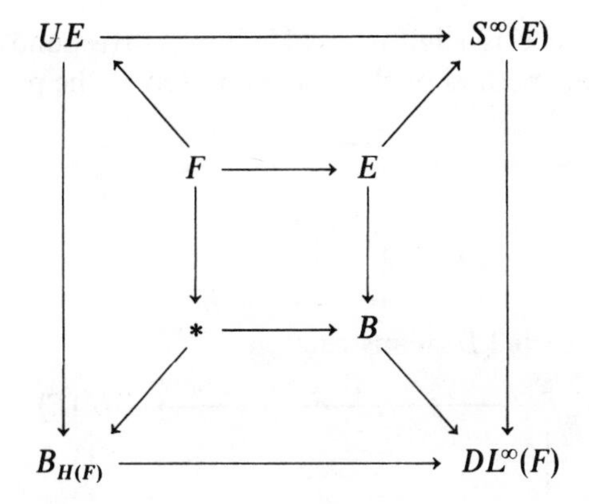

with $B_{H(F)} \to DL^\infty(F)$ being a weak homotopy equivalence. For CW B, we then have $B \to DL^\infty(F)$ factoring through $B_{H(F)}$, providing a classifying map.

(Strictly speaking, we must replace $UE \to B_{H(F)}$ by a Hurewicz fibering in the usual way or modify our construction following Fuchs [**14**], to obtain a Dold fibration at each stage.)

The reason we did not work entirely in terms of Prin E is that, at the time, there was no way to recover E from Prin E. We will return to this point later.

Theorem 3 was subsequently extended to arbitrary CW F by Allaud [**1**; 1966] and to arbitrary spaces F by Dold [**8**; 1966]. Their approach was quite different. They showed that $LF(B)$ is a representable functor, hence corresponds to $[B, Z]$ for some Z [**4**]. A crucial condition to be verified involves a solution of:

THE PASTING PROBLEM. Given $B = B_1 \cup B_2$ and fibrations $E_i \to B_i$ which are f.h.e. over $B_1 \cap B_2$, when does there exist a fibration $E \to B$ such that $E \mid B_i$ is f.h.e. E_i?

Several other people have solutions for various forms of this problem. In particular, Wirth [**34**], Fuchs [**14**], and Dyer have answers in terms of the mapping cylinder construction which are presented in the present context of classification.

It takes additional work to show the above Z is $B_{H(F)}$.

Finally, for this approach, consider the problem of classifying principal fibrations, i.e., fibrations $H \to E \to B$ with an associative action of H fibrewise on E. Porter [**24**; 1969] classifies such fibrations up to equivalence in terms of shm maps

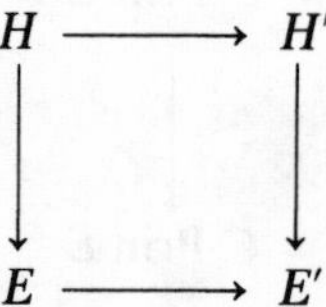

which, if we fix H, yields the following: Let $PH(B)$ denote the set of such equivalence classes.

THEOREM 4. *For* CW *spaces*, $PH(B)$ *is in* 1-1 *correspondence with* $[B, B_G]$.

Porter's equivalence is slightly weaker and corresponds to $[B, B_G]/H(B_G)$. He uses the following method of classification: Given the principal bundles

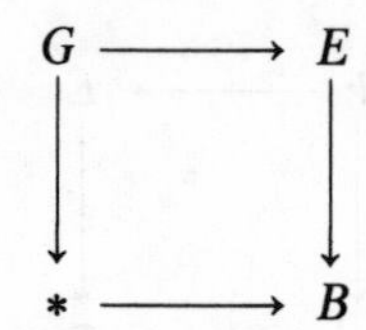

imbed them in the iterated DL-construction

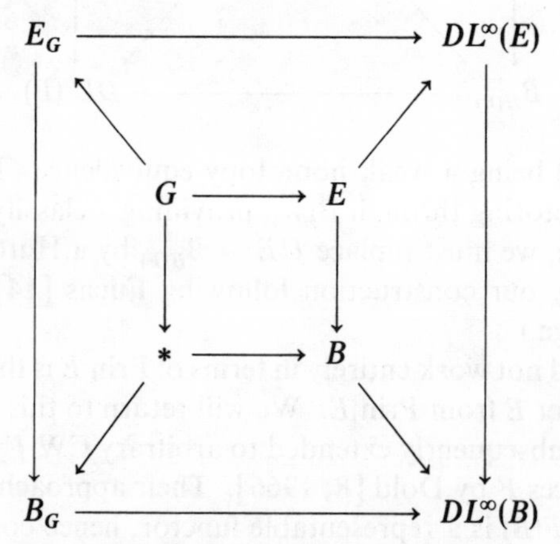

Again $B_G \simeq DL^\infty(B)$ (the total spaces are both acyclic) so we can factor through $B \to B_G$. It is easy to check that Prin $S(E) = DL(\text{Prin } E)$ and under this correspondence Porter's and my classifying maps correspond.

Both Porter's and my methods can be smoothed by the use of Fuchs' [**14**; 1969] modified Dold-Lashof construction which produces universal fibrations rather than quasifibrations.

Let us turn now to another viewpoint. Barratt, Gugenheim, and Moore [**3**; 1959] classified semisimplicial Kan fibrations $\mathscr{F} \to \mathscr{E} \to \mathscr{B}$ by $[\mathscr{B}, \overline{W}AM\mathscr{F}]$, where $AM\mathscr{F}$ is the automorphism group of the minimal complex $M\mathscr{F}$ of $\mathscr{F}$. The point of using minimal complexes is to pass to an associated *bundle* (rather than fibration) and to force homotopy equivalences to be actually automorphisms. Given fibrations of spaces $F \to E \to B$ and using the singular complex $\mathscr{S}$ and geometric realization $|\ \ |$, one might then hope to prove that the classes of such fibrations are in 1-1 correspondence with $[B, |\overline{W}AM\mathscr{S}F|]$. As far as I know, some of the many propositions necessary to carry out such a program had not then been written down, cf. [**2**].

Let us turn now to a more geometric *and* more algebraic way of classifying fibrations. Consider covering spaces, the simplest nontrivial examples of fibrations. Regular coverings are classified in terms of $\mathrm{Hom}(\pi_1(B), \mathrm{Aut}(F))$. Recently Teleman [**31**] has tried to generalize this relation, in part motivated by the situation in differential geometry. By using a suitable piecewise-differentiable model for $\hat{\Omega}B$, one can classify differentiable bundles in terms of holonomy (via connections or parallel transport) which is essentially in $\mathrm{Hom}(\hat{\Omega}B, G)$. This, I believe, is the background behind:

THEOREM 5. *For normal locally arc-connected, paracompact B, $LG(B)$ is in 1-1 correspondence with* {*conjugacy classes of* $\mathrm{Hom}(\bar{\Omega}B, G)$}.

Here $\bar{\Omega}X$ is the loop space with parameters $[0, 1]$ but a groupoid operation $\lambda * \mu$ which, when defined, traverses the first half of λ and the last half of μ. The point is to be able to lift paths consistently so as to have loops act associatively. This approach was extended to fibre homotopy equivalence by Curtis and Lashof [**6**]. The more usual ΩX can be used to explain the idea behind the map and it will turn out to be an shm map, this being good enough for the classification desired.

Let then $F \to E \to B$ be a Hurewicz fibration. (A Dold fibration would do, but would complicate the following formula slightly.) Consider

$$f_0 : \Omega B \times F \to E \quad \text{by } f_0(\lambda, f) = f \in F \subset E$$

and

$$g_t : \Omega B \times F \to B \quad \text{by } g_t(\lambda, f) = \lambda(t).$$

Applying the CHP we obtain $\tau = f_1 : \Omega B \times F \to F$ which tells us how F is transported around any loop in B. The homotopy class of τ is an invariant of $E \to B$ and, together with appropriate homotopies, a complete invariant. Hilton [**18**] observed that τ is "homotopy associative", i.e., $\tau(\lambda, \tau(\mu, f)) \simeq \tau(\mu + \lambda, f)$. I then showed [**28**; 1966] the existence of appropriate higher homotopies $\tau_i : I^{i-1} \times (\Omega B)^i \times F \to F$ such that the adjoints $\bar{\tau}_i : I^{i-1} \times (\Omega B)^i \to H(F)$ form an shm map. These form a complete invariant as, up to homotopy, E can be reconstructed from $\{\tau_i\}$ by a bar-type construction, i.e.,

$$E \cong \bigcup I^i \times (\Omega B)^i \times F/\sim,$$

where

$$\begin{aligned} &(t_1, \dots, t_{i-1}, \lambda_1, \dots, \lambda_i, f) \\ &\quad \sim (\dots, \hat{t}_j, \dots, \lambda_{j+1}\lambda_j, \dots) && \text{if } t_j = 0, \\ &\quad \sim (t_1, \dots, t_{j-1}, \lambda_1, \dots, \lambda_j, \tau_{i-j-1}(t_{j+1}, \dots, t_{i-1}, \lambda_{j+1}, \dots, f)) && \text{if } t_j = 1. \end{aligned}$$

THEOREM 6 [28]. *$LF(B)$ is in* 1-1 *correspondence with* $\{$*homotopy classes of transports* $\{\tau_i\}$ *up to action of* $H(F)$ *on* $F\}$.

Independently Drachman [11; 1966] in his thesis succeeded in classifying principal H-fibrations via shm maps.

THEOREM 7. *For connected H, $PH(B)$ is in* 1-1 *correspondence with* $\mathrm{Shm}_\pi(\Omega B, H)$.

The constructive part of the proof is similar to the construction of a transport. The CHP is used to construct $f_1 : \Omega B \to H$ covering

$$\begin{aligned} g_t : \Omega B &\to B \\ \lambda &\to \lambda(t). \end{aligned}$$

The CHP is then applied repeatedly to construct the rest of the shm map.

Drachman and I observed that for $H = H(F)$ and $P = \text{Prin } E$, transports were adjoint to his shm maps which in turn corresponded to classifying maps: $\mathrm{Shm}_\pi(\Omega B, H(F))/\pi_0(H(F)) \xleftrightarrow{1\text{-}1} [B, B_{H(F)}]$. The correspondence between these classifications can be clarified further by the following methods for recovering E from Prin E. Classically, confronted with a structural group G, we consider $\text{Prin}_G\, E$ and obtain E as $\text{Prin}_G \times_G F$ but using $H(F)$ instead of G we do not factor out as readily. One way around the difficulty is to generate an equivalence relation, so as to make sense of $\text{Prin } E \times_{H(F)} F$; that this procedure yields a quasifibration or fibration has been shown by Husseini [19; 1969] and Dyer [12]. Alternatively, we can reconstruct E by a bar-type construction, i.e.,

$$E \cong \bigcup \text{Prin } E \times \Delta^n \times H(F)^n \times F/\sim$$

with the usual identifications on all but the first and last faces where we have

$$\begin{aligned} (t_0, \ldots, t_n, \varphi, \varphi_1, \ldots, \varphi_n, f) &\sim (t_1, \ldots, t_n, \varphi\varphi_1, \ldots, \varphi_n, f) \quad \text{if } t_0 = 0, \\ &\sim (t_0, \ldots, t_{n-1}, \varphi_1, \ldots, \varphi_n(f)) \quad \text{if } t_n = 0. \end{aligned}$$

Thus we have $LF(B) \xleftrightarrow{1\text{-}1} PH(F)(B)$.

Notice that this construction gives a geometric realization of the E-M spectral sequence

$$\begin{aligned} E^2 &\cong \mathrm{Tor}^{H_*(G)}(H_*(\text{Prin } E), H_*(F)) \\ &\Downarrow \\ H_*(E)& \end{aligned}$$

We summarize the comparison of these various classifications in the following diagram:

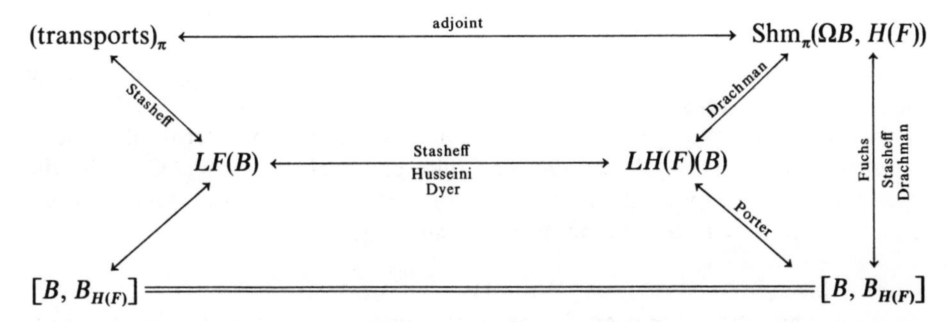

There are several ways of bringing homomorphisms into the picture. We have already mentioned Lashof's use of $\bar{\Omega}X$. A fairly obvious way which has occurred to many people is to replace F by the homotopy equivalent F' obtained by pulling the path space PB back over $p: E \to B$. The loop space ΩB then acts associatively on F'; we have a homomorphism $\Omega B \to H(F')$. There are difficulties with this approach to classification. First F' depends on E so to classify we must fix some F. We then have $H(F') \cong H(F)$ but the equivalence is at best shm.

Husseini [**19**; 1969] has a somewhat more structured solution. Functorially, he replaces ΩB by an associative complex M which is free enough to permit the construction of classifying homomorphisms $M \to H(F)$. Again, however, $\Omega B \to M$ is only shm.

Finally notice we have classified fibrations and fibrations with associative actions of the fibre. One might ask about actions which are not associative. Nowlan [**22**] shows that corresponding classification theorems exist, even for fibrations with A_n-actions of F on E. In particular he shows in the CW category:

THEOREM 8. *Let F be an associative H-space, then $E \to B$ is equivalent to a fibration induced from $E_F \to B_F$ iff E admits a fibrewise sha action of F.*

Having exhausted the classifying map approach, let us look at transition functions. Recall that a G-bundle can be described in terms of a cover $\{U_\alpha\}$ of B and homeomorphisms

$$(*) \qquad \begin{array}{ccc} p^{-1}(U_\alpha) & \underset{k_\alpha}{\overset{h_\alpha}{\rightleftarrows}} & U_\alpha \times F \\ & \searrow \quad \swarrow & \\ & U_\alpha & \end{array}$$

Over $U_\alpha \cap U_\beta$, we can make the following definition for $g_{\alpha\beta}: U_\alpha \cap U_\beta \to G$.

$$(U_\alpha \cap U_\beta) \times F \xrightarrow{h_\beta k_\alpha} (U_\alpha \cap U_\beta) \times F,$$
$$(x, f) \longrightarrow (x, g_{\alpha\beta}(x)f).$$

These satisfy $g_{\alpha\beta}g_{\beta\gamma} = g_{\alpha\gamma}$ over $U_\alpha \cap U_\beta \cap U_\gamma$. This is the 1-cocycle condition for sheaf cohomology so we obtain $[p] \in H^1(B; \mathscr{G})$ and indeed equivalent bundles give cohomologous cocycles.

It would seem reasonable to ask for a classifying map $B \to B_G$ in terms of $\{g_{\alpha\beta}\}$. For some reason this was done only recently: by Wirth [**35**; 1964] in a more general context, by tom Dieck [**32**; 1966] very explicitly in terms of Milnor's construction and by Segal [**25**; 1968] as follows: Recall $B \simeq B_{\{U_\alpha\}}$. Consider the map from the groupoid of the cover into G determined by $g_{\alpha\beta}$. The 1-cocycle condition shows it to be a homomorphism so we get

$$B \simeq B_{\{U_\alpha\}} \to B_G.$$

Now what about fibrations? Reasonable ones (e.g., having the WCHP over a weakly locally contractible paracompact base) are locally fibre homotopy equivalent to products [**13**], [**9**, Fuchs' thesis]. In fact Dyer [**12**] has reworked the foundations of the subject so as to emphasize the local approach and the similarities with the classical bundle theory. A variation on the pasting problem plays a crucial role: namely, showing the mapping cylinder of a fibre homotopy equivalence has one end as strong fibrewise deformation retract. This local approach permits the reconstruction of E from Prin E in a direct fashion. Classification then proceeds in terms of a universal example along classical lines.

What of transition functions? Since h_β is not inverse to k_β in (*) we have only $g_{\alpha\beta}g_{\beta\gamma} \simeq g_{\alpha\gamma}$ via some $g_{\alpha\beta\gamma}: I \times U_\alpha \cap U_\beta \cap U_\gamma \to H(F)$. This turns out to be part of the data necessary to classify $E \to B$. Wirth [**34**; 1964] defines transition families so as to prove:

THEOREM 9. *$LF(B)$ is in 1-1 correspondence with equivalence classes of transition families*

$$\{g_{\sigma^n}: I^{n-1} \times U_{\sigma^n} \to H(F)\},$$

where

$$\sigma^n = (\alpha_0, \dots, \alpha_n), \qquad U_{\sigma^n} = U_{\alpha_0} \cap \cdots \cap U_{\alpha_n}$$

and g_{σ^n} are suitably related on ∂I^{n-1}.

This is essentially the shm condition so again we have $B \simeq B_{\{U_\alpha\}} \to B_{H(F)}$. On the other hand, so far I have not seen a definition of "sheaf up to homotopy" so as to make $[g_\sigma] \in H^1(B; ?)$.

The theory of fibrations is thus fairly complete and well worked out on a conceptual level; the rest should be applications and computations.

REFERENCES

1. G. Allaud, *On the classification of fiber spaces*, Math. Z. **92** (1966), 110–125. MR **32** #6462.

2. M. G. Barratt, Lecture notes (to appear).

3. M. G. Barratt, V. K. A. M. Gugenheim and J. C. Moore, *On semisimplicial fibre-bundles*, Amer. J. Math. **81** (1959), 639–657. MR **22** #1895.

4. E. H. Brown, Jr., *Cohomology theories*, Ann. of Math. (2) **75** (1962), 467–484. MR **25** #1551.

5. S. S. Chern and Y.-F. Sun, *The imbedding theorem for fibre bundles*, Trans. Amer. Math. Soc. **67** (1949), 286–303. MR **11**, 378.

6. M. L. Curtis, *Classification spaces for a class of fibre spaces*, Ann. of Math. (2) **60** (1954), 304–316. MR **16,** 846.

7. M. L. Curtis and R. Lashof, *Homotopy equivalence of fibre bundles*, Proc. Amer. Math. Soc. **9** (1958), 178–182. MR **20** #4838.

8. A. Dold, *Halbexakte Homotopiefunktoren*, Lecture Notes in Math., no. 12, Springer-Verlag, Berlin and New York, 1966. MR **33** #6622.

9. ———, *Partitions of unity in the theory of fibrations*, Ann. of Math. (2) **78** (1963), 223–255. MR **27** #5264.

10. A. Dold and R. Lashof, *Principal quasi-fibrations and fibre homotopy equivalence of bundles*, Illinois J. Math. **3** (1959), 285–305. MR **21** #331.

11. B. Drachman, *A generalization of the Steenrod classification theorem to H-spaces*, Trans. Amer. Math. Soc. **153** (1971), 53–88.

12. E. Dyer, (to appear).

13. E. Fadell, *On fibre spaces*, Trans. Amer. Math. Soc. **90** (1959), 1–14. MR **21** #330.

14. M. Fuchs, *A modified Dold-Lashof construction that does classify H-principal fibrations* (to appear).

15. ———, *The section extension theorem and loop fibrations*, Michigan Math. J. **15** (1968), 401–406. MR **38** #3868.

16. ———, *Verallgemeinerte Homotopie-Homomorphismen und klassifizierende Räume*, Math. Ann. **161** (1965), 197–230. MR **33** #3295.

17. K. F. Gauss, *Disquisitiones generales circa superficies curvas*, Commentationes societas regiae scientarium Gottingensis recentiores VI (1828).

18. P. Hilton, *Homotopy theory and duality*, Gordon and Breach, New York, 1965. MR **33** #6624.

19. S. Y. Husseini, *The topology of classical groups and related topics*, Gordon and Breach, New York, 1969.

20. R. K. Lashof, *Classification of fibre bundles by the loop space of the base*, Ann. of Math. (2) **64** (1956), 436–446. MR **18,** 497.

21. J. W. Milnor, *Construction of universal bundles.* II, Ann. of Math. (2) **63** (1956), 430–436. MR **17,** 1120.

22. R. Nowlan, *The classification of A_n-fibrations*, Thesis, Notre Dame, Ind., 1969.

23. L. S. Pontrjagin, *Classification of some skew products*, C. R. (Dokl.) Acad. Sci. USSR **47** (1945), 322–325. MR **7,** 138.

24. G. J. Porter, *Homomorphisms of principle fibrations*, Illinois J. Math. (to appear).

25. G. Segal, *Classifying spaces and spectral sequences*, Inst. Hautes Études Sci. Publ. Math. No. 34 (1968), 105–112. MR **38** #718.

26. J. D. Stasheff, *A classification theorem for fibre spaces*, Topology **2** (1963), 239–246. MR **27** #4235.

27. ———, *Associated fibre spaces*, Michigan Math. J. **15** (1968), 457–470. MR **40** #2083.

28. ———, *"Parallel" transport in fibre spaces*, Bol. Soc. Mat. Mexicana (2) **11** (1966), 68–84. MR **38** #5219.

29. N. E. Steenrod, *The classification of sphere bundles*, Ann. of Math. (2) **45** (1944), 294–311. MR **5,** 214.

30. ———, *The topology of fibre bundles*, Princeton Math. Series, vol. 14, Princeton Univ. Press, Princeton, N.J., 1951. MR **12,** 522.

31. C. Teleman, *Généralisation du groupe fondamental d'un espace topologique*, C. R. Acad. Sci. Paris **248** (1959), 2845–2846. MR **23** #A1385a.

32. T. tom Dieck, *Klassifikation numerierbarer Bündel*, Arch. Math. (Basel) **17** (1966), 395–399. MR **34** #6776.

33. H. Whitney, *Topological properties of differentiable manifolds*, Bull. Amer. Math. Soc. **43** (1937), 785–805.

34. J. F. Wirth, *Fibre spaces and the higher homotopy cocycle relations*, Thesis, Notre Dame, Ind., 1965.

35. ———, Private communication.

III. AND IV. STATUS[2] REPORT ON FINITE H-COMPLEXES AND INFINITE LOOP SPACES

These reports will be concerned with recent developments in the theory of H-spaces with the hope of indicating the present state of the art with an eye toward where the field may be heading. There will be no attempt to discuss history or foundations, as was done in [13].

The small intersection of the two reports is covered by the result of Hubbuck (cf. [9]):

THEOREM. *Let X be a connected finite complex. If X is a homotopy commutative H-space, then $X \cong S^1 \times \cdots \times S^1$, n times, for some $0 \leqq n < \infty$.*

The proof makes use of the projective plane of X, i.e.,

$$XP(2) = SX \cup \Delta^2 \times X^2.$$

Consider the exact sequence in ordinary cohomology or K-theory of the pair $XP(2), SX$. The homotopy commutativity of X induces an automorphism of the exact sequence from which, with powerful machinery, Hubbuck obtains his answer.

III. FINITE H-COMPLEXES

The major problem here is that of classification, and more generally the study of analogues to results in the theory of compact Lie groups. Not long ago the following was a viable conjecture.

DEMOLISHED CONJECTURE. *If a finite complex X admits a multiplication, then there is a Lie group G_X such that $X \simeq G_X \times S^7 \times \cdots \times S^7$.*

Such spaces occur often enough for us to propose calling them "quasi-Lie" H-spaces. An analogous conjecture for sha spaces X might have omitted reference to S^7. I will generally not bother with sha or ha analogues of the statements I make.

The conjecture above was demolished by Hilton and Roitberg's [7] discovery of a multiplication on a 10-manifold M_7, the total space of the principal S^3-bundle over S^7 classified by 7ω, where ω classifies $Sp(2) \to S^7$. Since M_7 is not of the homotopy type of any Lie group, the classification will not reduce to that for Lie groups. To be precise, we want not only to classify the underlying spaces but we also want to classify up to H-equivalence.

---

[2] Comments in [ ] have been added in October 1970. See also the Proceedings of the Neuchâtel Conference on H-spaces, 1970, Lecture Notes in Maths, no. 196.

DEFINITION. Two H-spaces (X, m) and (Y, m) are H-equivalent if there is a homotopy equivalence $h: X \to Y$ which is an H-map, i.e., we have

$$\begin{array}{ccc} X \times X & \xrightarrow{h} & X \\ {\scriptstyle h \times h}\downarrow & \simeq & \downarrow{\scriptstyle h} \\ Y \times Y & \xrightarrow[h]{} & Y \end{array}$$

This *is* an equivalence relation.

THEOREM (CURJEL-DOUGLAS). *Up to H-equivalence, there are finitely many finite H-complexes of any given dimension.*

The proof involves estimating the number of choices in building the Postnikov system for such a space.

One approach to the classification problem is to investigate how much such a complex must look homologically like a Lie group. This is Hopf's original attack, showing for example that for connected *H-spaces* $H^*(X; Q) \approx E(x_1, \ldots, x_r)$, an X exterior algebra on odd dimensional generators. The rank of X is r.

For connected X of rank 1, Browder [2] gave a complete classification up to homotopy: i.e., $X \simeq S^1, S^3, S^7, RP^3$, or RP^7. There is only one structure possible on S^1, while on S^3 and S^7 there are respectively 12 and 120 homotopy classes of multiplications. However each is H-equivalent to its transpose or opposite, so the number of inequivalent structures is 6 or 60 respectively. The corresponding reduction for RP^3 or RP^7 has not been worked out (cf. [10], [12]).

For rank 2, Adams [1] studied X of the form $X \simeq S^p \cup e^q \cup e^{p+q}$ and, using $\mathscr{A}(2)$ in the projective plane of X, showed for $q > p + 1 > 2$ that (p, q) must be one of the pairs (3, 3), (3, 5), (3, 7), (7, 7), (7, 11), or (7, 15). Quasi-Lie examples exist except for (7, 11) and (7, 15). Thomas [15], [16] has carried out extensive computations of this sort, eliciting the structure of $H^*(XP(2); Z_2)$. He often must assume $H^*(X; Z_2)$ primitively generated, but Curjel has just shown that given some multiplication on X, there is one such that $H^*(X; Q)$ *is* primitively generated. Thus, in the absence of 2-torsion, $H^*(X; Z_2)$ will be primitively generated.

The cases (7, 11) and (7, 15) left open by Adams have been resolved by Douglas and Sigrist [5] and independently by Hubbuck [9]. The advance is made possible by using K-theory and, for example, playing ψ^2 off against ψ^3.

More generally there is the problem of determining the *type* of a finite H-complex; i.e., the set of dimensions in which rational cohomology ring generators occur. Hubbuck [9] has a machine (involving K-theory and the ψ^k-operations) which so far has analyzed the situation up to rank 5; the types all occur in quasi-Lie groups. [Further results for finite group complexes are being produced by Ewing [20].]

Given such restrictions on $H^*(X)$, how divergent can the homotopy types be? Using a technique of Zabrodsky [**17**] for "mixing homotopy type", I [**14**] was able to show the Hilton-Roitberg M_7 has in fact the homotopy type of a loop space. The same approach revealed that M_n is an H-space for $n \not\equiv 2\,(4)$. Zabrodsky [**18**] later showed M_n for $n \equiv 2\,(4)$ was indeed not an H-space. The proof involves tertiary cohomology operations.

Meanwhile Curtis and Mislin [**3**], [**4**] investigated $SU(3)$-bundles over S^7 and discovered two new homotopy types which are H-spaces.

Prior to his death, Jerome Harrison analyzed the known examples and produced the following result [**6**]:

Let $H \to G \to G/H = S^n$ be a representation of S^n as a (Lie) homogeneous space, classified by $\alpha \in \pi_{n-1}(H)$. Assume α of finite order and write $\alpha = \alpha_2 + \alpha_3 + \cdots + \alpha_p + \cdots + \alpha_q$, where α_p is of pth power order, p prime. Let E_β be classified by $\beta = \sum \varepsilon_p \alpha_p$. For $\varepsilon_p = 0, \pm 1$, E_β is an H-space iff $\varepsilon_2 = \pm 1$ or $n = 1, 3, 7$.

This complements a result of Curtis and Mislin [**3**] which, in the above setting, says:

If α belongs to a cyclic summand Z_q of $\pi_{n-1}(H)$ and $\beta = \lambda\alpha$ with λ prime to q, then E_β is an H-space (in fact sha). It also complements a very recent result of Zabrodsky [**19**] which says:

If $G, H = Sp(n+1), Sp(n)$, or $SU(n+1), SU(n)$, $n > 2$, then E_β is *not* an H-space if $\beta = \lambda\alpha$, λ even.

Notice E_β is the pull-back of G over a map $S^n \to S^n$ of degree λ.

In terms of examples of low dimension or rank, the first open[3] questions are [**6**]:

(1) $SU(6) \to SU(7) \to S^{13}$, $\lambda = 3$.

(2) $Sp(3) \to Sp(4) \to S^{15}$, $\lambda = 3$.

(3) $\mathrm{Spin}(7) \to \mathrm{Spin}(9) \to S^{15}$, $\lambda = 3$.

(4) $SU(3) \to G_2 \to S^6$, $\lambda = 2$. [No. See next comment.]

The exceptional group G_2 continues to be exceptional from our point of view. An interesting open question is: If X is a finite H-complex and $H^*(X; Q) \approx H^*(G_2; Q) \approx E(x_3, x_{11})$, how much like G_2 must X be? In particular must X have the same 2-torsion as G_2? [As expected, presenting this problem to Hubbuck produced a solution: $H^*(X; Z_2) \approx H^*(G_2; Z_2)$.]

In spite of this profusion of examples, the classification of homotopy types of finite H-complexes in low dimensions is solvable.

Given Zabrodsky's work and mine on M_n, he [**18**], Hilton and Roitberg [**8**] and Curtis and Mislin by three different methods classify simply connected torsion free finite H-complexes of rank 2:

$$S^3 \times S^3, \quad S^7 \times S^7, \quad SU(3) \text{ or } M_n, \quad n \not\equiv 2\,(4).$$

Curtis and Mislin have also carried out the classification of all finite H-complexes of dimension $\leqq 9$. Again the examples all look like quasi-Lie groups in a certain sense.

[3] Added in proof. Zabrodsky has now shown in the classical cases that $E_{\lambda\alpha}$ is an H-space iff λ is odd.

To be precise, all the evidence so far supports the following:

CONJECTURE. *Given a finite H-complex X, there is a family of quasi-Lie groups $G_X(p)$ and mod p equivalences $X \to G_X(p)$. (For any Lie group G and most primes G is mod p equivalent to a product of odd dimensional spheres.)*

[I have since shown this to be false; it may still be true for finite group complexes.]

There are other ways in which finite complexes might resemble compact Lie groups. The H-space problem session reviews some of the outstanding problems regarding torsion relevant to finite H-complexes. Rector **[11]** has recently initiated a program of studying subgroups in the finite H-complex setting:

A subfinite group complex H of a finite group complex G is a subgroup H such that the fibre of $B_H \to B_G$ is again of the homotopy type of a finite complex.

This turns out to be a workable definition, but produces some surprises: There are groups of the homotopy type of S^3 which admit no subfinite group complexes S^1. Part of Rector's program is the development of an appropriate analogue of the Weyl group.

For H-complexes which are not groups, I propose the following:

A subfinite H-complex H of a finite H-complex G is an H-map $H \to G$ such that there is a fibration

$$\begin{array}{ccccc} F & \to & E & \to & B \\ \wr\| & & \wr\| & & \\ H & \to & G & & \end{array}$$

of spaces of the homotopy type of finite complexes.

Many of the problems mentioned here and in the problem session have "group" or "sha" versions as well as a straight "H-space" version. Some would also be reasonable in a "homotopy associative" version. Homotopy commutativity in its most extreme form will be examined in the next part of this series.

REFERENCES

1. J. F. Adams, *H-spaces with few cells*, Topology **1** (1962), 67–72. MR **26** #5574.

2. W. Browder, *Fiberings of spheres and H-spaces which are rational homology spheres*, Bull. Amer. Math. Soc. **68** (1962), 202–203. MR **26** #5580.

3. M. Curtis and G. Mislin, *H-spaces which are bundles over S^7*, J. Pure Appl. Algebra (to appear).

4. ———, *Two new H-spaces*, Bull. Amer. Math. Soc. **76** (1970), 851–852.

5. R. R. Douglas and F. Sigrist, *Sphere bundles over spheres and H-spaces*, Topology **8** (1969), 115–118. MR **39** #6312.

6. J. Harrison and J. D. Stasheff, *Families of H-spaces*, Quart. J. Math. Oxford Ser. (2) (to appear).

7. P. J. Hilton and J. Roitberg, *On principal S^3-bundles over spheres*, Ann. of Math. (2) **90** (1969), 91–107. MR **39** #7624.

8. ———, *On the classification problem for H-spaces of rank two*, Comment. Math. Helv.

9. J. R. Hubbuck, *Generalized cohomology operations and H-spaces of low rank*, Trans. Amer. Math. Soc. **141** (1969), 335–360. MR **40** #2059.

10. C. M. Naylor, *Multiplications on* SO(3), Michigan Math. J. **13** (1966), 27–31. MR **32** #8342.

11. D. Rector, *Subgroups of finite dimensional topological groups* (preprint).

12. E. Rees, *Multiplications on projective spaces*, Michigan Math. J. **16** (1969), 297–302.

13. J. D. Stasheff, *H-spaces from a homotopy point of view*, Lecture Notes in Math., no. 161, Springer-Verlag, Berlin and New York, 1970.

14. ———, *Manifolds of the homotopy type of (non-Lie) groups*, Bull. Amer. Math. Soc. **75** (1969), 998–1000. MR **39** #7623.

15. E. Thomas, *Steenrod squares and H-spaces*, Ann. of Math. (2) **77** (1963), 306–317. MR **26** #3057.

16. ———, *Steenrod squares and H-spaces*. II, Ann. of Math. (2) **81** (1965), 473–495. MR **31** #1671.

17. A. Zabrodsky, *Homotopy associativity and finite* CW-*complexes*, Topology **9** (1970), 121–128.

18. ———, *The classification of simply connected H-spaces with three cells*. I, II (preprint).

19. ———, *On sphere extensions of classical Lie groups*, Proc. Sympos. Pure Math., vol. 22, Amer. Math. Soc., Providence, R.I., 1971.

20. J. Ewing, *On the type of associative H-spaces* (preprint).

IV. INFINITE LOOP SPACES (INCLUDING DYER-LASHOF OPERATIONS)[4]

Recently there has been renewed interest in infinite loop spaces, nourished especially by Boardman's proof of the conjecture that PL, Top, F, etc., are infinite loop spaces.

DEFINITION. X is an infinite loop space if there exists X_n, $n \geqq 1$, and maps $X_n \to \Omega X_{n+1}$ which are homotopy equivalences, $X_1 = X$.

Given an infinite loop space X, Dyer and Lashof [**8**] construct maps $W\Sigma_n \times_{\Sigma_n} X^n \to X$ from which they derive their operations, analogous to Steenrod operations in cohomology. These operations were very useful in computing $\Omega^n S^n X$, especially $H_*(\Omega^\infty S^\infty X)$ and $H_*(\Omega^\infty S^\infty)$ and recently, in the hands of Milgram mod 2 [**17**] and May mod p [**13**], [**14**], Tsuchiya [**22**], Kochman [**9**], and Madsen [**12**], these operations have played a crucial role in calculating the cohomology of $F = \lim H(S^n)$ where $H(\ )$ denotes auto-homotopy equivalences.

In the preprint version of [**8**], although not in the published version, Dyer and Lashof showed that the spaces $W\Sigma_n \times_{\Sigma_n} X^n$ go together to form good models for infinite loop spaces. Specifically, they study a construction

$$TX = \bigcup W\Sigma_n \times_{\Sigma_n} X^n/\sim$$

where the equivalence relation $\sim$ is reasonably related to $\Sigma_n \subset \Sigma_{n+1}$ corresponding to $X^n = X^n \times * \subset X^{n+1}$. They show TX is in certain cases a reasonable approximation to $QX = \Omega^\infty S^\infty X$.

Barratt [**3**] has recently considered, in a semisimplicial context, a construction ΓX which is the universal group of TX (this is equivalent to taking ΓX to be a free group on the generating set of cells on which TX is the free monoid or

[4] This part of the talk was revised considerably for distribution to the participants on the basis of subsequent talks by and discussions with J. M. Boardman and M. G. Barratt. The present version has been revised further in the light of subsequent developments due to Anderson [**1**], [**2**], Beck [**4**], and especially May [**15**].

equivalently defining $\Gamma X = \Omega BTX$). He has shown $\Gamma X \simeq QX = \Omega^\infty S^\infty X$ semisimplicially (cf. [**18**]).

One approach to infinite loop spaces, due to Beck [**5**] uses the fact that Q (as well as Γ or T) has very rigid internal structures. There is a canonical retraction $QQ(\ ) \xrightarrow{\varepsilon} Q(\ )$ given in terms of $S\Omega(\ ) \to (\ )$ by $(t, \lambda) \to \lambda(t)$ or of $\Gamma\Gamma(\ ) \xrightarrow{\varepsilon} \Gamma(\ )$ in terms of wreath product. These maps have the property that $\varepsilon(\varepsilon Q) = \varepsilon(Q\varepsilon)$, i.e.

$$\begin{array}{ccc} QQQ & \xrightarrow{\varepsilon Q} & QQ \\ {\scriptstyle Q\varepsilon}\downarrow & & \downarrow{\scriptstyle \varepsilon} \\ QQ & \xrightarrow[\varepsilon]{} & Q \end{array}$$

and similarly for Γ or T. In categorical language, we have a triple.

Beck proves: Given a retraction $r: QX \to X$ such that $r\varepsilon = r(Qr)$ (i.e. X is a Q-algebra), then X is an infinite loop space. The map ε and its particular definition in terms of $S\Omega$ are used to construct the iterated $B \cdots BX$ simultaneously.

Notice that for X to be a T-algebra would imply the existence of a very compatible family

$$W\Sigma_n \times_{\Sigma_n} X^n \to X,$$

which looks exactly like a lot of higher homotopies, the sort of structure one would expect to use to identify an infinite loop space in homotopy invariant terms.

Let us return briefly to the obvious approach to identifying infinite loop space indicated in our first lecture. Suppose X were an associative H-space. If $m: X \times X \to X$ were a homomorphism, then B_X would again be an associative H-space, but in fact m is a homomorphism (with respect to itself) iff it is commutative. The next approach, carried out by Sugawara [**21**] and myself [**20**] is to consider m being an shm map, in which case m or (X, m) is called shc(ommutative). This would yield a multiplication on B_X and further homotopies on X would be needed to construct B_{B_X}, and so on.

It was precisely to handle such families of higher homotopies, at least at the chain level, that Mac Lane, [**11**] working with Adams, introduced the notions of PACT's and PROP's.

Boardman and Vogt's [**7**] "categories of operators" are topological PROP's. Success was due to their ability to manipulate PROP's and above all, to Boardman's clever construction of PROP's based on the linear geometry of O, Top, and F.

For the purpose of this report I would like to present my own interpretation of Boardman and Vogt's work, being influenced as much as I can by features which appear in the work of Anderson [**1**], [**2**] and Segal [**19**], Beck [**4**], and May [**15**].

First "recall" the definition of a topological PROP [**11**], [**15**]:

A topological PROP P is a category with objects $0, 1, 2, \ldots$ satisfying the following conditions:

(a) The morphisms $P(j, k)$ constitute a topological space and composition is continuous.

(b) There is given a continuous associative bifunctor $\oplus: P \times P \to P$ such that on objects $j \oplus k = j + k$.

(c) There is given an inclusion of the symmetric group Σ_K in $P(k, k)$.

(d) For $\sigma \in \Sigma_j$, $\tau \in \Sigma_k$, the morphism $\sigma + \tau \in \Sigma_{j+k} P(j + k, j + k)$ and is that permutation which acts as σ on the first j elements, as τ on the last k.

(e) For $p_i \in P(j_i, k_i)$ and $\sigma \in \Sigma_m$, we have

$$\sigma(k_1, \ldots, k_m)(p_1 + \cdots + p_m) = (p_{\sigma^{-1}(1)} + \cdots + p_{\sigma^{-1}(m)})(j_1, \ldots, j_m)$$

where $\sigma(k_1, \ldots, k_m) \in \Sigma_k$, $k = \sum k_i$ and acts by permuting the blocks in the given partition of k elements.

EXAMPLE. End X is the PROP given by End $X(j; k) = \{X^j \to X^k\}$ with all the obvious structure.

Henceforth PROP will mean topological PROP.

If we have a morphism of PROP's $P \to \text{End } X$ we say that P acts on X or X is a P-space.

Consider the PROP which is generated by taking $W\Sigma(n, 1) = W\Sigma_n$. The PROP's P for the theory of infinite loop spaces are like $W\Sigma$ in that $P(n, 1)$ will be contractible and Σ_n-free. (May points out: if they were Σ_n-contractible, they could act only on spaces of the homotopy type of an abelian monoid.) Following May, we call such PROP's E_∞-PROP's.

Let E then be such an E_∞-PROP. Let X be an E-space. In particular X will be an H-space. Suppose in fact that X is associative and E acts via $\text{Hom}(X^n, X^r)$, then there would be an induced action of E on B_X. Realizing we are unlikely to stumble on such structure in nature, Boardman and Vogt [6], [7] develop the following approach: They generalize the W-construction so as to (1) apply to PROP's, (2) preserve homotopy type, (3) have the property:

For any PROP P, WP acts on X iff WP acts on any space of the homotopy type of X. If $\mathscr{A}$ is the PROP characterizing a topological monoid, $W\mathscr{A}(n, 1)$ is the complex K_n for my A_n-spaces.

They also define a tensor product of PROP's, and then use ordinary homotopy theory to define an equivariant map $W(\mathscr{A} \otimes P) \to P$ or, equally well $W(\mathscr{A} \otimes W\Sigma) \to P$.

Next, there is a construction (due independently to Frank Adams) which given a $W\mathscr{A}$ action on X produces a space $M_{\mathscr{A}}X \simeq X$ with an associative multiplication. Vogt generalizes this construction so that given an action of $W(\mathscr{A} \otimes P)$ on X he constructs $M_{\mathscr{A} \otimes P}X$ and an action of $\mathscr{A} \otimes P$ on $Y = M_{\mathscr{A} \otimes P}X$. In particular, this means Y is a monoid and P acts via $\text{Hom}(Y^n, Y)$ and hence on B_Y. Now they can iterate.

Boardman chooses P judiciously in terms of classical vector space structure to get P for $X = O$, Top, F and their classifying spaces [7]. The switch $W(\mathscr{A} \otimes W\Sigma) \to P$ allows the argument to be in terms of $W\Sigma$ after the first stage.

Attention has been called to this special case of Boardman and Vogt's theory by the work of Segal [19] and Quillen. There is a notion of "transfer" or "trace" for a representable functor which Quillen shows can be interpreted as a compatible family $W\Sigma_n \times_{\Sigma_n} X^n \to X$ for the representing X.

THEOREM (SEGAL). *A representable functor extends to a cohomology theory iff the functor admits a trace.*

Segal's proof involves showing that $\{W\Sigma_n \times_{\Sigma_n} X^n \to X\}$ yields iterated classifying spaces for ΩX, though the proof is formally different from that we have just sketched.

Segal's method (cf. Anderson's talk [1] for this conference) handles the higher homotopies by *not* using a single space X and its powers X^n but rather a family X_i such that $X_n \simeq (X_1)^n$. A direct comparison of the methods should be possible in terms of a generalized notion of PROP capable of handling such families of spaces.

It is perhaps worth remarking that at the present time the interest in infinite loop spaces is to obtain Dyer-Lashof operations. Note that the Boardman-Vogt approach can construct $W\Sigma_n \times_{\Sigma_n} (BX)^n \to BX$ before constructing the further classifying spaces.

Tsuchiya [22] was the first to describe such maps $W\Sigma_n \times_{\Sigma_n} X^n \to X$ directly for X being the monoids or classifying spaces of geometric interest, O, BO, F, BF, etc. This approach has been applied to some of the quotients such as G/Top by Madsen [12]. A somewhat different description of such maps has been given by Milgram at this conference. Although sufficient for computational purposes, such maps apparently are not compatible strictly enough to generate PROP's or to make X a T-algebra; perhaps homotopy PROP's or strong homotopy triple-algebras are needed.

Milgram [16] constructs models for $\Omega^q S^q X$ in terms of X^n and certain non-Σ_n-free cells $C(n)$. With hindsight, it may be possible to make his model look more like part of the Dyer-Lashof TX.

May [15] points out that the use above of precisely $W\Sigma_n$ is a convenience. The homology of the symmetric group does not depend on the choice of resolution; the Dyer-Lashof operations are equally well determined by $P(n, 1) \times_{\Sigma_n} X^n \to X$ if P is an E_∞-PROP acting on X.

May [15] has developed another approach utilizing the smoother parts of Boardman and Beck. Recall that Beck constructed $B \cdots BX$ for a Q-algebra X. May

(1) shows how a PROP can give rise to a triple,

(2) shows the triple C associated to Boardman's "little cubes" category maps by a *map of triples* into Q, preserving homotopy type,

(3) uses the direct product of PROP's to relate a space X with an E_∞-PROP action to Beck's construction using C as an intermediary and thus constructs $B \cdots BX$.

Beck has now [4] another approach. For a given E_∞-PROP P he defines "suspension" in the category of P-spaces (essentially as a quotient of the free P-space generated by the ordinary suspension). He then asserts that this gives a classifying space for X. The iterated classifying space is then the iterated P-suspension.

In addition to the "geometric" spaces listed above and their quotients, current interest includes certain products of $K(\pi, n)$'s with "exotic" infinite loop structures, e.g., $Y = \prod K(Z_2, 2^i)$ with $H_*(Y) \approx Z_2[\alpha_2]$ being studied by Kraines [**10**] and $Z = \prod K(Z_2, 2^i)$ with $H_*(Z) = Z_2[\alpha_2, \alpha_6, \ldots]$ being studied by D. S. Kahn. There apparently is a relevant PROP.

In conclusion, the branches of H-space theory devoted to finite complexes on the one hand and infinite loop spaces on the other are flourishing. One anticipates a survey of algebraic topology in the 70's will recount a variety of interesting developments.

References

1. D. W. Anderson, *Spectra and Γ-sets*, Proc. Sympos. Pure Math., vol. 22, Amer. Math. Soc., Providence, R.I., 1971.

2. ———, Internat. Congress Mathematicians, Nice, 1970.

3. M. G. Barratt, *A free group functor for stable homotopy*, Proc. Sympos. Pure Math., vol. 22, Amer. Math. Soc., Providence, R.I., 1971.

4. J. Beck, *Classifying spaces for homotopy-everything H-spaces* (preprint). (Hopefully to appear in the Proceedings of the Conference on H-spaces, Neuchatel, 1970.)

5. ———, *On H-spaces and infinite loop spaces*, Category Theory, Homology Theory and their Applications (Battelle Institute Conference, Seattle, Wash., 1968), vol. 3, Springer-Verlag, Berlin and New York, 1969, pp. 139–153. MR **40** #2079.

6. J. M. Boardman, *Homotopy structures and the language of trees*, Proc. Sympos. Pure Math., vol. 22, Amer. Math. Soc., Providence, R.I., 1971.

7. J. M. Boardman and R. M. Vogt, *Homotopy-everything H-spaces*, Bull. Amer. Math. Soc. **74** (1968), 1117–1122. MR **38** #5215.

8. E. Dyer and R. K. Lashof, *Homology of iterated loop spaces*, Amer. J. Math. **84** (1962), 35–88. MR **25** #4523.

9. S. O. Kochman, *The homology of classical groups over the Dyer-Lashof algebra, etc.*, Thesis, University of Chicago, Chicago, Ill., 1970.

10. D. Kraines, *The cohomology of some k-stage Postnikov systems*, Proc. Adv. Study Inst. Algebraic Topology, Aarhus, 1970.

11. S. Mac Lane, *Categorical algebra.* Chapter IV, Colloq. Lectures, Amer. Math. Soc., Providence, R.I., 1963, pp. 26–29.

12. I. Madsen, *On the action of the Dyer-Lashof algebra in $H_*(G)$ and $H_*(G/\text{Top})$*, Thesis, University of Chicago, Chicago, Ill., 1970.

13. J. P. May, *Categories of spectra and infinite loop spaces*, Battelle Institute Conference Category Theory, Homology Theory and Their Applications (Seattle, Wash., 1968), vol. 3, Springer-Verlag, Berlin and New York, 1969, pp. 448–479. MR **40** #2073.

14. J. P. May, *Homology operations in loop spaces*, Proc. Sympos. Pure Math., vol. 22, Amer. Math. Soc., Providence, R.I., 1971.

15. ———, *The geometry of iterated loop spaces* (preprint).

16. R. J. Milgram, *Iterated loop spaces*, Ann. of Math. (2) **84** (1966), 386–403. MR **34** #6767.

17. ———, *The mod two spherical characteristic classes*, Ann. of Math. (2) **92** (1970), 238–261.

18. S. B. Priddy, *On $\Omega^\infty S^\infty$ and the infinite symmetric group*, Proc. Sympos. Pure Math., vol. 22, Amer. Math. Soc., Providence, R.I., 1971.

19. G. Segal, *Homotopy-everything H-spaces* (preprint).

20. J. D. Stasheff, Thesis, Oxford University, 1960.

21. M. Sugawara, *On the homotopy-commutativity of groups and loop spaces*, Mem. Coll. Sci. Univ. Kyoto Ser. A Math. **33** (1960/61), 257–269. MR **22** #11394.

22. A. Tsuchiya, *Spherical characteristic classes mod p*, Proc. Japan Acad. **44** (1968).

TEMPLE UNIVERSITY

ON SPECTRA $V(n)$

HIROSI TODA

Let p be an odd prime. The Milnor's elements $Q_{i+1} = [\wp^{p^i}, Q_i]$, $Q_0 = \Delta$, generate an exterior subalgebra $E(Q_i)$ of the Steenrod algebra $\mathscr{A} = \wp \otimes E(Q_i)$, where $\wp$ is a subalgebra of $\mathscr{A}$ generated by the reduced powers $\wp^i$. We give $E(Q_i)$ an $\mathscr{A}$-module structure by the bijection $E(Q_i) \to \mathscr{A} /\!/ \wp$. Denote by $Q(m)$ a subalgebra of $E(Q_i)$ spanned by the first m bases and give it an $\mathscr{A}$-module structure as a quotient of $E(Q_i)$. Then we denote by

$$V(n + (k/2^{n+1})), \qquad 0 \leqq k \leqq 2^{n+1},$$

a spectrum having the cohomology $H^*(V(n + (k/2^{n+1})); Z_p) \cong Q(2^{n+1} + k)$. In particular $H^*(V(n); Z_p) \cong E(Q_0, Q_1, \ldots, Q_n)$.

THEOREM 1. *$V(3)$ exists for $p \geqq 7$, $V(2\frac{3}{4})$ exists for $p = 5$, $V(1\frac{1}{2})$ exists for $p = 3$.*

Note that for $p = 2$, $V(\frac{1}{2})$ exists but $V(1)$ does not exist since in $E(Sq^1, Sq^2Sq^1 + Sq^3)$ Adem relation $Sq^2Sq^2 = Sq^3Sq^1$ does not hold. We shall see that the above theorem is best possible for $p = 3$. The proof of the theorem is due to the following general

LEMMA. *Let M^* be a graded $\mathscr{A}$-module. Assume that $M^n \neq 0$ implies $\mathrm{Ext}_{\mathscr{A}}^{s+2,s+n}(M^*, Z_p) = 0$ for $s \geqq 1$. Then there exists a spectrum X such that $H^*(X; Z_p) \cong M^*$ as $\mathscr{A}$-modules.*

We have an isomorphism $\mathrm{Ext}_{\mathscr{A}}^{*,*}(E(Q_i), Z_p) \cong \mathrm{Ext}_{\wp}^{*,*}(Z_p, Z_p) = H^*(\wp)$ and we can estimate it by use of May's spectral sequences:

$$P(b_j^i) \otimes H^*(U(L)) \Rightarrow H^*(V(L)) \quad \text{and} \quad H^{*,*}(V(L)) \Rightarrow H^{*,*}(\wp),$$

where $U(L)$ is an exterior algebra $E(R_j^i)$ with the differential $\delta(R_j^i) = \sum_{k=1}^{j-1} R_{j-k}^{i+k} R_k^i$, $\deg(R_j^i) = (1, 2p^{i+j} - 2p^i)$, $\deg(b_j^i) = (2, 2p^{i+j+1} - 2p^{i+1})$ for $i \geqq 0$ and $j \geqq 1$. We have $H^*(U(L)) = \{1, h_0, h_1, g_0, k_0, k_0h_0, h_2, h_2h_0, g_1, l_1, l_2, l_1h_1, k_1, \ldots\}$ for

AMS 1970 *subject classifications*. Primary 55-02, 55E20, 55E45, 55G10, 55H15; Secondary 55G20.

$h_i = \{R_1^i\}$, $g_i = \{R_2^i R_1^i\}$, $k_i = \{R_2^i R_1^{i+1}\}$, $l_i = \{R_{4-i}^{i-1} R_2^0 R_1^{i-1}\}$. Then the theorem follows from the lemma just by checking the degrees.

A distinguished property of $V(n)$ is that its homotopy is relatively simple as well as its cohomology, i.e. the Adams spectral sequence

$$E_2 = \mathrm{Ext}_{\mathscr{A}}^{*,*}(H^*(V(n); Z_p), Z_p) \Rightarrow \pi_*(V(n))$$

has few nontrivial differentials by dimensional reasons. For $n = 3$ (thus $p \geqq 7$), $H^*(V(3); Z_p) \cong E(Q_i)$ hence $E_2 \cong H^*(\not{p})$ for degree $< 2p^4 - 2$. The same is true for $n = 2$ and for degree $< 2p^3 - 2$. Moreover it is true that ($p \geqq 5$)

$$E_2 = \mathrm{Ext}_{\mathscr{A}}^{*,*}(H^*(V(2); Z_p), Z_p) \cong Z_p[c] \otimes H^*(\not{p})$$

for degree $< 2p^4 - 2$, $\deg(c) = (1, 2p^3 - 1)$.

An application to the homotopy groups of spheres is given by considering the inclusion homomorphism

$$i_*: \pi_*(S^0) \to \pi_*(V(2)).$$

For example, $b = b_1^0$ is a permanent cycle and detects an element $\beta_1 \in \pi_{pq-2}(S^0)$, $q = 2(p-1)$, of order p. By checking the powers b^r in the spectral sequence for $V(2)$, we find that b^r is not a boundary for $r < p(p-1)$. Thus we have

$$(\beta_1)^r \neq 0 \quad \text{for } r < p(p-1).$$

Remark that this is true for the case $p = 3: \beta_1^5 \neq 0$, and that $\beta_1^6 = 0$. Also it is known $\beta_1^{p^2+1} = 0$ for general $p \geqq 3$.

Another example is a series of elements $\varepsilon^{(t)} \in \pi_{a(t)}(S^0)$, $a(t) = (tp^2 + 1)q - (2t + 1)$, given by $\varepsilon^{(t+1)} = \{\alpha_1, \beta_1^p, \varepsilon^{(t)}\}$ and $\varepsilon^{(0)} = \alpha_1$, where α_1 is detected by h_0. For $t < p$, $\varepsilon^{(t)}$ is detected by $h_0(b_1^1)^t$ and nontrivial.

To obtain more concrete results on $\pi_*(S^0)$, one may suggest computing, $\pi_*(V(2))$, $\pi_*(V(1))$ and then $\pi_*(V(0)) \cong \pi_*(S^0) \otimes Z_p + \mathrm{Tor}(\pi_{*-1}(S^0), Z_p)$. The above two spectral sequences of May and that of Adams collapse for degree $< p^2 q - 3$, $q = 2(p-1)$, and the first nontrivial differential is $d_{2p-1}: b_1^1 \to h_0(b)^p$ which corresponds to the relation $\alpha_1 \beta_1^p = 0$. We can choose a spectrum $V(2)$ as a mapping cone of a map

$$\beta: \Sigma^{pq+p} V(1) \to V(1),$$

then we have an exact sequence

$$\cdots \longrightarrow \pi_{*-pq-q}(V(1)) \xrightarrow{\beta_*} \pi_*(V(1)) \xrightarrow{i_{2*}} \pi_*(V(2)) \xrightarrow{\pi_{2*}} \cdots.$$

For degree $< p^2 q - 3$ we have by dimensional reasons

$$\pi_*(V(1)) \cong Z_p[\beta] \otimes \pi_*(V(2)),$$

where

$$\pi_*(V(2)) \cong Z_p[b] \otimes \{1, h_0, h_1, g_0, k_0, k_0 h_0\}.$$

Similarly $V(1)$ is a mapping cone of a map

$$\alpha : \Sigma^q V(0) \to V(0),$$

and we have a homotopy exact sequence associated with the cofibering

$$V(0) \xrightarrow{i_1} V(1) \xrightarrow{\pi_1} \Sigma^{q+1} V(0),$$

where $V(0) = S^0 \cup e^1$ is the Moore spectrum.

The computation of $\pi_*(V(0))$ from $\pi_*(V(1))$ is not simple and it can be done by the aid of the structure of automorphism algebra $\mathscr{A}_*(V(1)) = \sum \mathscr{A}_t(V(1))$, $\mathscr{A}_t(V(1)) = \{\Sigma^t V(1), V(1)\}$. The details are too long to describe here, the following relations are very interesting:

$$\beta^2\delta_1 - 2\beta\delta_1\beta + \delta_1\beta^2 = 0,$$

$$\beta^3\delta_0 - 3\beta^2\delta_0\beta + 3\beta\delta_0\beta^2 - \delta_0\beta^3 = 0,$$

where $\beta \in \mathscr{A}_{pq+q}(V(1))$, $\delta_1 = i_1\pi_1 \in \mathscr{A}_{-q-1}(V(1))$ and $\delta_0 = i_1 i\pi\pi_1 \in \mathscr{A}_{-q-2}(V(1))$ for the cofibering $S^0 \xrightarrow{i} V(0) \xrightarrow{\pi} S^1$. An analogous relation

$$\alpha^2\delta - 2\alpha\delta\alpha + \delta\alpha^2 = 0, \qquad \delta = i\pi,$$

in $\mathscr{A}_*(V(0))$ was obtained by Yamamoto (Osaka J. Math. **2** (1965)). These relations can be proved by use of higher cohomology operations or by use of slight modifications of J. Cohen's operations in his thesis.

The nontriviality of the elements

$$\alpha_r = \pi\alpha^r i \in \pi_{rq-1}(S^0)$$

was proved by Adams invariant e. Recently, L. Smith has shown the nontriviality of the elements

$$\beta_s = \pi\pi_1\beta^s i_1 i \in \pi_{(sp+s-1)q-2}(S^0).$$

β_s and $\alpha'_r = (1/r)\alpha_r$ generate multiplicatively the p-components of $\pi_*(S^0)$ for degree $< p^2q + q - 3$. As an application of the above relations we have

THEOREM 2 ($p \geqq 5$). (i) *$tu\beta_r\beta_s = rs\beta_t\beta_u$ $(t + u = r + s)$; (ii) $\beta_1^p\beta_s = \beta_s^{p+1} = 0$ for $s \geqq 2$.*

Now we go back to the spectra $V(n)$. We can restrict to choose $V(n + (k/2^{n+1}))$ such that it consists of $2^{n+1} + k$ cells $* \cup e_1 \cup e_2 \cup \cdots \cup e_b$, $b = 2^{n+1} + k$, each e_{2i} attaching to e_{2i-1} by degree p. Consider two such spectra $V(a)$, $V'(a)$, $a = n + (k/2^{n+1})$, and extend the identity map of the bottom sphere $S^0 = * \cup e_1$, then the obstruction to the extension lies in $\pi_*(V'(a))$ for $* = \dim e_i - 1$. Similarly we consider a multiplication $V(a) \wedge V(b) \to V(c)$ $(a, b \leqq c)$, i.e. a map extending the inclusion of $(V(a) \wedge S^0) \cup (S^0 \wedge V(b))$, then the obstruction belongs to $\pi_*(V(c))$ for $* = \dim e_i + \dim e_j - 1$, $e_i \subset V(a)$, $e_j \subset V(b)$. Then the following theorem is obtained as Theorem 1 by checking the degrees.

THEOREM 3. *The spectra given in Theorem* 1 *are unique up to homotopy equivalence, and the following multiplications exist*:

$$V(3) \wedge V(3) \to V(3) \quad \text{for } p \geqq 11,$$
$$V(2\tfrac{1}{4}) \wedge V(3) \to V(3) \quad \text{for } p = 7,$$
$$V(1\tfrac{1}{2}) \wedge V(2\tfrac{1}{4}) \to V(2\tfrac{3}{4}) \quad \text{for } p = 5,$$
$$V(0) \wedge V(2\tfrac{3}{4}) \to V(2\tfrac{3}{4}) \quad \text{for } p = 5,$$
$$V(\tfrac{1}{2}) \wedge V(\tfrac{1}{2}) \to V(1) \quad \text{for } p = 3,$$

and

$$V(0) \wedge V(1\tfrac{1}{4}) \to V(1\tfrac{1}{2}) \quad \text{for } p = 3.$$

For the case $p = 3$ we have the following negative statements.

THEOREM 4. *Let* $p = 3$.

(i) *The multiplication* $V(0) \wedge V(0) \to V(0)$ *is not homotopy associative.*

(ii) *Any multiplication* $V(\tfrac{1}{2}) \wedge V(1) \to V(1)$ (*thus* $V(1) \wedge V(1) \to V(1)$) *does not exist.*

(iii) *Any multiplication* $V(0) \wedge V(1\tfrac{1}{2}) \to V(1\tfrac{1}{2})$ *does not exist.*

(iv) $V(1\tfrac{3}{4})$ *does not exist.*

A similar situation appears in the Adams spectral sequence

$$E_2 = \mathrm{Ext}_{\mathscr{A}}^{*,*}(H^*(V(1); Z_p), Z_p) \Rightarrow \pi_*(V(1)),$$

i.e. for degree $< p^2q - 3$, it collapses if $p \geqq 5$ but it does not collapse if $p = 3$.

As indicated in Adams' speech the statement (ii) implies an exotic Cartan formula

$$\Phi(xy) = (\Phi x)y + x(\Phi y) + (\mathscr{P}^1 \Delta x)(\Delta \mathscr{P}^1 \Delta y) + (\Delta \mathscr{P}^1 \Delta x)(\mathscr{P}^1 \Delta y),$$

up to sign, for the secondary operation Φ associated with the relation $\mathscr{P}^2 \mathscr{P}^1 = 0$, since the obstruction to the existence of a multiplication in (ii) is detected by Φ, i.e. $b = b_1^0$.

The nonassociativity (i) is proved as follows. Let $T: V(0) \wedge V(0) \to V(0) \wedge V(0)$ be the switching map and $P = (T \wedge 1)(1 \wedge T)$ a cyclic permutation of $V(0)^{(3)} = V(0) \wedge V(0) \wedge V(0)$. Let $\mu: V(0) \wedge V(0) \to V(0)$ be a multiplication. Since $\pi_i(V(0)) = 0$ for $0 < i < 3$ and $\pi_3(V(0))$ is generated by $i\alpha_1$, μ is unique, commutative: $\mu = \mu \circ T$ and the associator is

$$\mu \circ (1 \wedge \mu) - \mu \circ (\mu \wedge 1) = x \cdot i\alpha_1(\pi \wedge \pi \wedge \pi)$$

for some $x \in Z_3$ (up to homotopy). Since $T(1 \wedge \mu)P = \mu \wedge 1$ we have $\mu(1 \wedge \mu) - \mu(1 \wedge \mu)P = x \cdot i_1\alpha(\pi \wedge \pi \wedge \pi)$. Now assume $x = 0$, then we have a homotopy $h: V(0)^{(3)} \times I \to V(0)$ between $\mu(1 \wedge \mu)$ and $\mu(1 \wedge \mu)P$ and it induces a map $\bar{h}: X \to V(0)$ with $\bar{h} \mid V(0)^{(3)} = \mu(1 \wedge \mu)$, where X is obtained from $V(0)^{(3)} \times 1$ by identifying $V(0)^{(3)} \times 1$ with $V(0)^{(3)} \times 0$ by P, i.e.

$$X = V(0)^{(3)} \times_\pi S^1 = ep^1(V(0))$$

for the cyclic group π generated by P. Nishida's formula shows that $\mathscr{P}^1 H^0(X; Z_3) = H^4(X; Z_3)$ but this contradicts the naturality of $\mathscr{P}^1$ since $H^0(X; Z_3) = \bar{h}^* H^0(V(0); Z_3)$ and $H^4(V(0); Z_3) = 0$. Thus we conclude $x \neq 0$, and the nonassociativity (i).

Next we see how to get (ii) from (i). By Theorem 3 we have two multiplications

$$\varphi_1 : V(\tfrac{1}{2}) \wedge V(\tfrac{1}{2}) \to V(1) \qquad \text{and} \qquad \varphi_2 : V(1) \wedge V(0) \to V(1)$$

as extensions of μ. We assume the existence of a multiplication $\varphi : V(\frac{1}{2}) \wedge V(1) \to V(1)$ and consider the difference

$$d = \varphi(1 \wedge \varphi_0) - \varphi_2(\varphi_1 \wedge 1) : K \to V(1),$$

where $K = V(\frac{1}{2}) \wedge V(\frac{1}{2}) \wedge V(0)$ and $\varphi_0 = \varphi \mid V(\frac{1}{2}) \wedge V(0)$. K consists of cells of dimensions 0, 1, 2, 3, 5, 6, 10, 11 and the cells of dimensions 0, 1, 2, 5, 6, 7 form a subcomplex L of K such that

$$K/L = (S^3 \vee S^{10}) \cup e_1^7 \cup e_2^7 \cup e^{11}$$

and $\mathscr{P}^2 H^3(K/L; Z_3) = H^{11}(K/L; Z_3)$. Since $\pi_i(V(1)) = 0$ for $0 < i < 10$ and $i \neq 3$, the difference d is factored through K/L:

$$d = f \circ \pi' : K \to K/L \to V(1).$$

By (i), $f \mid S^3 = \pm i_1 i \alpha_1$, that is, the functional operation

$$\mathscr{P}_f^1 : H^0(V(1); Z_3) \to H^3(K/L; Z_3)$$

is an isomorphism. In the mapping cone of f we have $\mathscr{P}^2 \mathscr{P}^1 H^0(e_f; Z_3) \neq 0$ but this contradicts Adem relation $\mathscr{P}^2 \mathscr{P}^1 = 0$. Thus (ii) is proved. Since the obstruction to the existence of φ is in $\pi_{10}(V(1)) \approx Z_3$ the generator of which is detected by the secondary operation Φ associated with $\mathscr{P}^2 \mathscr{P}^1 = 0$, we see that for the fundamental classes $u \in H^0(V(\frac{1}{2}); Z_3)$, $v \in H^0(V(1); Z_3)$ the relation

$$\Phi(u \wedge v) = \pm(\mathscr{P}^1 \Delta u) \wedge (\Delta \mathscr{P}^1 \Delta v)$$

holds.

Similarly (iii) is proved by constructing a difference

$$V(0) \wedge V(0) \wedge V(1\tfrac{1}{4}) \to V(1\tfrac{1}{2})$$

the mapping cone of which contradicts Adem relation $\mathscr{P}^p \mathscr{P}^1 \mathscr{P}^1 - 2\mathscr{P}^1 \mathscr{P}^p \mathscr{P}^1 + \mathscr{P}^1 \mathscr{P}^1 \mathscr{P}^p = 0$. Hence, for the secondary operation Φ' associated with the relation we have

$$\Phi'(u \wedge v) = \pm(\Delta u) \wedge (\Delta \mathscr{P}^3 \mathscr{P}^1 \Delta v) \quad \text{in } V(0) \wedge V(1\tfrac{1}{2}).$$

The idea to prove (iv) is the following. By (ii) we cannot find a multiplication as in (ii), but by attaching 11-cell to $V(1)$ by a generator $i_1 i \beta_1$ of $\pi_{10}(V(1))$ we have a multiplication of a type

$$\bar{\varphi} : V(\tfrac{1}{2}) \wedge V(1) \to V(1) \cup e^{11}$$

which extends $V(\frac{1}{2}) \wedge V(\frac{1}{2}) \cup V(0) \wedge V(1) \to V(1)$ and has degree ± 1 on the top cells e^{11}. Define a complex (spectrum)

$$V(1\tfrac{3}{4})^+ = V(1\tfrac{1}{2}) \cup e^{11} \cup e^{22}$$

from the disjoint union of $V(\tfrac{1}{2}) \wedge V(1\tfrac{1}{4})$ and $V(1) \cup e^{11}$ by identifying the corresponding points under $\bar{\varphi}$. Consider the boundary homomorphisms

$$\partial : \pi_{22}(V(1\tfrac{3}{4})^+, V(1\tfrac{1}{2}) \cup e^{11}) \to \pi_{21}(V(1\tfrac{1}{2}) \cup e^{11}, V(1\tfrac{1}{4})) \approx \pi_{21}(S^{11}) \oplus \pi_{21}(S^{18})$$

and

$$\partial' : \pi_{21}(V(1\tfrac{1}{2}) \cup e^{11}, V(1\tfrac{1}{4})) \to \pi_{20}(V(1\tfrac{1}{4})).$$

For the class $\varepsilon \in \pi_{22}$ of e^{22}, we have $\partial(\varepsilon) = \beta_1 \oplus \alpha_1$ up to sign. The image $\partial'(\alpha_1)$ just gives the obstruction to the existence of $V(1\tfrac{3}{4})$. Then $0 = \partial'\partial(\varepsilon) = \partial'(\beta_1 \oplus \alpha_1) = \beta_1^2 + \partial'(\alpha_1)$. Now checking $\beta_1^2 \neq 0$ in $V(1\tfrac{1}{4})$ we have $\partial'(\alpha_1) \neq 0$ and the nonexistence (iv) of $V(1\tfrac{3}{4})$.

This idea may also apply to prove that for $p = 5$, $V(2\tfrac{7}{8})$ exists iff a multiplication $V(1\tfrac{3}{4}) \wedge V(2) \to V(2)$ exists. Finally we propose the following

CONJECTURE. *For odd prime* $p = 2k - 1$, $V(k - (1/2^{k-1}))$ *exists but* $V(k - (1/2^k))$ *does not exist*?

NORTHWESTERN UNIVERSITY AND
KYOTO UNIVERSITY

ON SPHERE EXTENSIONS OF CLASSICAL LIE GROUPS

ALEXANDER ZABRODSKY

1. **A short survey and summary of results.** In [4] Hilton and Roitberg discovered the first known example of a finite CW-complex which admits an H-structure, and a long drought was over. In [2], [6], and [9] other examples followed. Most of these examples were of one type. One may refer to it as a sphere extension of a classical group.

Let $(G_n, d) = (SU(n), 2)$ or $(Sp(n), 4)$. One has the classical fibration

$$G_{n-1} \to G_n \to S^{dn-1}.$$

Thus, one considers G_n as an S^{dn-1} extension of G_{n-1}. If $f_\lambda : S^{dn-1} \to S^{dn-1}$ is a map of degree λ then most of the newly discovered finite CW-H-spaces were of the $M(n, \lambda)$ type obtained by the induced fibration

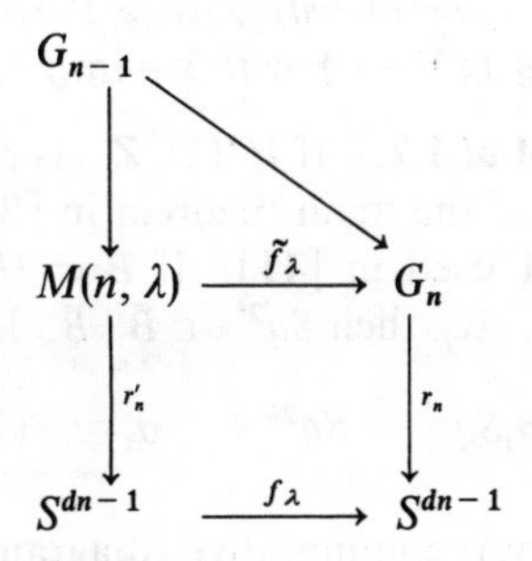

The following question naturally arises:

1.0. For what values of λ does $M(n, \lambda)$ admit an H-structure?

In [3], [5], [6], and [10], this question was given a complete answer for

---

AMS 1970 *subject classifications*. Primary 55D45.

$dn \leqq 8$. The general problem was studied in [**7**], and some conjectures were suggested in [**10**].

However, the complexity and tediousness of the proofs in [**10**] that helped settle the question for G_n, $d = Sp(2), 4$ were not encouraging. There was little hope that these proofs can be generalized to the case $dn > 8$.

Fortunately (though not surprisingly) the case $dn = 8$ proved to be exceptional and for $dn > 8$, one can use the decomposability of Sq^{dn} via primary ($dn \neq 2^j$) or secondary ($dn = 2^j$) operations.

Here we sketch an outline of the proof of the following:

1.1. THEOREM. *If $M(n, \lambda)$ admits an H-structure $dn \neq 2^j$, then λ is odd. If $M(n, \lambda)$ admits a homotopy associative H-structure ($dn = 2^j$, $j > 3$) then λ is odd.*

The restriction in the case $dn = 2^j$ may turn out to be unnecessary. As a matter of fact, here one can already relax the homotopy associativity condition by the assumption that $H^*(M(n, \lambda), Z_2)$ is primitively generated.

The proof of 1.1 is based on the following two theorems:

1.2. THEOREM. *Let X, μ be an H-space, $H^*(X, Z_2) = \Lambda(x_1, x_2, \ldots, x)$, $\dim x_i = odd$, $\dim x_i \leqq \dim x_{i+1} \leqq \dim x = 2k - 1$, $k \neq 2^t$. Then*

$$Qx \in \overline{\mathscr{A}(2)} \cdot QH^*(X, Z_2).$$

($QH^*(X, Z_2)$ is the module of indecomposables in $H^*(X, Z_2)$ and $Qx \in QH^*(X, Z_2)$ is the image of x.)

1.3. THEOREM. *Let X, μ be an H-space, $H^*(X, Z_2) = \Lambda(x_1, x_2, \ldots, x_j, x)$, $\dim x_i = odd$, $\dim x_i \leqq \dim x_{i+1} \leqq \dim x = 2^t - 1$, $t > 3$, $x_i, x \in PH^*(X, Z_2)$. Then x is detected by a secondary operation defined on elements of dimension $< 2^t - 1$, i.e. there exists a (generalized) 2-stage Postnikov system Y, $\Pi_m(Y) = 0$ if $m \geqq 2^t - 1$ and a mapping $f: X \to Y$ with $x \in \operatorname{im} f^*$.*

2. **An outline of the proof of 1.2.** If $H^*(X, Z_2)$ is primitively generated, 1.2 is an immediate consequence of the main theorem in [**8**]. In the general case, the technique is similar to that used in [**11**]: If $B \subset H^*(X, Z_2)$ is the $\mathscr{A}(2)$-subalgebra generated by $x_1, \ldots, x_j$, then $Sq^{2^t} x \in \bar{B} \cdot \bar{B}$. If

$$\sum_t \alpha_t Sq^{2^t} = Sq^{2k}, \qquad \alpha_t \in \mathscr{A}(2),$$

then one has a (homotopy) commutative diagram (see top of page 281) where $f: X \to X_0 = \prod_{i \leqq j} K(Z_2, |x_i|)$ $(f^* \iota_{|x_i|} = x_i)$ is an H-mapping with $\operatorname{im} f^* = B$,

$$f_1 : X_0 \to X_1 = \prod_s K(Z_2, m_s)$$

satisfies $\operatorname{im} f_1^* = \operatorname{subker} f^* : \ker f^* = \overline{\operatorname{im} f_1^*} \cdot H^*(X_0, Z_2)$, $f_1 \circ f \sim *$, $g^* \iota = x$; g_1 exists as $\sum_t \alpha_t g_0^* \iota_{2k-1+2^t}$ is odd dimensional decomposable in $\ker f^*$ and hence

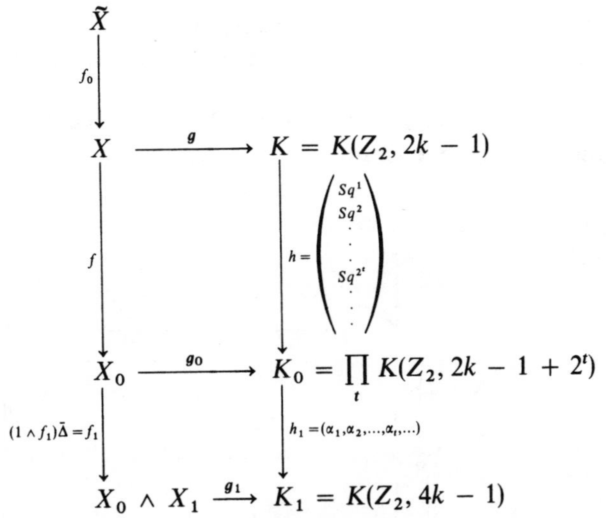

$$
\begin{array}{ccc}
\tilde{X} & & \\
\downarrow{\scriptstyle f_0} & & \\
X & \xrightarrow{g} & K = K(Z_2, 2k-1) \\
\downarrow{\scriptstyle f} & & \downarrow{\scriptstyle h = \begin{pmatrix} Sq^1 \\ Sq^2 \\ \vdots \\ Sq^{2^t} \\ \vdots \end{pmatrix}} \\
X_0 & \xrightarrow{g_0} & K_0 = \prod_t K(Z_2, 2k-1+2^t) \\
\downarrow{\scriptstyle (1 \wedge f_1)\bar{\Delta} = f_1} & & \downarrow{\scriptstyle h_1 = (\alpha_1, \alpha_2, \ldots, \alpha_t, \ldots)} \\
X_0 \wedge X_1 & \xrightarrow{g_1} & K_1 = K(Z_2, 4k-1)
\end{array}
$$

in $\overline{\ker f^* \cdot \bar{H}^*(X_0, Z_2)}$. $f_0 : \tilde{X} \to X$ is the fibration induced by f from $\Omega X_0 \to \mathscr{L}X_0 \to X_0$. As in [11], homotopies $l_1 : X \to \mathscr{L}(X_0 \wedge X_1)$, $l_0 : \tilde{X} \to \mathscr{L}K_0$ and $l_2 : K \to \mathscr{L}K_1$ realizing $* \sim \bar{f}_1 \circ f, * \sim f \circ f_0$ and $* \sim h_1 \circ h$ can be chosen so that:

2.1. $(\mathscr{L}g_1 \circ l_1 - l_2 \circ g) \circ f_0 \sim \mathscr{L}h_1(\mathscr{L}g_0 \circ l_0 - w \circ f_0) - l_2 \circ g \circ f$. ($w : X \to PK_0$ is the homotopy $h \circ g \sim g_0 \circ f$.)

The homotopy class z of the map in 2.1 lies in $[\tilde{X}, \Omega K_1] = H^{4k-2}(\tilde{X}, Z_2)$ and represents an element in $\phi(f_0^*(x))$, where ϕ is the secondary operation associated with $e(\sum \alpha_t Sq^{2^t}) > 2k - 1$. One can see that

$$\bar{\tilde{\mu}}^* z \equiv f_0^* x \otimes f_0^* x \bmod \sum_t \alpha_t(\operatorname{im} f_0^* \otimes \operatorname{im} f_0^*).$$

But, by 2.1, $z \in \operatorname{im} f_0^*$, hence, z is decomposable, $f_0^* x \in \overline{\mathscr{A}(2)} \operatorname{im} f_0^*$ and consequently, $x + b \in \overline{\mathscr{A}(2)} \cdot H(X, Z_2)$, $b \in B$, and 1.2 follows.

3. **An outline of the proof of 1.3.** By [1], given the 2-stage Postnikov system E_m constructed by the fibration

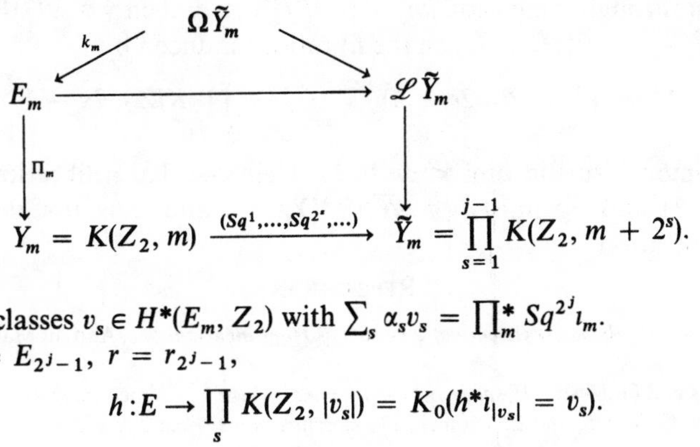

$$
\begin{array}{ccccc}
 & & \Omega\tilde{Y}_m & & \\
 & \swarrow{\scriptstyle k_m} & & \searrow & \\
E_m & & \longrightarrow & & \mathscr{L}\tilde{Y}_m \\
\downarrow{\scriptstyle \Pi_m} & & & & \downarrow \\
Y_m = K(Z_2, m) & & \xrightarrow{(Sq^1, \ldots, Sq^{2^s}, \ldots)} & & \tilde{Y}_m = \prod_{s=1}^{j-1} K(Z_2, m + 2^s).
\end{array}
$$

There exist classes $v_s \in H^*(E_m, Z_2)$ with $\sum_s \alpha_s v_s = \prod_m^* Sq^{2^j} \iota_m$.

Let $E = E_{2^j-1}$, $r = r_{2^j-1}$,

$$h : E \to \prod_s K(Z_2, |v_s|) = K_0(h^* \iota_{|v_s|} = v_s).$$

Then $h^* \sum_s \alpha_s \iota_{|v_s|} = 0$. If $h_0 : \tilde{E} \to E$ is the ΩK_0-principal fibration induced by h, then there exists a class $v \in H^{2^j-2}(\tilde{E}, Z_2)$, v restricts to $\sum_s \alpha_s \sigma^* \iota_{|v_s|}$ in $H^*(\Omega K_0, Z_2)$ and $\bar{\mu}_{\tilde{E}}^* v = h_0^* r^* \iota \otimes h_0^* r^* \iota$.

Now, $Sq^{2^t} x = 0$ implies that there exists $g : X \to E$ with $g^* r^* \iota = x$. As in the proof of 1.2, there exists the diagram

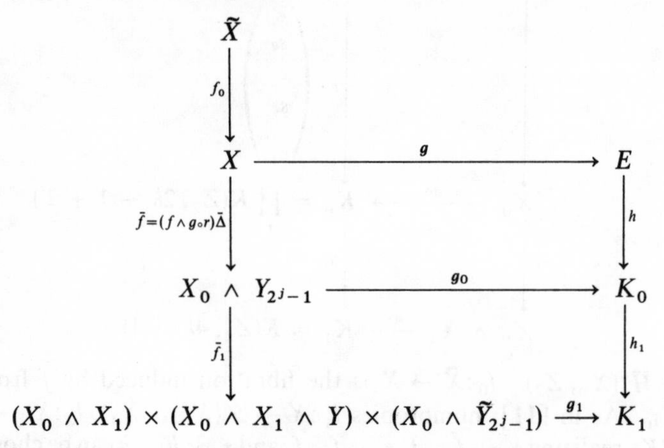

Now, if $l_0 : \tilde{X} \to \mathscr{L}K_0$ is the homotopy $* \sim h \circ g \circ f_0$ induced by $* \sim \bar{f} \circ f_0$ then the homotopy class z of $\mathscr{L}h_1 l_0 - l_2 \circ g \circ f_0 : \tilde{X} \to \Omega K_1$ represents a class in some third order operation defined on $f_0^* x$. As in 2.1, one has:

3.1. $z \in \operatorname{im} f_0^*$, and hence, z is decomposable.

Now, $g \circ f_0$ is an H-mapping, l_0 induces a lifting $\tilde{g} : \tilde{X} \to \tilde{E}$, $\tilde{g}^* v = z$, and consequently, the deviation of $\tilde{g}$ from being an H-mapping is represented by a mapping $\tilde{X} \wedge \tilde{X} \to \Omega K_0$ and $\tilde{\mu}^* z \equiv f_0^* x \otimes f_0^* x \bmod \sum_s \alpha_s H^*(\tilde{X}, Z_2)$. It follows that $f_0^* x \in \mathscr{A}(2) H^*(X, Z_2)$ which is equivalent to 1.3.

4. **An outline of the proof of 1.1.** $H^*(M(n, \lambda), Z_2) \approx H^*(G_n, Z_2)$ as an algebra and $\tilde{f}_\lambda^*$ induces an isomorphism in dim $< dn - 1$. Moreover, if $Y_\lambda \to G_n$ is the fibration induced by $G_n \to K(Z_{2^s}, dn - 1)$ (representing $r_n^* \iota_{dn-1}$), $\lambda = 2^s(2\lambda' + 1)$, then $\tilde{f}_\lambda$ can be lifted to $\tilde{f}'_\lambda : M(n, \lambda) \to Y_\lambda$, $\tilde{f}'^*_\lambda : H^*(Y_\lambda, Z_2) \to H^*(M(n, \lambda), Z_2)$ is an isomorphism through dimension $dn - 1$. If $\tilde{f}'^*_\lambda y = x$ then $y \notin \mathscr{A}(2) H^*(Y_\lambda, Z_2)$.

Let $f'_0 : \tilde{Y}_\lambda \to Y_\lambda$, $f''_0 : \tilde{G} \to G_n$ be the fibrations induced by

$$Y_\lambda \to \prod_{2s < dn} K(Z_2, 2s - 1), \qquad G_n \to \prod_{2s < dn} K(Z_2, 2s - 1)$$

inducing epimorphisms in dim $< dn - 1$. Then one has a fibration $\tilde{Y}_\lambda \to \tilde{G} \to K(Z_{2^s}, dn - 1)$ and again, $f'^*_0 y \notin \overline{\mathscr{A}(2)} H^*(\tilde{Y}_\lambda, Z_2)$ and x is not detected by a secondary operation.

References

1. J. F. Adams, *On the non-existence of elements of Hopf invariant one*, Ann. of Math. (2) **72** (1960), 20–104. MR **25** #4530.

2. M. Curtis and G. Mislin, *H-spaces which are bundles over* S^7 (mimeographed).

3. M. Curtis, G. Mislin, and E. Thomas, *The classification of rank* 2 *H-spaces*, unpublished.

4. P. Hilton and J. Roitberg, *On principal S^3 bundles over spheres*, Ann. of Math. (2) **90** (1969), 91–107. MR **39** #7624.

5. ———, *On the classification problem for H-spaces of rank two* (mimeographed).

6. J. Stasheff, *Manifolds of the homotopy type of (non-Lie) groups*, Bull. Amer. Math. Soc. **75** (1969), 998–1000.

7. J. Stasheff and J. Harrison, *Families of H-spaces* (mimeographed).

8. E. Thomas, *Steenrod squares and H-spaces*, Ann. of Math. (2) **77** (1963), 306–317. MR **26** #3057.

9. A. Zabrodsky, *Homotopy associativity and finite* CW *complexes*, Topology **9** (1970), 121–128.

10. ———, *The classification of H-spaces with three cells.* I, II, Math. Scand. (to appear).

11. ———, *Secondary operations in the module of indecomposables*, Summer Institute in Algebraic Topology, Aarhus, 1970.

UNIVERSITY OF ILLINOIS AT CHICAGO CIRCLE

AUTHOR INDEX

Roman numbers refer to pages on which a reference is made to an author or a work of an author. *Italic numbers* refer to pages on which a complete reference to a work by the author is given. **Boldface numbers** indicate the first page of the articles in the book.

SUBJECT INDEX